Fuel Cell and Hydrogen Technologies in Maritime Transportation

C. Ozgur Colpan • Suleyman Aykut Korkmaz •
Olgun Konur
Editors

Fuel Cell and Hydrogen Technologies in Maritime Transportation

Springer

Editors
C. Ozgur Colpan
Department of Mechanical Engineering
Dokuz Eylül Üniversitesi
Buca, Izmir, Türkiye

Suleyman Aykut Korkmaz
Department of Marine Engineering
Dokuz Eylül University
Buca, Izmir, Türkiye

Olgun Konur
Department of Marine Engineering
Dokuz Eylul University
Buca, Izmir, Türkiye

ISBN 978-3-031-86109-3 ISBN 978-3-031-86110-9 (eBook)
https://doi.org/10.1007/978-3-031-86110-9

This Springer imprint is published by the registered company Springer Nature Switzerland AG
The registered company address is: Gewerbestrasse 11, 6330 Cham, Switzerland

Preface

The maritime transportation sector is undergoing a significant transformation as the urgency to address climate change and reduce greenhouse gas emissions intensifies. As global trade continues to expand, so does the environmental footprint of the shipping industry, which has prompted growing interest in alternative, sustainable energy solutions to reduce its environmental impact. Hydrogen and fuel cell technologies have emerged as sustainable solutions to decarbonize marine transport, offering a promising path for marine vessels toward reducing emissions while maintaining the operational demands of the maritime sector. This book provides an in-depth exploration of the current state and future potential of fuel cell and hydrogen technologies in maritime transportation. It brings together cutting-edge research, innovative technologies, and practical applications, showcasing the role hydrogen can play in reshaping the maritime industry toward a sustainable future.

The book covers a wide range of topics starting with the challenges and solutions related to hydrogen storage and handling on ships, a critical aspect of ensuring safety and efficiency at sea. It also delves into the use of various hydrogen-based fuels, including ammonia, methanol, and both liquid and gaseous hydrogen, analyzing their suitability for maritime applications and their potential to reduce emissions. The role of green hydrogen in sustainable transportation is explored, identifying opportunities and barriers that lie ahead in achieving widespread adoption of this clean fuel. An overview of advanced technologies and recent applications of fuel cell-powered marine vessels are discussed in detail, including fuel cells for autonomous underwater and surface vehicles, which offer promising new opportunities for cleaner operations in specialized areas. Energy management strategies for hybrid fuel cell systems are also represented, focusing on optimizing performance and reducing costs for ships using a combination of fuel cells and conventional technologies. An innovative case study on onboard hydrogen production highlights the potential of producing green hydrogen through electrolysis powered by waste heat recovery, showcasing a sustainable energy loop within maritime operations. The book also assesses the impact of hydrogen injection methods on engine performance and emissions, providing insights into how hydrogen can be integrated into existing internal combustion engine-powered vessels. Recognizing the importance of infrastructure, the emerging role of seaports as hubs for hydrogen fuel is analyzed, along with the potential of hydrogen-powered vessels to transform waterborne transportation from an ecosystem perspective. The environmental impacts of hydrogen-powered ships are further examined through life cycle assessments, offering a comprehensive view of the sustainability of these technologies over the lifespan of a vessel. Finally, the book explores the current status and future pathways for hydrogen and fuel cell adoption in the maritime industry, highlighting ongoing technological advancements and the challenges ahead.

This book aims to serve as a valuable resource for researchers, engineers, policymakers, and industry stakeholders involved in maritime transportation and sustainability. It offers a comprehensive view of the latest advancements in hydrogen and fuel cell technologies while addressing the practical considerations required to implement these innovations in the real world. We extend our sincere gratitude to all the chapter authors and reviewers for their invaluable contributions to the success of this book, and to the editorial team at Springer for their continued support for helping and guiding us in each step in bringing this book to life. We hope this book inspires ongoing efforts on further innovation and collaboration in the pursuit of a cleaner, greener, and more sustainable maritime industry.

Buca, Izmir, Türkiye

C. Ozgur Colpan
S. Aykut Korkmaz
Olgun Konur

Contents

Hydrogen Storage and Handling on Ships

Nikolaos Souglakos, Alexandros Komianos, Xue Xu, and Evangelos Boulougouris

Abbreviations

ABS	American Bureau Of Shipping
AiP	Approval in Principle
ALARP	As Low As Reasonable Possible
BV	Bureau Veritas
CH_2	Compressed Hydrogen
CNG	Compressed Natural Gas
DAL	Design Accidental Load
DDT	Deflagration to Detonation Transition
EMS	Emergency Shutdown System
ERA	Explosion Risk Analysis
ESD	Emergency Shutdown Valves
FPR	Fuel Preparation Room
FSA	Formal Safety Assessment
GHG	Greenhouse Gas Emissions
HISC	Hydrogen Induced Stress Cracking
IMO	International Maritime Organization
ITP	Inspection and Test Plan
LEL	Lower Explosive Limit
LH_2	Liquid Hydrogen
LNG	Liquified Natural Gas
LR	Lloyds Register
NDT	Non-Destructive Testing

N. Souglakos (✉) · A. Komianos · X. Xu · E. Boulougouris
Department of Naval Architecture, Ocean and Marine Engineering, University of Strathclyde, Glasgow, UK
e-mail: nikolaos.souglakos.2020@uni.strath.ac.uk; alexandros.komianos@strath.ac.uk; xue.xu@strath.ac.uk; evangelos.boulougouris@strath.ac.uk

C. O. Colpan et al. (eds.), *Fuel Cell and Hydrogen Technologies in Maritime Transportation*, https://doi.org/10.1007/978-3-031-86110-9_1

P&IDs	Piping and Instrumentation Drawings
PRV	Pressure Relief Valves
PSD	Process Shutdown Valves
PSV	Pressure Safety Valve
QCP	Quality Control Plan
SOLAS	Safety of Life at Sea
TCS	Tank Connection Space

1 Introduction

Climate change applies pressure to decarbonize all industries, including the maritime. Recognizing this need, international organizations, governments, and stakeholders are developing strategies to reduce Greenhouse Gas Emissions (GHG). The International Maritime Organization (IMO) set an ambitious target for the full decarbonization of the maritime industry by 2050 [1]. To achieve this goal, the development of new technologies and fuels capable of reducing the GHG footprint of maritime assets and processes is undeniably crucial. The challenge is to ensure that the production of these fuels will not shift the emissions to other sectors. The milestones introduced call for a 20% reduction by 2030, a 70% reduction by 2040, and full decarbonization of the maritime industry by 2050, as presented in Fig. 1.

Besides hydrogen, there is a variety of alternative fuels such as methane (as liquefied natural gas, LNG), methanol, ammonia, and ethanol. However, for the best selection of alternative fuel, many different factors must be considered, such as chemical properties, environmental effects, safety and health issues, availability and supply chain, cost and competition with other market sectors, infrastructure, and their potential for use in marine propulsion [3]. Therefore, choosing the future marine fuel is a rather laborious and uncertain decision, as it is a complex situation that requires consideration of many factors.

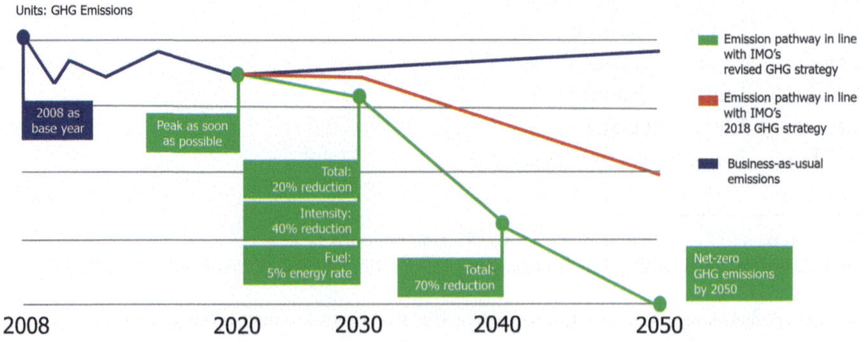

Fig. 1 Revised IMO GHG Strategy. (Adapted from [2])

Hydrogen is one of the most promising fuels for replacing carbon-based fuels due to the absence of pollutants, its abundance in nature, and the prospect of producing it from sustainable energy sources. Many cutting-edge projects are currently trying to address issues related to hydrogen production, transport, and storage while maximizing ocean renewable energy, such as the "Ocean-Fuel—Ocean Renewable Energy FuelsRE" EPSRC project. However, the choice of hydrogen as the fuel with wide use absolutely necessitates thorough research on its safety. The physical properties of hydrogen, its low density, chemical reactivity, and its wide range of flammability and explosiveness create the need for a safe design of its storage and handling facilities, as well as provisions for pressure relief in high-pressure installations and piping under various operational and emergency conditions [4]. Based on the aforementioned, it becomes clear that it is necessary to comprehend these particular characteristics of hydrogen, as well as safety issues in an uneventful case of dispersion, which could occur rapidly and is particularly likely to lead to a possible ignition or explosion provided that a heat source is present.

A key issue with hydrogen is its small molecular size, which allows it to leak even from airtight systems. In the case of high-pressure storage, its ability to cause blistering and cracking in steel is a significant safety risk. When stored in low temperatures, it can cause solidification/freeze of oxygen and nitrogen, which leads to significant mechanical stress and embrittlement of alloys that are rated for hydrogen. Finally, its high evaporation rate combined with the broad flammable limits make it a high risk of ignition and explosion in case of dispersion [5].

In this chapter, the current state-of-the-art regarding hydrogen storage and handling is explored, focusing on the safety aspects due to its unique physical properties. Different types of hydrogen fuel containment and the related regulatory framework, the design considerations for safer handling, and the status of research projects in the maritime sector are discussed, showcasing that hydrogen uptake in the maritime industry is not straightforward, always compared against conventional hydrocarbon fuels.

2 Available Types of Hydrogen Storage Across All Industries

The purpose of this section is to provide knowledge and insight into the available types of storage of hydrogen across all industries, either in compressed form or as a liquid [6]. Other industries, such as automotive, have considerable experience in hydrogen storage and, subsequently, handling compared to maritime, which is currently in the infancy stage. In general, there are currently five types of compressed hydrogen tanks, Types I-V, but these are pertinent to other industries and not the maritime. Each type of tank has a varying degree of material and design complexity, which also leads to different storage pressures and, therefore, capacities. A common aspect amongst Type I-V, or pressurized, tanks is their cylindrical shape, allowing for high-pressure storage.

Type I tanks are entirely made of alloy and can withstand pressures of 200–300 bar, whilst these allow 1.7 wt% gravimetric storage. Type II tanks are constructed out of a polymer liner, which is wrapped around a metallic cylinder and allows for up to 200 bar pressure with a 2.1 wt% gravimetric storage [7]. Type III tanks have a metallic liner that is entirely wrapped with composite fiber and allows higher storage pressures (compared to Type I and II) of up to 700 bar. Type IV tanks are constructed out of a polymeric liner, which is fully wrapped with composite fibers, being a plastic bladder in essence, and also allow for up to 700 bar storage pressure. Lastly, Type V tanks are fully composite with no liner or metallic parts [7] and allow for much greater storage pressures while having a 5.7 wt% gravimetric capacity.

Hydrogen storage tanks in the maritime industry are not entirely straightforward in the current status of legislation. These are in an immature stage, mainly due to the absence of relevant prescriptive rules. The most applicable to the hydrogen set of rules, the International Code of Safety for Ships Using Gases of Other Low-Flashpoint Fuels, the IGF Code offers three distinct types of storage [8]. Although the IGF Code covers mainly LNG as fuel and does not explicitly have prescriptions pertinent to hydrogen, it is used as the basis for the storage tank philosophy and design.

Chapter 6.4.15 of the IGF Code elaborates on the tank types and the design of those [1]. Currently, there are four different tank types, namely Type A to C and membrane tanks (integral). Type A tanks are independent and designed using traditional plane surfaces which conform to ship geometries and, as per the current status of the IGF, have a design vapor pressure of less than 0.7 MPa/bar. Type A tanks are not pressurized; therefore, hydrogen is expected to be in a liquid state, specifically below −163 °C.

Moreover, Type A tanks necessitate a complete secondary barrier, enclosing the tank so that it can contain the LNG for 15 days. Type B tanks are independent tanks of prismatic shapes and have the same storage environment as Type A. Model testing and appropriate rigorous analysis, including relevant analytical tools, are employed to design those, being the sole difference between those and Type A tanks, except for materials. Furthermore, unlike type A, Type B tanks require only partial secondary barriers, owing to the sophistication and effort of their design, but this is reflected in a higher design cost.

Type C tanks are pressure vessels storing the storage medium in magnitudes greater than 2 bar. The usual shapes of Type C tanks are cylindrical, bi-lobe, and tri-lobe. Since LNG is being pressurized, there is no need for secondary barriers. Lastly, another option is integral membrane tanks, having the advantage of being built into the supporting ship structures. Storage conditions are cryogenic, while these have two barriers, able to contain the liquid LNG for 15 days.

Table 1 presents and summarizes the available types of storage tanks as per the IGF Code. The above aim is to inform you of the available types and relevant conditions for the storage of low-flashpoint fuels as per the IGF Code. The code itself does not include hydrogen as a fuel, but these types of storage can surely be used for hydrogen.

Table 1 Available tanks for LNG

	Independent Tanks			Integral Tank
Feature	IMO Type A	IMO Type B	IMO Type C	Membrane
Geometry	Self-supporting independent prismatic tank with option of inclined boundaries		Pressure vessel (cylindrical, bi-lobe, tri-lobe design)	Prismatic tank, built into the supporting ship structure
Space utilization	Moderate to Good		Low to Moderate	Good
Temperature/ Pressure	−163 °C/<0.7 bar		NA/>2 bar (overpressure possible)	−163 °C/<0.7 bar
Barriers	Two barriers The second barrier enclosing the tank is able to contain liquid gas for 15 days.	Two barriers Partial second barriers, designed to contain the liquid phase of LNG fuel for 15 days	One barrier	Two barriers A secondary barrier designed to contain the liquid phase of LNG fuel for 15 days
Design complexity	Moderate	High	Low to moderate	Moderate
Manufacturing	Prefabrication—Independent ship structure		Prefabrication—Independent ship structure, may be located on deck	Construction of the tank inside the pre-manufactured tank compartment
Material	9% nickel steel, stainless steel, aluminum	9% nickel steel, high-manganese steel, aluminum, and SUS	9% nickel steel (SUS)	Stainless steel, nickel steel (Invar)
Sloshing risk	Small due to swash bulkhead		Small due to shape, volume, and swash bulkheads	Significant
Main challenges	Secondary barrier	Design complexity	Weight and space utilization	Sloshing and series construction

Source: [9]

3 Regulatory Framework of a Hydrogen-Fueled Vessel

3.1 Common Regulatory Elements in IGF Code and Classification Society Rules

This specific section provides a summation of the regulations as per the International Code of Safety for Ships Using Gases or Other Low-Flashpoint Fuels [8] and the rules issued by Classification Societies regarding hydrogen-fueled vessels. Focus was given on the rules regarding the storage and handling of hydrogen; otherwise, this section would be an exhaustive enumeration and narration of overlapping immature regulations. Hence, focus was given to the available types of enclosed spaces, not as per SOLAS categorization, but according to the function of the room and its relevant elements (such as piping and relevant fittings, bunkering station, tank connection space, fuel preparation rooms, etc.). Nevertheless, the aspects of fire safety and hazardous zones are common between the IGF and the classification societies that were perused for the purpose of this section.

3.2 *Bunkering Station*

According to the IGF code [8], the bunkering station must be located on an open deck to ensure adequate ventilation (Part A-1, Sec.8, Reg. 8.3.1.1) and to have an arrangement that limits any spilled fuel (Part A-1, Sec.8, Reg. 8.3.1.3). It is obligatory for the bunkering station either it's situated in an enclosed or semi-enclosed space, a risk assessment to be conducted (Part A-1, Sec.8, Reg. 8.3.1.1). The bunkering station must be separated by A-60 class partitions from machinery spaces (Category A), accommodation spaces, control spaces, and other areas with high fire risk (Part A-1, Sec.11, Reg. 11.3.6). The control of the bunkering station must be possible both remotely and from the bunkering station itself, using remote-controlled valves (Part A-1, Sec.5, Reg. 5.5.1). A manual and a remote operated valve in series must be installed at each connection point of all bunkering lines (Part A-1, Sec.8, Reg. 8.5.3). In both cases, the pressure, temperature, and level of the tank must be monitorable, and in case of high-level alarm the pressure safety valve (PSV) will be activated (Part A-1, Sec.15, Reg. 15.4.2.2), or in case overfill, an alarm should be triggered, activating the automatic Emergency Shutdown System (EMS) (Part A-1, Sec.5, Reg. 5.5.1). In any case of release, or stoppage of the ventilation, during bunkering operation, an audible and visible alarm must be triggered, and EMS activated (Part A-1, Sec.5, Reg. 5.5.2 & 3). Necessary means must be provided to allow the safe release of pressure from the bunker suction pumps and lines into the fuel tank or to another safe location (Part A-1, Sec.8, Reg. 8.3.1.4). Following the same mentality, any potential damage to the piping system should not result in damage to the fuel containment system (Part A-1, Sec.8, Reg. 8.3.1.2), hull, or deck structure, as the low-temperature fuel can be released in the form of a cold jet (Part A-1, Sec.8, Reg. 8.3.1.5 & 6) and cause structural damage to the vessel. The bunker lines should be designed in such a way that they can be purged with inert gas (Part A-1, Sec.8, Reg. 8.5.1), drained after the end of bunkering operation (Part A-1, Sec.8, Reg. 8.5.4), gas-free (Part A-1, Sec.8, Reg. 8.5.5) and isolated appropriately (in case of cross-over lines) (Part A-1, Sec.8, Reg. 8.5.6).

As for the flexible hoses, these need to be compatible with the temperature of the fuel (Part A-1, Sec. 8, Reg. 8.3.2.1) and capable of withstanding a bursting pressure equal to five times the pressure during bunkering (Part A-1, Sec. 8, Reg. 8.3.2.2). The bunkering manifold should be capable of enduring loads and be equipped with dry breakaway/ quick release mechanism (Part A-1, Sec. 8, Reg. 8.4.1). The code also states the necessity of installing a dry chemical powder fire extinguisher system, capable of discharging 3.5 kg/sec for 45 seconds, as a safety measure (Part A-1, Sec.11, Reg.11.6.1).

In the rule set of Hydrogen-fueled vessels of Bureau Veritas (BV) [10], it is additionally specified that entrances and openings should not face bunkering stations (Ch.1 Sec.3 Reg.1.6.2), and the flexible hoses need to be supported to meet bending radius requirements, such as avoiding ground/deck contact, and allow for vessel's and bunkering unit movements (Ch.1 Sec.6 Reg. 3.1.1). Lloyd's Register (LR) [11] categorizes the requirement for hydrogen transfer hoses to be double-walled and vacuum insulated (Sec. 8, Reg. LR 8.3.2–02), and bunkering manifolds should be insulated to prevent gas condensation/solidification (Sec. 8, Reg. LR 8.4–02).

3.3 Tank Connection Space

As per IGF [8], Tank Connection Space (TCS) characterizes all the spaces, which are the peripheral areas of all tank connections and the fuel tank (Part A, Sec.2, Reg.2.2.15). The code specifies that the ventilation system must be of the extraction type, with 30 air changes per hour capacity, and fail-safe fire dampers must be installed. The Bureau Veritas rules [10] indicate that the enclosed space should be gastight, and it may still require enclosure within a tank connection space (Ch.1 Sec.3 Reg. 1.4.1). The tank connection space's leakage resilience must undergo thorough assessment based on defined leakage scenarios, evaluating released/leaked hydrogen quantity, pressure build-up, hydrogen concentration, and temperature variations in the containment and hold spaces (Ch.1 Sec.3 Reg.1.4.2) and must prevent hull exposure to extreme cooling; bulkhead materials shall match lowest probable temperature of the fuel (Ch.1 Sec.3 Reg.1.4.3). The enclosed space must withstand maximum pressure from leaks while venting to a mast is necessary for relief (Ch.1 Sec.3 Reg.1.4.4). Access to the TCS is restricted, unless directly from open deck; secured with bolted hatch or gastight door with airlock (Ch.1 Sec.3 Reg.1.4.5). As far as the required adequate forced ventilation is concerned, it is necessary to maintain gas concentrations below 25% of the Lower Explosive Limit (LEL) in worst-case scenarios (Ch.1 Sec.4 Reg.1.4.1). Fans must ensure continuous ventilation, with redundancy in case of fan failure, including the catastrophic ventilation fans (Ch.1 Sec.4 Reg.1.4.2).

LR rules [11] categorize the regulations concerning the tank connection space according to the fuel's state, whether in a gaseous or liquid form. For both gas and liquid systems, LR indicates that the TCS should be designed to include only the appropriate equipment (Sec. 5 Reg. LR 5.7-02, Sec. 5 Reg. LR 5.8-02), and the volume of the space should be minimized (Sec, 5 Reg. LR 5.7-03, Sec, 5 Reg. LR 5.8-03). The rules also provide an alternative to inert the space instead of having ventilation (Sec. 13 Reg. LR 13.4-02 & -03). For gaseous hydrogen systems, the space should be located on the main deck to ensure unobstructed flow in the event of a dispersion scenario; otherwise, an alternative design safety assessment should be conducted (Sec, 5 Reg. LR 5.7-01). Regarding the piping system of liquid hydrogen, it is stated that the pipes connecting the master isolation valve to downstream components should be double-walled and either inserted or vacuum insulated (Sec, 5 Reg. LR 5.8-04). The piping sections most prone to trapping liquefied hydrogen should be equipped with pressure relief valves (Sec. 5 Reg. LR 5.8-06).

3.4 Fuel Preparation Room

In accordance with IGF [8] the Fuel Preparation Room (FPR) is the enclosed space which include all the necessary equipment (pumps, compressors and vaporizers) for conditioning the fuel before transferring to the consumers (Part A, Sec. 2, Reg. 2.2.17) and shall be located on the deck (Part A-1, Sec.5, Reg. 5.8) and considered as a Type A machinery space (Part A-1, Sec.11, Reg. 11.3.1). If the FPR is located below deck, its entrance should have direct access from the deck or alternatively, an airlock should be provided (Part A-1, Sec. 5, Reg. 5.11.2). The airlock must be designed with a simple geometrical shape, have an area of not less than 1.5 square meters (Part A-1, Sec.5, Regulation 5.12.4), and be able to withstand critical scenarios without releasing any gas into any safe zone (Part A-1, Sec. 5, Reg.5.12.3). It should feature gastight bulkheads and doors spaced between 1.5 and 2.5 meters apart and at least 300 mm high, with self-closing mechanisms (Part A-1, Sec. 5, Reg.5.12.1), and be equipped with over-pressured ventilation (Part A-1, Sec. 5, Reg.5.12.2). The fuel preparation should be equipped with mechanical ventilation, should operate continuously whenever the equipment within the room is in operation (Part A-1, Sec. 13, Reg. 13.6.3) and be capable of achieving 30 air changes per hour (Part A-1, Sec. 13, Reg. 13.6.1) Furthermore, the system should be designed to ensure that the redundancy of the ventilation system is not reduced by more than 50% in the event of a fan or group of fans becoming inoperable (Part A-1, Sec. 13, Reg. 13.6.2), and in the event of a loss of under pressure, access to the space must not be restricted (Part A-1, Sec. 5, Reg. 5.12.6). The fuel preparation room must be equipped with gas detectors rated at 20% LEL (Part A-1, Sec. 15, Reg. 15.8.1.4).

As per BV [10], fuel preparation rooms must contain cryogenic leaks safely, with materials chosen for the lowest probable temperature. The FPR must prevent hull exposure to extreme cooling and endure maximum pressure from leaks, with optional venting to a safe area. Shafting penetrations through bulkheads or decks must be gastight (Ch.1 Sec.3 Reg.1.5. 3–5). Also, the mechanical ventilation system shall maintain a gas concentration below 25% LEL in all probable leakage scenarios (Ch.1 Sec.4 Reg.1.5.1).

The American Bureau of Shipping (ABS) rule for hydrogen-fueled vessels [12] specifies that in Fuel Preparation Rooms where a potential explosion due to leakage is highly likely, consideration should be given to the possible installation of explosion-resistant structural protection. This could include reinforced steel bulkheads and decks, energy-absorbing structural arrangements, or the integration of barricade structures (Sec.4, Reg. 3). Drip trays and spray shields, or equivalent measures, should be installed where cryogenic hydrogen leakage may occur. These arrangements should prevent unacceptable cooling in the event of cryogenic leakages, considering the probable maximum leakage scenario (Sec. 5 Reg. 6.3).

An additional automatic shutdown valve may serve as the fuel tank outlet valve required by the Marine Vessel Rules if it contains the master fuel valve. Both valves should be positioned outside the machinery space, with the FPR having an additional automatic shutdown valve (Sec. 9 Reg. 5.2). The ventilation system for the FPR must be independent of other ventilation systems (Sec. 13 Reg. 6.1.1). Air inlet openings should be positioned as low as possible in the space, while exhaust openings should be at the highest point and opposite sides to the air inlet openings to prevent hydrogen accumulation. Ventilation should initiate circulation from the bottom and exhaust from the top (Sec. 13 Reg. 6.1.3).

3.5 Piping, Fittings, and Materials

As per IGF [8] and specifically Section 5, the fuel piping regulations specify that pipes should be positioned at least 800 mm away from the ship's side and that they need to be avoided to pass through accommodation spaces, service areas, and electrical rooms. Additionally, piping in ro-ro spaces and on open decks must be shielded from mechanical damage. Gas fuel piping within the emergency shutdown system (EMS) protected machinery spaces must be placed away from electrical installations and tanks holding flammable liquids (Part A-1, Sec. 5).

Fuel pipes passing through ship enclosure spaces require secondary protection, such as ventilated ducts or double-wall piping systems, meeting ventilation standards and equipped with gas detection (Part A-1, Sec. 9, Reg. 9.5.1). In gas-safe machinery spaces, fuel piping must be enclosed within either pressurized with inert gas double pipes or under- pressure ventilated ducts (Part A-1, Sec. 9, Reg. 9.6.1). Gas fuel supply to consumers must not exceed 1.0 MPa in EMS -protected machinery spaces (Part A-1, Sec. 9, Reg. 9.7.2). The design pressure of outer pipes or ducts in fuel systems must be equal or exceed the maximum working pressure of inner pipes (Part A-1, Sec. 9, Reg. 9.8.1), taking into consideration the maximum built-up pressure and the local instantaneous peak pressure.

The gas supply systems must feature fail-to-close and fail-to-open ventilation valves to ensure safety during emergencies. Moreover, these systems must be capable of utilizing double block and bleed valves not only for emergencies but also for routine engine shutdowns (Part A-1, Sec. 9, Reg. 9.4.6). The automatic ventilation must be activated downstream of the double block and bleed valve upon shutdown of the master gas fuel valve (Part A-1, Sec. 9, Reg. 9.4.7). To guarantee safety during maintenance, each engine's gas supply line must be equipped with a manually operated shutdown valve (Part A-1, Sec. 9, Reg. 9.4.8). In installations with multiple engines, integration of the master gas fuel valve and double block and

Table 2 Hydrogen piping system [10]

Piping system		Design conditions		Single wall arrangement	Double wall arrangement	
		Design pressure	Design temperature		Inner pipe	Outer pipe
Gaseous H_2	Low-pressure lines	$P = 10$ bar	Any	I	II	II
	High-pressure lines	$P > 10$ bar	Any	I	I	II
Liquefied H_2		Any	Any	I	I	II

bleed valve functions is permitted (Part A-1, Sec. 9, Reg. 9.4.9). Furthermore, main gas supply lines entering EMS protected spaces must incorporate rupture detection systems with automatic shutoff valves (Part A-1, Sec. 9, Reg. 9.4.10).

Piping carrying low-temperature fuel must be insulated to prevent moisture condensation (Part A-1, Sec. 9, Reg. 7.3.1.4). High-pressure fuel piping systems must be structurally strong, confirmed through stress analysis considering factors like weight, acceleration, internal pressure, and ship movement (Part A-1, Sec. 9, Reg. 7.3.4.4). For temperatures below −110 °C, a comprehensive stress analysis is required (Part A-1, Sec. 9, Reg. 7.3.4.5). Piping from tanks to release valves must meet safety standards equivalent to type C tanks, with dynamic stress limitations (Part A-1, Sec. 9, Reg. 6.3.6). Inert gas supply lines must feature double block and bleed valves to prevent gas return, with additional non-return valves, all placed outside non-hazardous spaces (Part A-1, Sec. 9, Reg. 6.13.2). Pressure Relief Valves (PRV) must be able to discharge an amount determined by either the maximum capacity of the inerting system or vapors generated under fire exposure, with no more than a 20% rise in tank pressure (Part A-1, Sec. 9, Reg. 6.7.3.1.1).

As per BV rules [10] for hydrogen-fueled vessels, piping systems are categorized into classes I, II, and III. These classifications are crucial for determining material acceptance, joint selection, heat treatment, welding procedures, pressure testing protocols, and certification of fittings (Table 2).

BV's rulebook (Ch.1, Sec.9, Reg. 4.15) specifies that handling liquefied hydrogen and cold hydrogen vapor piping systems must include thermal insulation. This requirement aims to minimize heat leakage and ensure personnel safety by preventing accidental contact with cold surfaces.

As per DNV, the materials selected for either LH_2 or CH_2 shall be able to withstand both the physical and chemical properties of H_2, especially in terms of Hydrogen Induced Stress Cracking (HISC) [6]. As an initiation point for the proper design of the piping system can be used the below:

- ISO 15156 Part 1, Part 2, and Part 3. Petroleum and natural gas industries—Materials for use in H_2S-containing environments in oil and gas production.
- DNV-RUSHIP Pt.2 Ch.1 to Ch.4 Materials & welding.

The standards and rule sets shall be used for reference when designing them:

- IMO Resolution MSC 420(97): Interim recommendations for the carriage of liquefied hydrogen in bulk.
- IGF Code.
- ISO/TR 15916: Basic considerations for the safety of hydrogen systems.
- EN 13480 Metallic industrial piping.
- ASME B31.3 Process Piping.
- ASME B31.12 Hydrogen Piping and Pipelines.
- EN ISO 5817 Welding—Fusion-welded joints in steel, nickel, titanium, and their alloys (beam welding excluded)—Quality levels for imperfections.
- ISO 10675 Non-destructive testing of welds—Acceptance levels for radiographic testing.
- ISO 11666 Non-destructive testing of welds—Ultrasonic testing—Acceptance levels.
- EN 1779 Non-destructive testing—Leak testing—Criteria for method and technique selection.
- EN 13184 Non-destructive testing—Leak testing—Pressure change method.
- EN ISO 20485 Non-destructive testing—Leak testing—Tracer gas method.
- RUSHIP Pt.2 Ch.1 to Ch.4 Materials & welding.
- DNV-RU-SHIP Pt.5 Ch.7 Liquefied gas tankers.
- DNV-RU-SHIP Pt.4 Ch.6 Piping systems.
- DNV-RU-SHIP Pt.4 Ch.7 Pressure equipment.
- DNV-RP-D101 Structural analysis of piping systems (DNV, 2017a).

Attention shall always be given to the design loads of the piping system, irrespectively of LH_2 or CH_2. Improper design (stress) loads will lead to failures and extremely hazardous scenarios. The prescriptions within the IGF Code provide a solid starting point for both LH_2 and CH_2, which could be further supplemented via other Codes and Standards, always in lieu of updated rules (by the Administration—IMO). Relevant P&IDs (piping and instrumentation drawings) shall include all components (valves, pressure vessels, disconnection couplings, and instrumentation), along with specifications of design pressure, temperatures, and type (phase) of the operating medium. The evaluation of the stress loads and the piping system's flexibility can be supplemented through the recommended practices of RP-D101 Structural analysis of piping systems.

Particular attention shall also be paid to thermal expansion under all loads, especially cryogenic ones, including all green loads. Double wall pipes are necessary. Furthermore, the outer piping shall be evaluated through stress analysis, including expansion in case of accidental release from the primary piping (Fig. 2).

All operations mentioned above shall be risk-based, necessitating relevant hazard identification and risk assessment(s). Regarding fabrication and welding, DNV Rules Part 2 Ch. 1 to 4 provide adequate information but only as an initiation point. Additionally, the IGF Code might also provide requirements for welding, post-welded heat treatment, and Non-Destructive Testing (NDT). All relevant tests and inspections shall also be part of an approved Inspection and Test Plan (ITP),

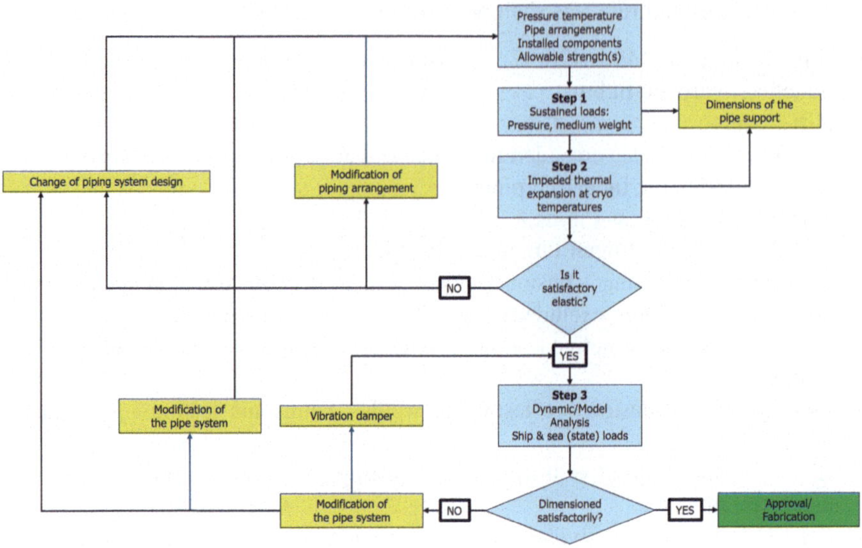

Fig. 2 Recommended approach for the analysis of pipe stress. (Adapted from: [6])

while the quality control of all components could be part of a Quality Control Plan (QCP).

Particular attention shall be given to critical components of the piping, such as emergence/process shutdown valves (ESD/PSD) and pressure relief valves (PRV). Failure of those components would surely lead to hazardous scenarios.

Applicable existing standards and DNV rules for all valve types are provided in:

- EN 12516–1/−2/−3/−4 Industrial valves.
- Part 1—Shell design strength.
- Part 2—Calculation method for steel valve shells.
- Part 3—Experimental method.
- Part 4—Calculation method for valve shells manufactured in metallic materials other than steel.
- EN 13445 Unfired pressure vessels.
- ASME B 16.34 Valves Flanged, Threaded, and Welding End.
- DNV-CP-0186 Class programme DNV-CP-0186, Valves.

Concerning PRVs, the following apply:

- EN ISO 21028-1 Cryogenic vessels—Toughness requirements for materials at cryogenic temperatures—Part 1: Temperatures below −80 °C.
- ISO 4126 Safety devices for protection against excessive pressure—Part 1: Safety valves & Part 4: Pilot-operated safety valves.
- EN 13648-1 Cryogenic vessels—Safety devices for protection against excessive pressure—Part 1: Safety valves for cryogenic service.

- ISO 11114-1. Gas cylinders Compatibility of cylinder and valve materials with gas contents—Part 1: Metallic materials.
- ISO 11114-2. Gas cylinders Compatibility of cylinder and valve materials with gas contents—Part 2: Non-metallic materials.
- ISO 21013-1. Cryogenic vessels Pressure relief accessories for cryogenic service—Part 1: Re-closable pressure relief valves.
- ISO 21013-3. Cryogenic vessels Pressure relief accessories for cryogenic service—Part 3: Sizing and capacity determination.
- ASME VIII-1, Div.1 ASME Boiler and Pressure Vessel Code.
- ASME B 16.34 Valves Flanged, Threaded, and Welding End.
- API 520 Sizing, Selection, and Installation of Pressure-relieving Devices.

3.6 Hydrogen Containment System(s)

As mentioned earlier, the IGF Code does not include hydrogen as a fuel; therefore, there are no relevant prescriptions for safe storage on board ships. To that end, this section elaborates on available prescriptions from the classification societies that were investigated. Specifically, for DNV, there are currently no rules on the safe storage of hydrogen, but as per [6], some valuable guidelines are provided. These are allocated according to the storage conditions of hydrogen, namely liquified (LH_2) and compressed (CH_2). Before delving into the two kinds, IGF prescribes some general requirements, regardless of the state of the storage medium.

As per the IGF Chap. 6: Fuel Containment System, tank containment systems situated below the deck are required to be gastight toward adjacent spaces. The fuel containment system is an integral part of the risk assessment, as prescribed in Chap. 4. Tank connections and relevant fittings (valves, flanges, etc.) are also required to be in enclosed gat-tight TCSs, except in cases where the tank is located on an open deck. TCS bulkhead materials shall be able to tolerate the properties of the stored medium and also any pressure build-ups during leakage. Needless to say, the explosive properties of hydrogen are much worse than those of LNG. A common factor across both storage arrangements and all tank types is the means for pressure release. These are not cited nor elaborated as these pertain to LNG and Compressed Natural Gas (CNG) only, which have different hazard levels, if not lower, than LH_2 and CH_2; therefore, these have been omitted from the review.

In terms of underdeck position placement, Reg. 5.3.3 provides prescriptions for the protection of the storage tanks. These are offered in two modes, a prescriptive one and a probabilistic one, presented in Table 3. The prescriptive ones are straightforward, necessitating minimum distances from the side, the bottom, and the shell plate. These are B/5 or 11.5 m (whichever is less), B/15 or 2 m (whichever is less), and a distance from the shell plate "d," where $d = 0.8–2.0$ m (for cargo ships), respectively. The probabilistic approach follows the mentality of the probabilistic damage stability regulation, allowing for lesser distances if risk allows so by considering the probability of damage.

Table 3 Comparison of the minimum tank to shell plating distance requirements

Deterministic approach	*Probabilistic approach*
Minimum distance from side: B/5 or 11.5 m (whichever is less)	The probability of tank damage for cargo ships $f_{CN} < 0.04$, where $f_{CN} = f_l \times f_t \times f_v$
Minimum distance from bottom: B/15 or 2.0 m (whichever is less)	f_l = probability of collision damage in the longitudinal direction
Minimum distance from shell plate: "d"	f_t = probability of inboard penetration accounting for available side protection
$d = 0.8$–2.0 m for cargo ships, depending on tank size	f_v = probability accounting for vertical extent of damage
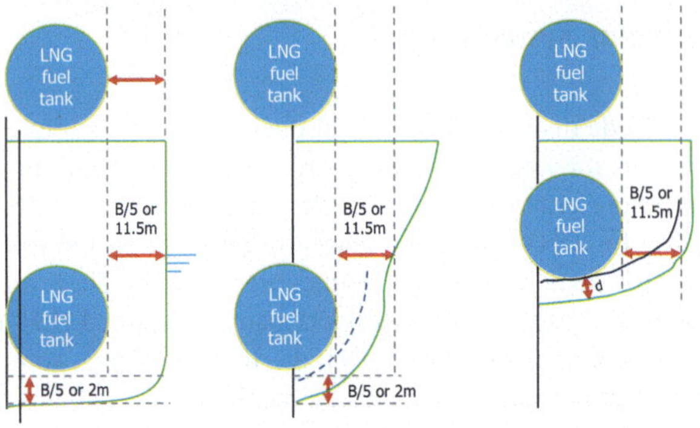	Formulations are based on and refer to SOLAS Regulations II-1/7-1 and 7-2.6, used for damage stability calculations of cargo and passenger ships.
	Factor f_l is calculated using the formulations of the p-factor in SOLAS Regulation II-1/7-1.
	Factor f_t is a function of the distance between the tank and the side shell.
	Factor f_v is a function of the distance from the deepest draught to the lower-most extreme boundary of the LNG tank.

Adapted from: [9]

According to DNV, which has not yet released any rules for the safe storage and handling of hydrogen, the tank location distances, as prescribed in the IGF, shall not be used directly. All hydrogen fuel tanks fitted on board a vessel require approval in principle and shall be dealt with on a case-by-case basis. However, DNV provides general guidance towards the placement of those and, in turn, the obtainment of AiP. Specifically, DNV advises that the following rules/circulars/documents provide a solid basis for the AiP:

- IGF requirements for methanol (apply those for H_2).
- IMO Res MSC.420(97): Interim recommendations for carriage of liquefied hydrogen in bulk.
- Other standards (pertinent to LH_2 and CH_2—cited further down below).

3.6.1 Liquified Hydrogen Storage (LH_2)

Liquified gas fuel shall have arrangements such that it can be safely emptied from the storage tanks (Reg. 6.3.11). In addition, the fuel tanks shall be able to be interested, vented, and purged if required. Furthermore, the worst environmental conditions that should be taken into account during the design is that of the North Atlantic.

Adequate access must be available to inspect the containment system, while types A and B are required to have a complete secondary barrier. Table 4 presents the secondary barrier necessities according to the type of tank.

In terms of the safety principles of the liquified storage tanks, cited in Ch. 6.4.2, these pertain to the hazards of LNG. The secondary barrier must be able to withstand all leakages that may develop, including the thermal insulation/lagging that is required to keep the LNG in storage conditions. Prescriptions are also available to reduce or omit the secondary barrier, subject to alternative design. Specifically, the secondary barrier can be partial, where small leaks may develop, while these are, in theory, handled and disposed of by a safety system. Extensions of this rule to hydrogen cannot be foreseen by the Authors of this chapter, keeping in mind its volatility. As presented in Table 1, the secondary barrier itself shall be able to withstand leaked of the (liquified) gas fuel for 15 days. Mechanical damages shall not jeopardize the integrity of the secondary barrier, and failure of supporting elements on the hull structure shall not harm the function of the tank.

DNV advises that inspiration can be drawn from the IGF and the International Code of the Construction and Equipment of Ships Carrying Liquefied Gases in Bulk (IGC Code) [13]. Currently, there is a pilot project which allows the BULK carriage of hydrogen between Australia and Japan, but these are not part of the IGC; therefore, there are no prescriptions to use LH_2 as fuel on IGC, although the ship might be carrying it as a cargo, with no potential to be used a boil-off gas. LH_2 vented from cargo spaces is just lost to the atmosphere.

3.6.2 Compressed Hydrogen Storage (CH_2)

Compressed storage in the IGF is described in Ch. 6.4.15.3, obviously having in mind compressed natural gas (CNG), covering design pressures, loads, structural arrangements, etc. Chapter 6.6 elaborates on the pertinent regulations for CNG. The most important prescription is that compressed CNG tanks shall not be located in enclosed spaces except for cases where depressurization and inertion means are

Table 4 Requirements for the secondary barrier

<table>
<tr><th>Basic tank type</th><th>Secondary barrier requirement</th></tr>
<tr><td>Membrane</td><td rowspan="2">Complete secondary barrier</td></tr>
<tr><td>Type A</td></tr>
<tr><td>Type B</td><td>Partial secondary barrier</td></tr>
<tr><td>Type C</td><td>No secondary barrier required</td></tr>
</table>

Adapted from: [8]

provided, surrounding surfaces and insulation (TCS) are able to withstand releases from the tank itself, and fixed fire-fighting systems are installed, applicable and able to fight CNG fires, including jet ones.

DNV segregates different types of compressed storage in terms of construction materials. Namely, DNV provides guidance, as in the LH_2, about steel and composite cylinders.

Steel cylinders shall be designed according to Class 1 cylinder prescriptions in Rules for Ships Pt4 Ch7 [14] or CNG cylinders as per Rules for Ships Pt5 Ch8 [15]. Also, the material used shall not be able to react or be permeable with hydrogen. Composite cylinders can be designed as per Rules for Ships Pt.5 Ch.8, and alternative standards may be accepted provided that these are designed conservatively against the "standard."

Furthermore, DNV suggests that guidance can be drawn from generic land-based rules covering CH_2 for existing pressure vessels, such as:

- Road transport of CH_2 governed by UN Model Regulation.
- EU Agreement on International Carriage of Dangerous Goods by Road (ADR).
- European Transportable Pressure Equipment Directive 2010/35/EU-"TPED."
- Seveso III Directive 2012/18/EU.
- EU standards for pressure vessels more than 0.5 bar on Pressure Equipment Directive (PED).
- EN 1252–1:1998—storage tank materials.
- EN 1797:2001 on the compatibility between the material and the gas.
- EN 13648 parts 1, 2, and 3 on safety devices concerning the protection against excessive pressure.
- ASME and NFPA standards.

4 Current Status of Hydrogen Storage and Handling in the Maritime Industry

4.1 Alternative Design and Arrangements (AD&A) and Risk Acceptance Criteria

According to the IGF Code Part A, a low-flashpoint fuel like hydrogen is allowed as long as the Alternative Design approach demonstrates that the hydrogen-specific systems are as safe, reliable, and dependable, being new and comparable to conventional oil-fueled ships. SOLAS regulation II-1/55 specifies how this risk equivalence shall be demonstrated, and this needs to be approved by the Administration. SOLAS regulation II-1/55 points to the method specified in MSC.1/Circ 1455 [16].

On the other hand, DNV, in lieu of regulations relevant to hydrogen-fueled ships, offers some guidance in the form of an additional notation, namely FC, in Section 3 of DNV-RU-SHIP Pt.6 Ch.2 [17]. The scope of Section 3 of the aforementioned set of rules is to provide requirements regarding the design and arrangement of FC power installation and the FC space itself. Items such as fuel reformers, converting liquid or gaseous primary fuels to hydrogen-rich gas, relevant control, monitoring, and safety systems, as well as materials, manufacturing, workmanship, and testing, are covered. Moreover, it clearly specifies that since these systems have not yet been approved by the IMO, additional acceptance by Flag authorities must be sought, whilst the fuel cell rooms will require acceptance on a case-by-case basis, both showcasing the maturity of the systems. There is no segregation on what the primary source of hydrogen for the fuel cell is, for example, pure hydrogen, ammonia, or other, and the majority of the prescriptions are focused on the fuel cell room itself. Piping requirements, ventilation (inlet, exhaust), and design principles are all focused on the enclosure containing the fuel cell(s).

The current status of the AD&A, as required in the IGF, is rather challenging to implement and may lead to very comprehensive but also expensive and potentially overdesigned solutions. An overview of the approval process is summarized in Figs. 3 and 4. The former presents an overview of the approval process for the preliminary design, showcasing the role of the submitter and the Administration, while the latter showcases the process for the final design. The procedure as a whole is not as short or simple as depicted in the figures. However, they offer considerable aid in understanding the complex process of AD&A. Needless to say, the submitter should approach the Administration as early as possible. On the other hand, the requirements are not exact, and these are treated on a case-by-case basis. Also, hydrogen is not explicitly mentioned in the AD&A, but it is not excluded either, paving the way forward for its safe uptake.

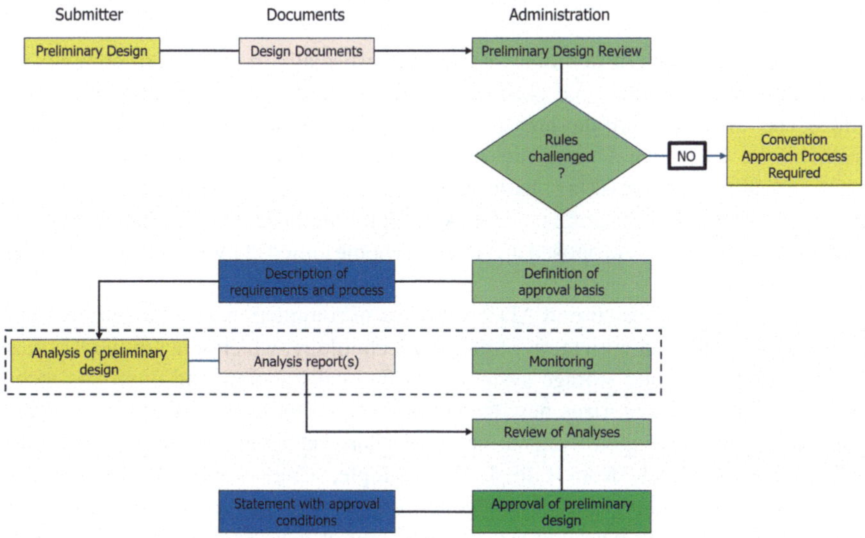

Fig. 3 Overview of the approval procedure for the preliminary design as per the AD&A, showcasing the roles of the stakeholders. (Adapted from: [6])

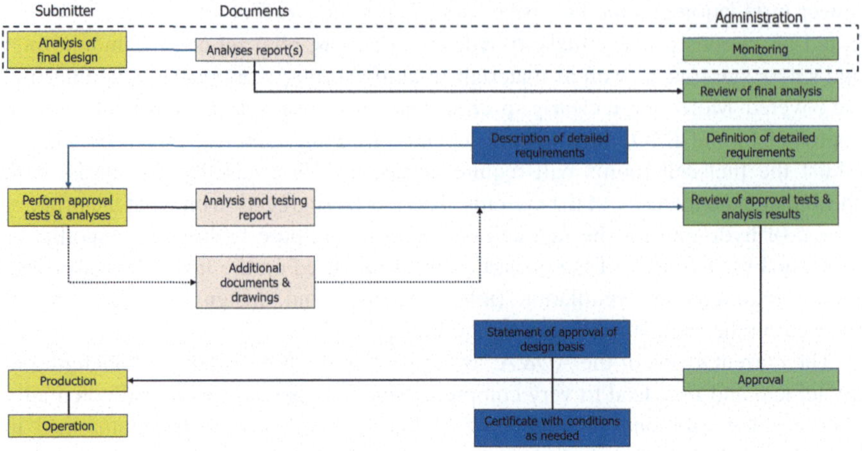

Fig. 4 Overview of the approval procedure for the final design as per the AD&A, showcasing the roles of the stakeholders. (Adapted from: [6])

Usually, the AD&A procedure is distinguished into two phases, namely, phase 1, incorporating the preliminary design with two distinct milestones (1 and 2), and phase 2, which incorporates the final design with another two pertinent milestones (3 and 4). The milestones are as follows:

- Development of preliminary design,
- Approval of preliminary design,
- Development of final design, and
- Approval.

Usually, a plethora of iterations is required, especially in the case of complex or novel designs, while the goal is to prove safety equivalence with the prescriptive rules. The latter is where the greatest challenge lies, hence why governing bodies, such as the European Union, are funding quite a few research projects, cited later on, aimed at derisking hydrogen. One of those projects, led by DNV, namely MarHySafe [6], aiming to derisk and provide insight on the safe uptake of hydrogen on board, has proposed an AD&A scheme aimed at hydrogen, presented in Figs. 5 and 6.

Another pitfall of the current AD&A process in conjunction with the introduction of hydrogen is the categorization of the new technology, which inherently classifies hydrogen handling and storage systems and fuel cells as new and unproven technologies. Table 5 categorizes new technology according to the AD&A procedure, explaining why hydrogen systems are particularly challenging, with a difficulty rating of 4 on a scale from 1 to 4. For example, a new application of existing technology, such as borrowing knowledge from other industries with experience in hydrogen systems and handling (i.e., automotive) or new technology itself, is also categorized as 4.

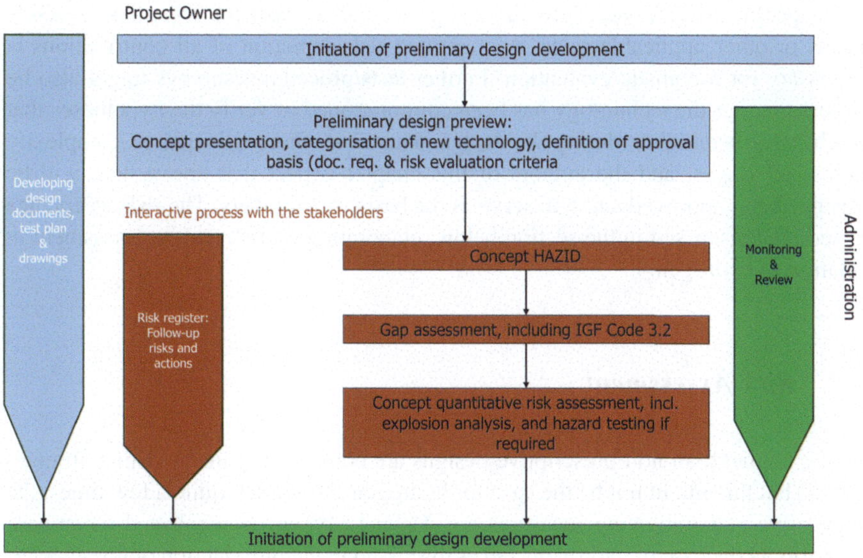

Fig. 5 Proposed steps towards the approval of the preliminary design as per the AD&A procedure aimed at hydrogen. (Adapted from: [6])

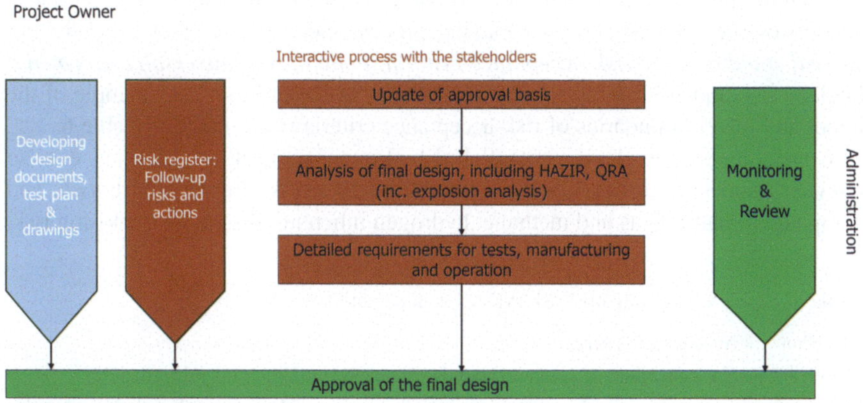

Fig. 6 Proposed steps towards the approval of the final design as per the AD&A procedure aimed at hydrogen. (Adapted from: [6])

Table 5 Categorization of new technology as per the AD&A

	Technology status		
Application area	Proven	Limited field history	New or unproven
Known	1	2	3
New	2	3	4

Source: [16]

As per the AD&A procedure, these require a risk assessment, regardless of IGF's needs or other applicable rules. A quantified risk depiction of all contributions is necessary for a realistic evaluation. Further tests/procedures/surveys might also be necessary after the technology has been commissioned to verify the hypotheses that might have been taken during the design. Therefore, there is increased complexity across all stages, and the authors of this chapter expect that one system will be designed and approved and will serve as the basis for all others. The risk acceptance criteria are discussed in the section below, elaborating on risk and its assessment in relation to hydrogen.

4.2 *Risk Assessment*

At the moment, all non-prescriptive designs are compared against existing arrangements, but this might not be the case for hydrogen. Moreover, quite a few times, the preventive and mitigating measures are designed based on a selected worst-case scenario. In the case of hydrogen, and as discussed in the introduction of this chapter, a small leak may lead to devastating and totally unacceptable scenarios. Risk-based approaches are expected to be at the heart of the AD&A process, meaning that the Formal Safety Assessment (FSA) is still relevant and extremely useful, providing a structured, systematic procedure for assessing the safety of design [18]. The latest version of the FSA for the rule-making process clearly states that "*…there are several standards for risk acceptance criteria, none as yet universally accepted.*" Societal risk and FN curves are still at the heart of the process. An example of the upper and lower boundaries of risk-acceptance criteria is presented in Table 6.

Considering the explosive potential of hydrogen, societal risk becomes ever so relevant, especially in cases of passenger ships. In terms of assessing its risk, and compared to natural gas and methane, hydrogen inherently has more explosion risk.

Table 6 Example of quantitative risk evaluation upper and lower bounds

Decision parameter	Acceptance criteria		
	Crew member	Lower bound—ALARP	Upper bound—ALARP
		Negligible (broadly acceptable) fatality risk/year	Maximum tolerable fatality risk/year
Individual risk		10^{-6}	10^{-3}
	Passenger	10^{-6}	10^{-4}
	Third parties, public ashore	10^{-6}	10^{-4}
	Target values for new ships	10^{-6}	The above values to be reduced by one order of magnitude
Societal	To groups of the above persons	To be derived by using economic parameters ref.: (MSC 72/16, 2000)	

Source: [18]

Also, there is a lack of leak frequency data concerning hydrogen. When hydrogen leaks out of a high-pressure vessel/system, it exhibits sonic behavior, while the speed of sound and density may cause different behavior. This is owed to hydrogen having four times the speed of sound and eight times lower density than natural gas if one considers releases from identical piping and leak diameters. Despite that, it exhibits a lower mass release rate. Concerning large hydrogen storage vessels, the cloud realized by a possible dispersion will be greater, bearing in mind the total gas volume at a concentration larger than the flammable one.

Under a hypothetical scenario of a potential high-pressure release in a confined space, a plume will be produced via a partly impinged jet zone, and a passive cloud will disperse before it is further spread, depending on the confined space. Due to the higher flammability range of hydrogen, compared to natural gas, the possibility of forming a larger flammable cloud is always higher, provided that the concentration is lower than 5–10% and that it does not come into contact with a hot surface or ignitor. Smaller hydrogen leaks do not always have the potential to explode, but they may lead to a fire, which is highly dependent on the ventilation arrangements and atmosphere conditions inside the confined space. In contrast, a larger cloud may also have the potential to reach hot surfaces or ignitors further away from the point of release.

Regarding potential explosion severity, a large hydrogen gas cloud with high concentration may give way to an explosion that is 5–10 times larger than that of natural gas. Even a small cloud, or a small enclosure, may result in a Deflagration to Detonation Transition (DDT). Moreover, hydrogen explosions are realized in a quicker manner. If blast walls or blast panels are to be used as a risk mitigation measure, protecting against a potential explosion, these should be rated according to the peak pressure as well as the time required to achieve it; otherwise, these are useless.

Concerning functional requirement criteria, Annex G of NORSOK standard Z013 provides a detailed guide on the explosion risk analysis (ERA), utilizing CFD simulations [19]. QRA through ERA allows for the calculation of the F/N risk curve. Results obtained from such analysis, usually done in the offshore industry, are the maximum pressures experienced, which leads to the selection of blast walls and panels, in the way of the Design Accidental Load (DAL). Fatalities possible from the realization of the scenarios are able to be calculated and relevant safety barriers can be stipulated as such. Borrowing from the offshore industry, a frequency of 1.0E-4 could be used for such events. Ships carrying large amounts of passengers may choose to have an even smaller frequency.

4.3 Current Research Projects in the Maritime Industry

As a zero-carbon emission fuel, hydrogen could be a favorable option for the decarbonization of the maritime industry. However, due to its explosive and volatile properties, ensuring large quantities of hydrogen safely stored on-board has proved to be a challenge, which is well reflected by a plethora of EU-funded research projects, aiming at exactly that. Project HyShip aims to design and construct a new ro-ro demonstration vessel using liquid green hydrogen as its fuel and develop an LH_2 supply chain and bunkering platform. Project LH2CRAFT aims to revolutionize LH_2 storage on a massive scale, catering to commercial vessels and emerging applications, all while enhancing safety, modularity, scalability, and environmental sustainability. By introducing harmonized standard vent sizing requirement, the HySEA project aims to conduct pre-normative research on vented deflagrations in enclosures and containers for hydrogen energy applications, to facilitate the safe and successful introduction of hydrogen energy systems (https://cordis.europa.eu/project/id/671461). Another project that aims to develop, validate, and demonstrate a new zero-emission passenger ship powered by hydrogen and the associated hydrogen distribution, storage, and bunkering solution is the project ZEAS.

All the EU-funded projects relevant to hydrogen handling and storage in the maritime industry are presented in Table 7.

Table 7 Hydrogen storage and handling projects currently funded by the European Union

Project	Funder	Objectives	Duration	Website
HyShip	H2020—SOCIETAL CHALLENGES—Smart, Green And Integrated Transport	Build a cargo vessel for commercial operation running on liquid hydrogen, establish a viable liquid hydrogen supply chain and a bunkering platform	January 2021 – December 2025	https://hyship.eu/
LH2CRAFT	Horizon Europe—Climate, Energy and Mobility	Design an innovative membrane-type containment system that is able to store LH_2 at frigid temperatures of −253 °C on a massive scale, catering to commercial vessels and emerging applications, all while enhancing safety, modularity, scalability, and environmental sustainability	June 2023 – May 2027	https://lh2craft.eu/
ZEAS	Horizon Europe—Climate, Energy and Mobility	The project will showcase a system powered by hydrogen fuel cells designed for maritime applications. This system encompasses a zero-emission passenger ship tailored for the Adriatic Sea, along with solutions for hydrogen distribution, storage, and bunkering	January 2024 – December 2027	https://teco2030.no/projects/zeas/

(continued)

Table 7 (continued)

Project	Funder	Objectives	Duration	Website
ELVHYS	Horizon Europe—Climate, Energy and Mobility	International standards related to cryogenic hydrogen transferring technologies for mobile applications (filling of truck, ship, stationary tanks) are missing, and the experience is limited or non-existent. ELVHYS aims to provide indications on inherently safer and efficient cryogenic hydrogen technologies and protocols in mobile applications by proposing innovative safety strategies including selection of effective safety barriers and hazard zoning strategies, which are the results of a detailed risk analysis	January 2023 – December 2025	https://elvhys.eu/
MARANDA	H2020—SOCIETAL CHALLENGES—Smart, Green And Integrated Transport	A mobile hydrogen storage container, refillable in any 350 bar hydrogen refueling station will be developed in this project, to increase hydrogen availability to marine sector as well as many other sectors	March 2017 – March 2022	https://cordis.europa.eu/project/id/735717
RESHIP	Horizon Europe—Climate, Energy and Mobility	Aims at enhancing energy efficiency performance and at addressing the current challenges for hydrogen usage on-board, including high energy demand, abrupt power spikes, demanding energy storage requirements	September 2022 – August 2025	https://www.reship-project.com/
sHYpS	Horizon Europe—Climate, Energy and Mobility	The project will develop a novel LH_2 swappable storage solution, which can be adapted to multiple types of vessels and speed up achieving IMO's target for 2030 and 2050	June 2022 – May 2026	https://www.shyps.eu/

(continued)

Table 7 (continued)

Project	Funder	Objectives	Duration	Website
e-SHyIPS	H2020—SOCIETAL CHALLENGES—Smart, Green And Integrated Transport	The project aims to conduct a regulatory framework review and assess experimental data on hydrogen ship design, safety systems, material and components, and bunkering procedures	January 2021 – December 2024	https://e-shyips.com/
Ocean REFuel	Engineering and Physical Sciences Research Council	Ocean REFuel is a five-year research program, funded with £10 M, focused on converting Ocean Renewable Energy into liquid and gaseous fuels	December 2021 – November 2026	Oceanrefuel-project@strath.ac.uk

References

1. I.M.O. (2023). Resolution MEPC.377(80): The 2023 IMO Strategy for the reduction of greenhouse gas emissions from ships.
2. DNV. (2023). Alternative fuels for containerships.
3. Hansson, J., Månsson, S., Brynolf, S., & Grahn, M. (2019). Alternative marine fuels: Prospects based on multi-criteria decision analysis involving Swedish stakeholders. *Biomass and Bioenergy, 126*, 159–173. https://doi.org/10.1016/j.biombioe.2019.05.008
4. Zhang, J., Delichatsios, M. A., & Venetsanos, A. G. (2010). Numerical studies of dispersion and flammable volume of hydrogen in enclosures. *International Journal of Hydrogen Energy, 35*(12), 6431–6437. https://doi.org/10.1016/j.ijhydene.2010.03.107
5. Astbury, G. R. (2008). A review of the properties and hazards of some alternative fuels. *Process Safety and Environmental Protection, 86*(6), 397–414. https://doi.org/10.1016/j.psep.2008.05.001
6. DNV. (2021). Handbook for Hydrogen Fuelled Vessels.
7. Van Hoecke, L., Laffineur, L., Campe, R., Perreault, P., Verbruggen, S. W., & Lenaerts, S. (2021). Challenges in the use of hydrogen for maritime applications. *Energy and Environmental Science, 14*(2), 815–843. https://doi.org/10.1039/d0ee01545h. Royal Society of Chemistry.
8. IMO. (2015). Resolution MSC.391(95): Adoption of the International Code of Safety for Ships Using Gases or Other Low-Flashpoint Fuels (IGF Code.
9. DNV. Alternative Fuels for Containerships – LNG, Methanol Extended, Ammonia
10. Bureau Veritas. (2023, November). Hydrogen Fuelled Ships – NR678 – November 2023.
11. Lloyds Register. (2023). LR-RU-012 Rules and Regulations for the Classification for Ships Using Hydrogen Fuelled Vessels – Appendices – Appendix LR3 – Requirements for S – LR Part A-1.
12. ABS. (2023). Requirements for Hydrogen Fueled Vessels.
13. IMO. (2014). Resolution MSC.370(93) (Adopted on 22 May 2014) Amendments to the International Code for the Construction and Equipment of Ships Carrying Liquefied Gases In Bulk (IGC Code).
14. DNV. (2021). Part 4 Systems and components Chapter 7 Pressure equipment.
15. DNV. (2021). Part 5 Ship types, Chapter 8 Compressed natural gas tankers.

16. IMO. (2013). MSC.1/Circ. 1455: Guidelines for the approval of alternatives and equivalents as provided for in various IMO instruments.
17. DNV. (2023). Part 6 Additional Class Notations, Chapter 2 Propulsion, Power Generation and Auxiliary Systems.
18. IMO. (2007). Consolidated text of the Guidelines for Formal Safety Assessment (FSA) for use in the IMO rule-making process MSC/Circ.1023–MEPC/Circ.392).
19. Standards Norway. (2010). NORSOK Standard Z-013, Risk and Emergency Preparedness Assessment.

Ammonia, Methanol, Liquid Hydrogen and Gas Hydrogen Utilization on Ships

Yuanqing Zhu, Jie Shi, Chong Xia, Yongming Feng, and Song Zhou

1 Introduction

In 2018, the International Maritime Organization (IMO) introduced a greenhouse gas (GHG) reduction strategy, aiming to reduce GHG emissions from the shipping industry by 50% compared to 2008 levels by 2050. However, in 2023, the IMO revised this target to achieve zero carbon emissions by 2050 [1]. This change underscores the urgent need for GHG emissions reduction. In fact, the shipping industry is facing dual pressures: ship exhaust pollution control and greenhouse gas reduction. The natural disasters caused by the greenhouse effect, the harm of air pollution to human health, and the vision of zero emissions in the shipping industry inspire the maritime industry to take action to protect the Earth's environment.

From the perspective of GHG emissions reduction, although measures such as ship type optimization, waste heat recovery, deceleration navigation, and installation of carbon capture devices have achieved effective carbon reduction effects [2], only by using zero-carbon/low-carbon alternative fuels can IMO's GHG emission reduction strategy be achieved [3–5]. In fact, the shipping industry is actively exploring the feasibility of alternative fuels, and currently, the most promising fuels are hydrogen, ammonia, and methanol [6–10]. Among these three alternative fuels, methanol is the most likely one to achieve significant application in the near future and achieve effective carbon reduction in the coming years. Two perspectives exist regarding methanol. One argues for completely eco-friendly production, ensuring carbon neutrality and minimal environmental impact. The opposite view suggests that a small amount of carbon content in methanol is still effective for decarbonization but can significantly reduce the cost of methanol production. Of

Y. Zhu (✉) · J. Shi · C. Xia · Y. Feng · S. Zhou
College of Power and Energy Engineering, Harbin Engineering University, Harbin, China
e-mail: zhuyuanqing@hrbeu.edu.cn; shi@hrbeu.edu.cn; xiachong@hrbeu.edu.cn; fengyongming@hrbeu.edu.cn; songzhou@hrbeu.edu.cn

C. O. Colpan et al. (eds.), *Fuel Cell and Hydrogen Technologies in Maritime Transportation*, https://doi.org/10.1007/978-3-031-86110-9_2

course, there is no doubt that even in 2050, methanol fuel is expected to play an important role in the maritime industry. The application of hydrogen and ammonia as fuels on ships is considered to be at the forefront of current technology, as both substances are carbon-free. Nevertheless, each of them presents its own set of challenges that restrict their use; ammonia is toxic, and hydrogen gas is difficult to store on ships.

Nonetheless, when considering the issue of pollutant emissions, it appears that alternative fuels like ammonia, hydrogen, and methanol seem to have become more complex and may generate new pollutants, which may require new pollutant emission regulations. For instance, when ammonia, methanol, or hydrogen are used as fuels for internal combustion engines, NO_x pollutants are still emitted, with NH_3 potentially even emitting more NO_x pollutants than conventional fuels. Taking ammonia as an example, zero-carbon ammonia eliminates CO_2 emissions. However, it may emit N_2O, a gas that is stronger than the greenhouse effect of CO_2 and may offset the carbon-free advantage of NH_3. The use of methanol as a fuel may cause emissions of non-conventional pollutants like formaldehyde.

The above analysis is based on the use of internal combustion engines as ship power; fuel cells may be more energy efficient than internal combustion engines. In addition to potential efficiency improvements, fuel cells offer several other advantages, including reduced noise levels, lower maintenance demands, and modular and flexible configurations. However, besides technological challenges, the cost of fuel cells is much higher than that of internal combustion engines. Therefore, its proportion in the shipbuilding market is small.

In summary, although alternative fuels such as ammonia may introduce other pollutants, their carbon reduction benefits are significant. Therefore, it is necessary to have a comprehensive understanding of the current application status of alternative fuels such as ammonia, methanol, and hydrogen on ships. The first section of this chapter is the introduction, the second section is the application of ammonia on ships, the third section is the application of methanol on ships, the fourth section is the application of gaseous hydrogen and liquid hydrogen on ships, and the fifth section is the conclusion.

2 Application of Ammonia on Ships

2.1 Ammonia Fuel

Ammonia is an important basic chemical with extensive applications in the fertilizer industry. Under normal atmospheric conditions, ammonia is present as a transparent gas with a potent, sharp odor. When subjected to higher pressures, it becomes a liquid, which simplifies storage and transportation. Ammonia has the capability to undergo cooling to a temperature as low as −33 °C when subjected to a pressure of 1 bar, enabling its storage in a container in the form of a liquid. Conversely, the liquid state of ammonia can be achieved by applying a pressure of 10 bar to ammonia under conditions of 25 °C. Ammonia's high octane number, which is approximately 130, is ideal for engines that have high compression ratios. This characteristic helps to decrease the likelihood of knock occurring. In theory, the high energy density and thermal efficiency of NH_3 reciprocating engines can be improved to over 50% by adopting a compression ratio of up to 35 and stoichiometric combustion. In addition, NH_3 is an efficient hydrogen carrier, containing 17.6 wt% H_2. Notably, its H_2 density per unit volume often surpasses that of liquid H_2, rendering it a viable alternative to traditional hydrogen carriers. At 293 K and 8.6 bar conditions, the H_2 density in liquid NH_3 (108 kg H_2/m^3) is four times higher than the metal hydride storage method (25 kg H_2/m^3) [11].

Ammonia, on the other hand, exhibits poisonous characteristics. It has the capacity to cause irritation to the eyes, lungs, and skin, even at low levels. At elevated levels of concentration or upon direct contact, it possesses the potential to pose an immediate risk to life. Furthermore, apart from posing a threat to human well-being, ammonia has the potential to cause corrosion in metal substances. When exposed to moisture, ammonia can chemically interact with copper, zinc, and various alloys, resulting in the deterioration of these materials. Consequently, the utilization of ammonia on ships may encounter difficulties due to this corrosive nature.

From the analysis of the physical and chemical properties of fuel, the advantage of ammonia fuel is that it is relatively easy to store as a liquid on board. Its disadvantage is that ammonia is toxic, so research on its protective measures is also important.

2.2 *Marine Ammonia-Fueled Internal Combustion Engine*

The utilization of ammonia as a fuel in engines has been proven to be environmentally friendly when subjected to a comprehensive life cycle assessment [12]. However, the combustion characteristics of ammonia hinder its application in internal combustion engines, and fortunately, there are various solutions available. This section will introduce various ammonia fuel engine technologies discovered in literature, such as pilot diesel ignited ammonia dual-fuel engines and pre-chamber ignition technology.

2.2.1 Ammonia Combustion

Ammonia has a low calorific value (18.6 MJ/kg) and a low maximum laminar burning velocity (0.07 m/s) [13]. Therefore, ammonia combustion generates less energy and is less intense than other fuels, resulting in a slower diffusion rate. Additionally, ammonia has a narrow combustion range (0.63–1.4) and is difficult to ignite, so it is not very suitable for use in engines. Even with the use of high-energy ignition systems, whether in reciprocating engines or gas turbines, ammonia often requires other auxiliary measures to enhance and improve combustion speed when used as engine fuel to ensure engine performance. Many measures to enhance combustion have been studied in the past, such as adding hydrogen, ammonia dissociation before combustion, high compression ratios, and using pilot oil.

During engine combustion, ammonia is more likely to form NO and NO_2 than other fuels. In addition, it is possible to generate N_2O, a potent greenhouse effect substance. The primary objective of utilizing NH_3 in engines is to mitigate CO_2 emissions, as NH_3 does not contain any carbon. Nevertheless, it is essential to analyze the release of N_2O, considering its high global warming potential (GWP). Gill et al. [14] revealed that ammonia increases N_2O emissions of about 1 g/kWh (equivalent to approximately 0.4% of the nitrogen present in the NH_3) on a dual-fuel engine. This increase completely negated the CO_2 emission reduction benefits during combustion. In addition, NH_3 combustion will emit unburned NH_3.

2.2.2 Pilot Diesel-Ignited Ammonia Dual-Fuel Engine

Due to the unfavorable combustion properties of NH_3, ensuring stable engine operation becomes problematic when using it as the sole fuel source. Consequently, researchers have concentrated their efforts on igniting combustion in ammonia fuel engines using pilot oil fuel. Typically, diesel is generally used as the pilot oil fuel, and there are two operational modes:

1. Low-pressure dual fuel (LPDF) mode. In the LPDF mode, NH_3 is mixed with air before combustion, forming a uniform mixture, and ignition is achieved by injecting high-reactivity diesel. LPDF can be achieved through modifications to traditional diesel engines, which have relatively low technical requirements. However, due to quenching effects, valve overlap, and the low flame speed of NH_3 combustion, a large amount of unburned NH_3 is emitted into the atmosphere [15]. We know from the properties of the fuel that ammonia is toxic, so although this mode has the advantage of being simple and easy to implement, it requires additional measures to eliminate ammonia slip.
2. High-pressure dual fuel (HPDF) mode. In the HPDF mode, the cylinder is injected with pilot fuel and liquid ammonia near the top dead center. Combustion takes place through diffusion, similar to conventional diesel engines. Due to the reduced knock risk, HPDF engines can achieve a higher compression ratio. Contrary to the LPDF mode, direct injection of liquid ammonia can substantially decrease ammonia slip [16]. Scharl et al. [17] determined three stages of the HPDF combustion process based on spatiotemporal integration spectroscopy and time-resolved spectroscopy: (i) First, diesel pilot ignition, where the flame is a typical diesel flame; (ii) After the pilot oil interacts with ammonia spray, the flame changes. During this process, flame emissions from NH_3 and diesel combustion are observed. (iii) Finally, it is mainly the pure ammonia combustion stage.

In terms of indicated thermal efficiency, the LPDF mode outperforms both pure diesel and HPDF modes mainly because it reduces in-cylinder transfer losses.

However, this advantage is accompanied by increased emissions of NH_3 and NO_x. On the other hand, the HPDF mode can attain an indicated thermal efficiency comparable to diesel mode and simultaneously decrease NO_x in the exhaust via in-cylinder thermal denitrification. According to the simulation results presented by Zhou et al. [18], both modes have the potential to achieve an indicated thermal efficiency of more than 50% in large, large, low-speed, two-stroke diesel engines. In the optimized ammonia-premixed combustion scenario, the LPDF mode is adopted, resulting in a 5.1% increase in thermal efficiency. However, in contrast with diesel mode, NO_x emissions have risen notably by 140.6%. Conversely, the optimized HPDF mode can attain an indicated thermal efficiency comparable to diesel mode while reducing NO_x emissions by approximately 47%. Li et al. [19] compared LPDF and HPDF, and the results showed that the recommended maximum ammonia ratio for LPDF mode was around 80%. If we continue to increase the diesel replacement rate, there is a higher chance of experiencing misfires. However, by utilizing HPDF mode, it is possible to achieve a diesel replacement rate as high as 97%.

2.2.3 Pre-chamber Ignition Technology

Pre-chamber ignition technology is generally applied to dual-fuel gas engines, especially ammonia-natural gas engines. Wu et al. [20, 21] demonstrated that to optimize the ammonia-natural gas engine's thermal efficiency, it is recommended to employ an NH_3 volume fraction ranging from 50% to 80%. The pre-chamber ignition technology is as follows.

1. During the initial stage of the combustion, the engine utilizes an intake manifold injection technique, whereby fuels are introduced into the intake port from the intake manifold. Subsequently, the fuels are combined with air at the intake port. In the subsequent compression phase, the premixed air-fuel mixture is injected into the cylinder.
2. In order to guarantee the complete injection of fuels into the engine cylinders during each cycle, the intake manifold opening period commences immediately after the closure of the exhaust valve. This precautionary measure effectively prevents any potential occurrence of ammonia slip at the exhaust port.
3. When the engine is operating to compression stroke, a limited quantity of the gas mixture within the primary combustion chamber undergoes compression into the pre-chamber. The natural gas blend becomes thinned to an almost stoichiometric blend before ignition. Upon the discharge of the spark plug electrode, a flame core is initiated. After nucleation, the flame commences, spreading within the pre-chamber, contingent on the local flame evolution at a laminar pace. Consequently, a high-pressure stream is engendered inside the nozzle, creating a flame jet that penetrates the primary combustion chamber and augments flame propagation under rare circumstances.
4. Furthermore, while undergoing the compression stroke, a portion of ammonia is pushed into the prechamber, where it undergoes dissociation into H_2 and N_2.

2.2.4 Liquid NH_3 Stratified Injection

The research field of direct injection of liquid ammonia has shown tremendous vitality. Lewandowski et al. [16] evaluated the evaporation process and cooling effect of ammonia injection and gained a deeper understanding of the mixing of ammonia in the cylinders under different operating conditions.

Liu et al. [22] introduced the ammonia-diesel stratified injection mode, depicted in Fig. 1a. Ammonia and diesel are co-axially injected, so the ammonia jet easily reaches the flame zone of diesel combustion during spray development. Ammonia-diesel stratified injection is a highly effective method for improving NH_3 combustion in large-diameter low-speed marine engines, with an energy replacement ratio of up to 99%. Another advantage of using this method is that ammonia and diesel use the same injector, which can reduce the number of injectors.

Ichikawa et al. [23] introduced the idea of a three-layer NH_3 stratified injection system. This system allows for the sequential injection of three layers from a single nozzle, resulting in the formation of three distinct layers of spray, as depicted in Fig. 1b. In this spray concept, liquid NH_3 is sandwiched between the supporting fuel. Supporting fuels refer to liquid fuels that possess the characteristics of easy ignition and combustion. Examples of such fuels include marine gas oil (MGO), diesel, and biodiesel. The experimental results show that this method provides an effective way to solve the challenge caused by the slow combustion speed of NH_3 in large two-stroke engines [24].

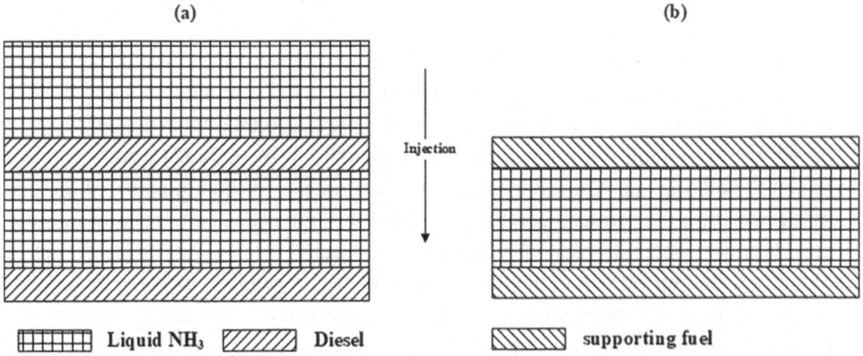

Fig. 1 Diagram of liquid ammonia stratified injection strategy: (**a**) ammonia/diesel mode; (**b**) supporting fuel/liquid NH_3/supporting fuel mode

2.2.5 In-Cylinder Gas Recirculation Technology

According to the analysis of the properties of ammonia combustion in Sect. 2.2.1, it is difficult to use ammonia alone to drive an internal combustion engine. One of the challenges is that ammonia has a high auto-ignition temperature, which means it needs a higher temperature to initiate ignition compared to conventional fuels like diesel. Additionally, ammonia exhibits a prolonged ignition delay, which can result in incomplete combustion or misfires, ultimately impacting the efficiency and performance of the engine. Lastly, the low laminar flame speed of NH_3 combustion further complicates its use in internal combustion engines. A slower flame propagation rate can lead to less efficient combustion processes and increase the emissions of unburned ammonia and other pollutants. Two approaches to attain pure ammonia marine combustion engines have been identified in scholarly works: high-temperature cylinder gas recirculation (HTCGR) and in-cylinder reforming gas recirculation (IRGR). Both methods involve the circulation of high-temperature gas or reformed gas into the cylinder, constituting in-cylinder circulation technology.

HTCGR was utilized to achieve pure NH_3-driven marine low-speed diesel engines in an article published in *Energy* [25]. The process of exhaust gas recirculation entails the redirection of a fraction of the engine's exhaust back into the intake manifold, where it mixes with fresh air-fuel mixture before re-entering the cylinder, thereby lowering the maximum combustion temperature within the cylinder. In contrast to EGR, the concept of HTCGR is to circulate the high-temperature gas from the engine's expansion stroke to the compression stroke in order to heat the ammonia in the compression stroke to reach its self-ignition temperature. As shown in Fig. 2, the principle of the HTCGR system is as follows: a constant pressure tank (CPT) is used to store and release high-temperature gas. There are two sources of high-temperature gas: one is from the expansion stroke, and the other is the heated air. During the expansion stroke of the engine, open the CGR valve to store some of the high-temperature gas in the CPT. At the compression stroke, the CGR out valve opens, and the high-temperature gas from the constant pressure tank enters the cylinder to heat the ammonia. Numerical simulation results indicate that with the

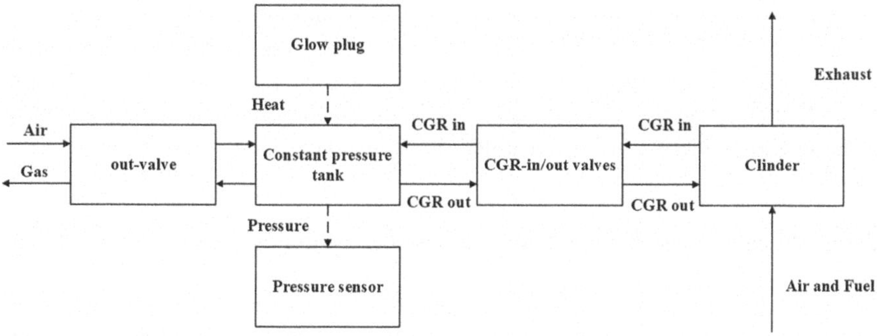

Fig. 2 The depiction of the HTCGR system

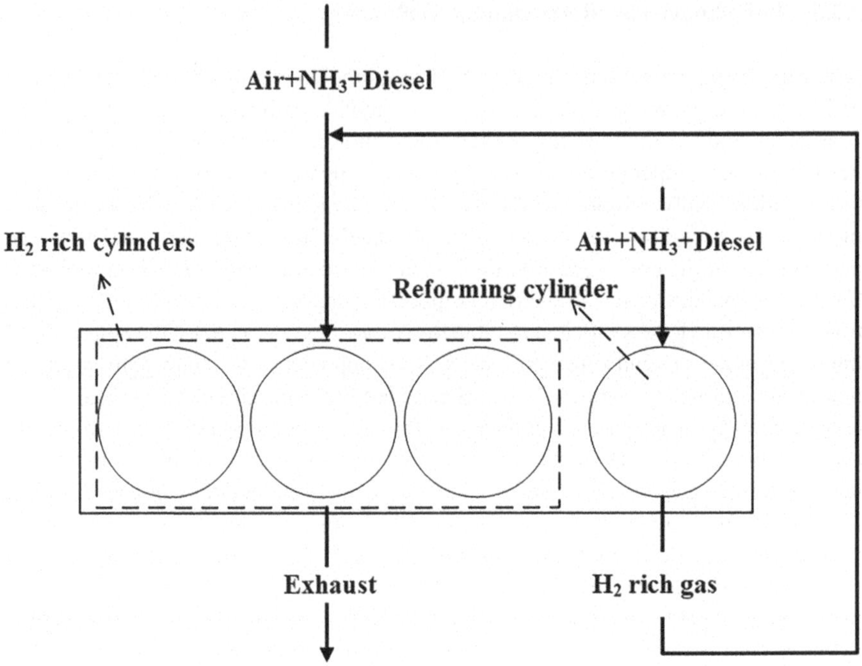

Fig. 3 The depiction of the IRGR

use of the HTCGR system, the temperature inside the cylinder before ammonia injection can reach 1200 K. Furthermore, the thermal efficiency can potentially reach 50%.

A paper published in *Nature Communications* proposes the concept of in-cylinder reforming gas recirculation (IRGR) for pilot-diesel-ignition ammonia combustion engines [26], which improves the thermal efficiency of the engine while reducing unburned NH_3, NO_x, N_2O, and GHG emissions. Figure 3 illustrates that in a high-temperature environment, the reforming cylinder contains air, ammonia, and diesel, undergoing a reforming reaction to produce H_2-rich gas. The H_2-rich gas from this specific reforming cylinder is then circulated back into other cylinders, effectively merging the benefits of H_2-rich combustion with EGR. This approach offers several advantages over installing a reformer externally on the engine: (i) A 23% reduction in fuel energy loss due to heat transfer; (ii) solved the problems of short catalyst lifespan and high cost; and (iii) avoided the complexity of engine structure.

2.2.6 Formation of N_2O and NO_x in Ammonia Fuel Engines

While it is true that ammonia can effectively eliminate CO_2 emissions, its usage does come with a potential downside. The increased utilization of ammonia may inadvertently result in a rise in NO_x emissions alongside the production of non-conventional pollutants such as N_2O. There are generally two ways for ammonia combustion to produce NO_x. The first involves the oxidation of ammonia, which produces fuel NO_x. The second pathway is through combustion, which raises the flame temperature and consequently leads to significant thermal NO_x formation. During the expansion process, as the cylinder temperature decreases, NO can be decomposed through the thermal de-oxidation process, which does not exist in pure diesel models. Due to the mutually exclusive temperature conditions required for the production of NO_x and N_2O, a trade-off exists between the emissions of NO_x and N_2O. Xu et al. [27] increased the injection pressure of pilot diesel and found that unburned NH_3 and N_2O emissions decreased while NO_x emissions increased. According to Wolfram et al. [28], the widespread adoption of ammonia fuel in the global shipping industry for decarbonization could potentially disrupt the global nitrogen cycle. As a result, it is essential to understand the processes that lead to the generation of N_2O and NO_x in ammonia engines. Wu et al. [29] used isotope labeling to label N in N_2, thereby labeling the nitrogen element in the thermal N_2O and NO_x generation pathways in ammonia diesel dual-fuel engines and clarifying the chemical reaction pathways of fuel and thermal nitrogen oxide generation evolution.

2.3 *Ammonia Fuel Cell*

2.3.1 Application of Marine Ammonia Fuel Cells

Ammonia fuel cells can be divided into two categories: direct and indirect. Direct ammonia fuel cells (DAFCs), which directly use ammonia, were first reported in the 1960s to operate at temperatures ranging from 50 °C to 200 °C, utilizing potassium hydroxide as an aqueous electrolyte [30]. Typically, the conversion of ammonia into hydrogen is necessary for its utilization in fuel cells. However, DAFC has the unique capability to directly employ ammonia without the need for this conversion process. Zhao et al. [31] presented a prototype of a DAFC with up to 135 mW cm^{-2} peak power density. This DAFC utilized cathode catalysts that were resistant to ammonia and did not contain any noble metals. Additionally, it employed high-temperature hydroxide exchange membranes that exhibited excellent stability. Indirect ammonia fuel cell is the process of reforming ammonia into hydrogen gas and then using it.

The focus of ongoing research on NH_3 fuel cells is to analyze the feasibility of their application in ships through simulation methods. Perčić et al. [32] demonstrated the use of solid oxide fuel cell (SOFC) ammonia fuel cells in short-distance sea transportation, showing that green ammonia can reduce CO_2 equivalent emissions by up to 84%. Wu et al. [33] studied the feasibility of using ammonia SOFC on container ships. The research results indicate that ammonia SOFC systems have distinct advantages compared to other fuel and propulsion technologies, making them well-suited for use in maritime applications. In a 20-year economic analysis of various marine propulsion systems, it was found that the overall cost of ammonia

SOFC propulsion systems is similar to that of internal combustion engine systems using Liquefied Natural Gas (LNG) and heavy fuel oil (HFO).

Duong et al. [34] evaluated a specialized ammonia SOFC combination system designed for use on ships. Their research found that the proposed multi-generation power generation system achieved energy efficiency and exergy efficiency levels of 60.4% and 57.3%, respectively. However, a drawback of this system is its slow response time to meet the power demands of the ship. The start-up and operation mechanisms are crucial and need to be able to respond quickly to the power system. Since SOFCs operate at high temperatures of 700–900 °C, they require time to preheat and start up the system. The potential challenge of meeting urgent power needs during critical vessel operations may arise due to delays in response time. Furthermore, Duong et al. [35] designed a fuel cell system using ammonia as a fuel for a regular cargo ship with a power of 3800 kW. The system integrates SOFC, proton exchange membrane fuel cell (PEMFC), and waste heat recovery systems. Ammonia is used as fuel for SOFC, and the waste heat in SOFC exhaust gas is reused to decompose ammonia into PEMFC and produce pure hydrogen. The integration of SOFC, PEMFC, and waste heat recovery systems is particularly beneficial for ships that require a quick response to start-up. By leveraging the unique features of both fuel cell types, this hybrid solution can optimize power generation for various operational stages. The PEMFC system, which operates at lower temperatures, can promptly deliver power for propulsion devices, ensuring swift start-up and maneuverability. In contrast, high-temperature SOFCs are better suited for stable maritime navigation conditions, generating a steady power supply for sustained cruising. This dual-fuel cell approach ensures a reliable and efficient energy source for diverse marine demands.

The literature review suggests that SOFCs are a promising fuel cell technology for maritime applications. Please refer to the literature review by Machaj et al. [36] for detailed information on SOFC technologies. Due to the high operating temperatures of SOFC, NH_3 decomposes into H_2 upon entering the fuel cell, eliminating the need for a separate ammonia decomposition reactor. However, the dynamic response of ammonia SOFC is slow. To address this challenge, a combination of high-temperature SOFC and low-temperature PEMFC may be a viable solution. Nevertheless, this approach will inevitably increase system complexity and cost. Another potential solution is the alkaline anion exchange membrane ammonia fuel cell, which operates at low temperatures, offers fast system start-up, and can directly utilize ammonia. These features make it suitable for applications on mobile devices. However, no relevant research on ship applications has been found to date.

2.4 Summary

Ammonia is considered a safer and more cost-effective fuel compared to hydrogen, with fewer storage challenges and greater sustainability. Despite its importance in the scientific and industrial sectors, its application in marine internal combustion engines is being limited by various challenges. Issues such as toxicity and subpar combustion performance hinder its direct substitution for traditional fuels. Additionally, the increase in NO_x and unburned NH_3 in exhaust emissions necessitates additional after-treatment systems. The emissions of N_2O from ammonia combustion also raise concerns about the overall greenhouse gas reduction benefits of utilizing an ammonia-based fuel.

Ammonia-fueled internal combustion engines require measures such as dual fuel to operate. The most common are ammonia-diesel dual-fuel engines and ammonia-natural gas dual-fuel engines. Ammonia-diesel dual-fuel engines have two modes of operation: LPDF mode and HPDF mode. Both modes have their own advantages and disadvantages, among which the HPDF mode can achieve an ammonia substitution rate of up to 97%. From this perspective, the HPDF mode may be cleaner. It is recommended to use prechamber ignition technology for ammonia natural gas dual fuel engines. Compared with traditional ignition methods, this method can achieve a higher ammonia substitution rate. In addition, the emergence of new technologies such as stratified injection of liquid NH_3 and high-temperature cylinder gas recirculation, as well as the analysis of pollutant mechanisms such as NO_x, indicate the progress of ammonia fuel internal combustion engine technology. However, current research on large two-stroke marine ammonia-fueled internal combustion engines remains inadequate. In the future, it is still necessary to expand research and development efforts for large two-stroke marine ammonia internal combustion engines, with the aim of fully exploiting the carbon-free benefits of NH_3.

In the discourse on selecting alternative fuels in the shipping industry, ammonia has consistently been considered the ultimate solution for decarbonization in shipping. Numerous studies have reported the significant emission reduction potential of ammonia applications on ships [37–39]. The emergence of the first batch of actual ship orders for ammonia in the second half of 2023 marks a significant breakthrough in the field of shipbuilding. Ammonia can be converted into electricity in fuel cells, and the cost-effectiveness of internal combustion engines provides advantages for its use on ships. However, it appears that ammonia internal combustion engines are becoming increasingly advanced. Progress in the use of ammonia to power internal combustion engines is accelerating [40]. Going forward, ammonia utilization as an internal combustion engine fuel will necessitate the development of appropriate combustion strategies to improve combustion efficiency and emissions. At the same time, the development of marine ammonia fuel cells should also be carried out.

3 Application of Methanol on Ships

3.1 Methanol

Methanol (CH_3OH) is the simplest form of alcohol. Under the temperature of the room and the pressure of the atmosphere, it is an odoriferous, transparent, highly flammable liquid that is both light and volatile. As a chemical raw material, methanol has a wide range of applications in industry and daily life. In 2023, methanol has become increasingly recognized as an alternative fuel for reducing GHG and pollutant emissions. Methanol offers a significant benefit as a fuel due to its ability to generate reduced levels of SO_x and NO_x emissions, as well as reduce CO_2 emissions. Furthermore, methanol leakage will not cause global warming.

Nevertheless, methanol has a low flash point of 11 °C and a low boiling point of 65 °C, which means it has a higher fire risk than diesel. Therefore, compared to traditional fuels, there are no major obstacles in transportation, storage, and refueling, except for small modifications made to adapt to low flash points. The new infrastructure does not require large-scale investment. However, there are also some drawbacks to using methanol as a fuel. Firstly, owing to the lower energy density of methanol compared to conventional marine fuels, it necessitates roughly double the storage capacity. Secondly, due to the incompatibility between methanol and certain materials, there are potential corrosion issues. Moreover, methanol is characterized by its acute toxicity toward the human body. It possesses a median lethal dose of 56.2 grams per individual and an inhalation lethal concentration ranging from 4000 to 13,000 ppm, which is deemed fatal. To prevent potential harm caused by internal ingestion, skin contact, or respiratory inhalation, it is imperative to proactively devise and implement necessary protective measures and emergency protocols.

To summarize, methanol presents numerous advantages as a marine fuel. These include favorable economic benefits, emission characteristics that are on par with LNG, enhanced system compatibility, and broader applicability.

3.2 Marine Methanol Internal Combustion Engine

3.2.1 Working Mode of Methanol Internal Combustion Engine

Methanol is not suitable for direct use in conventional diesel internal combustion engines intended for heavy fuel oil (HFO) or marine diesel due to its low cetane number, which renders it unsuitable for compression and ignition in such engines. Currently, methanol is primarily utilized in the dual fuel mode (DMDF) in marine diesel engines, where it is combined with diesel fuel for combustion [41, 42]. Simply, DMDF is accomplished by introducing methanol at low pressure into the intake manifold. This methanol is then mixed with air, resulting in a methanol-air mixture. Subsequently, it is ignited by diesel fuel near the top dead center. Furthermore, an alternative approach involves incorporating ignition improvers into pure methanol [43].

The direct dual fuel stratification (DDFS) method is an advanced combustion strategy designed for marine engines, offering the potential for high performance across various operational conditions. This approach involves the use of two direct injection injectors within the combustion chamber to precisely control the combustion process and heat release rate through adaptable injection methods. The larger cylinder diameter of marine engines facilitates the integration of dual injectors, making them well-suited for implementing the DDFS technique. Methanol is identified as a suitable fuel for DDFS engines due to its proven efficacy in automotive applications. Consequently, methanol and diesel DDFS marine engines present potential advantages, although research in this specific area remains limited. Sun et al. [44] investigated the impact of a performance optimization-focused fuel injection strategy on a marine methanol/diesel direct dual fuel stratification (DDFS) engine with a substantial methanol substitution rate of 95%. The findings suggest that the combustion state switching of methanol/diesel DDFS engines is significantly influenced by the fuel mixing process.

3.2.2 Methanol as a Hydrogen Carrier

Indeed, using methanol as a hydrogen carrier is a promising approach for hydrogen energy utilization on board ships. The HyMethShip project [45], funded by the EU Horizon 2020 Research and Innovation Program and supported by the EU Horizon 2020 Research and Innovation Program, offers a hopeful resolution for utilizing methanol in maritime vessels. The project aims to create an almost closed CO_2 cycle, reducing CO_2 emissions by 97% and NO_x emissions by over 80% while eliminating SO_x and PM emissions. As illustrated in Fig. 4, through the utilization of hydrogen derived from renewable sources and captured carbon, green methanol is produced and utilized as a carrier for hydrogen storage on ships. There exist two distinct methods for engine propulsion: the first involves the direct utilization of methanol, while the second entails the injection of methanol into the pre-combustion system, also known as the reforming system. In the membrane reformer system, a chemical process occurs in which methanol and water react to produce H_2 and CO_2. The H_2 generated serves as a fuel source for the ship's propulsion, while the CO_2 is captured, converted into liquid form, and then stored on board the vessels. The CO_2 that has been captured is subsequently conveyed to a port for storage and is utilized in the production of environmentally friendly methanol.

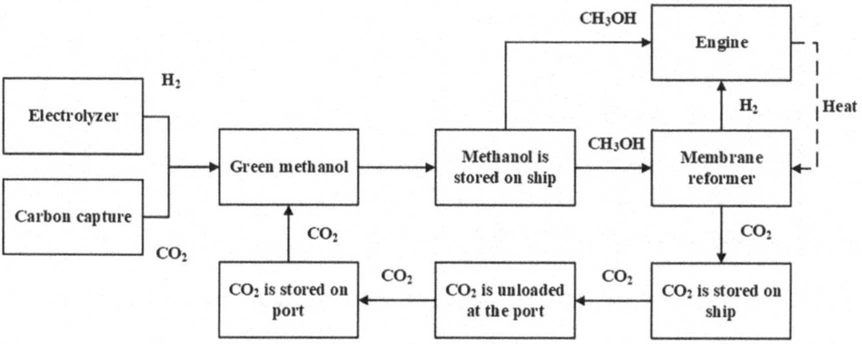

Fig. 4 Process diagram of methanol as a hydrogen carrier closed carbon cycle

3.2.3 Pollutant Emissions

To achieve near-zero emissions in marine methanol engines, it is crucial to efficiently remove various pollutants from the exhaust, which have different physical and chemical properties. These pollutants include unburned CH_3OH, HCHO, NO_x, CO, and HC, as well as CO_2. Oxidation catalytic converters (DOCs) are mainly used to reduce unburned CH_3OH, HCHO, CO, and HC emissions, while SCR, or water injection, is used to reduce NO_x emissions. However, there are few studies that efficiently remove all of these pollutants simultaneously. Gong et al. [46] significantly reduced HC and unburned CH_3OH emissions by optimizing injection parameters, but HCHO emissions showed an opposite trend to unburned CH_3OH emissions. The experiments conducted by Güdden et al. [47] suggest that marine large-bore high-speed methanol engines emit relatively high levels of formaldehyde (~1 g/kWh), and it is recommended to use DOC to remove formaldehyde from the exhaust. A paper published in *Energy* indicates that the implementation of water injection in engines resulted in a significant reduction of up to 16% in HC emissions and up to 12% in NO_x emissions when compared to the emissions data of standard engines [48].

The main solutions for CO_2 emissions are post-combustion carbon capture, pre-combustion carbon capture, and oxygen-rich combustion, as shown in Fig. 5 [49]. The relatively low operational temperature range of methanol steam reforming (~150–300 °C) makes pre-combustion carbon capture a more viable option compared to post-combustion carbon capture systems, which are more complex and have higher energy demands. However, pre-combustion carbon capture technology is still under development. The cost of producing green methanol, a carbon-neutral fuel, is

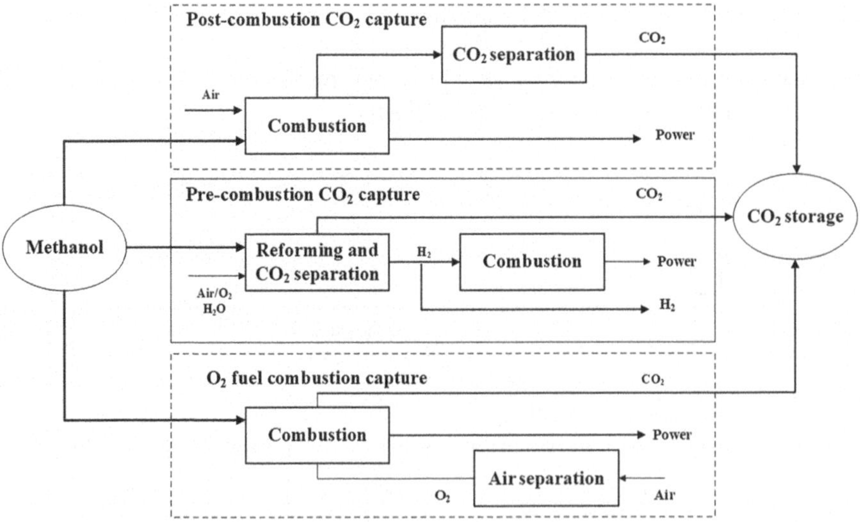

Fig. 5 The working principle of carbon capture

significantly higher compared to methanol derived from natural gas. Specifically, the production expenses of bio-methanol are approximately three times greater than those of natural gas-derived methanol, while the manufacturing costs of electric methanol are roughly eight times higher than those of methanol produced from natural gas [50].

Therefore, there is a need for further research and development to efficiently remove all pollutants from the exhaust of marine methanol engines and achieve near-zero emissions. Moreover, it is imperative to decrease the manufacturing expenses associated with green methanol in order to enhance its economic feasibility as a fuel alternative.

3.2.4 Commercial Application of Methanol Fuel Engine

Compared to ammonia, the research on methanol is more mature. Methanol dual-fuel two-stroke engines, developed by ship engine manufacturers like MAN Energy Solutions, have been successfully introduced into the market. These engines have been in operation since 2016, showcasing their reliability and efficiency. In addition to this, MAN Energy Solutions plans to offer methanol retrofitting for four-stroke engines starting in 2024. This retrofitting will address the challenges associated with fuel systems and injection technology, enabling the utilization of methanol in four-stroke engines. The company envisions the application of methanol four-stroke engines in various types of vessels, including container ships, ferries, fishing boats, and cruise ships. On the other hand, two-stroke engines powered by methanol are considered the optimal choice for oil tankers, container ships, and other specific ship applications. This distinction highlights the versatility and adaptability of methanol as a viable fuel option for the maritime industry. Anglo Belgian Corporation (ABC) focuses on developing medium-speed methanol dual fuel engines [51], which include 6-cylinder and 8-cylinder inline engines as well as 12-cylinder and 16-cylinder V-type engines, with a power range of 955 kW to 3.5 MW. According to the operating conditions and load, a ratio (energy ratio) of 70% methanol and 30% traditional fuel or biofuel can be achieved. Another advantage of these dual-fuel engines is that when there is no methanol available, operators can effortlessly switch to 100% traditional fuel or biofuel. China State Shipbuilding Corporation has developed a 6M320DM methanol fuel engine [52]. The cylinder diameter of this engine is 320 mm, the stroke is 420 mm, the single-cylinder power is 500 kW, and the speed is 750 rpm. It can adapt to various ship types, including deck ships, bulk carriers with a total tonnage of less than 20,000, chemical tankers, finished oil tankers, and ocean engineering ships. In summary, methanol as a marine fuel has cost-effectiveness, superior environmental performance, and technological feasibility in commercial applications, and it is gradually becoming one of the mainstream choices for ship power systems.

3.3 *Marine Methanol Fuel Cell*

Furthermore, apart from its application as a primary energy source in marine propulsion systems, methanol has demonstrated efficacy as a viable fuel option for fuel cell technology.

3.3.1 Indirect Methanol Fuel Cell

Methanol must undergo reform in order to produce hydrogen, which can subsequently be utilized in PEMFC or SOFC. Compared with internal combustion engines, PEMFC and SOFC have higher efficiency, ranging from 50 to 60% and 60%, respectively. However, without the recovery of adequate waste heat, the conversion of methanol to hydrogen will necessitate additional energy, thereby augmenting the overall energy consumption of this fuel production process. Presently, the development and testing of fuel cells that utilize methanol in the maritime industry are ongoing.

3.3.2 Direct Methanol Fuel Cell

Direct methanol fuel cells (DMFCs) offer several advantages, such as not requiring reforming processes, operating over a wide range of temperatures, emitting minimal pollutants, having a simple device structure, and enabling easy storage of liquid methanol fuel. These characteristics make DMFCs a sustainable and efficient power source. In comparison to lithium batteries, DMFCs exhibit lower operating temperatures, higher energy density, and improved safety and portability when compared to proton exchange membrane fuel cells (PEMFCs). However, the slow kinetics of methanol oxidation inherent in DMFCs significantly restrict their total energy and power density. As a result, DMFCs are still in the developmental phase [53, 54]. Leo et al. [55] conducted a study comparing the performance of PEMFC and DMFC for marine applications from an exergy perspective. The research revealed that DMFCs are associated with greater fire damage and loss, while the exergy efficiency of both systems is comparable.

3.4 *Summary*

The use of methanol in maritime applications primarily involves two methods: methanol-diesel dual-fuel engines and methanol fuel cells. At present, the methanol-diesel dual-fuel engine stands as the most extensively researched and applied method, boasting ample technical capabilities. Nevertheless, the primary obstacle to adopting methanol as an environmentally friendly marine fuel lies not in its onboard functionality but rather in the attainment of widespread green methanol production. Methanol is also an excellent hydrogen carrier that can produce hydrogen gas by reforming and supplying it to fuel cells. In fact, hydrogen obtained through methanol reforming can not only be used for fuel cells but may also be more suitable for internal combustion engines. Because internal combustion engines do not require high hydrogen concentrations like fuel cells. Nevertheless, the commonly used methanol-to-hydrogen production is an endothermic process that requires additional energy. Moreover, the combination of methanol-reforming for hydrogen production and carbon capture technology on board provides some interesting solutions for the application of methanol on ships. With its significant advantages and feasibility as a marine fuel, it is expected that methanol engines will significantly contribute to carbon reduction in the shipbuilding industry in the next 20 years.

4 Application of Gaseous Hydrogen and Liquid Hydrogen on Ships

4.1 Hydrogen

Hydrogen is regarded as a promising future fuel for reducing GHG emissions in the maritime sector. Nevertheless, in contrast to more conventional fuels like natural gas, hydrogen possesses unique traits that have a notable impact on its viability as a fuel source [56]. Hydrogen is flammable and explosive in a broader range of ways. Unlike other fuels, the interaction between hydrogen and materials is different. The molecule is small, allowing for easy diffusion and making it more susceptible to leakage compared to other fuels, despite its quicker dispersal from the leak. Hydrogen systems necessitate the use of special steel or alternative materials that are resistant to embrittlement caused by the diffusion of hydrogen into metals and specific types of steel. Therefore, while hydrogen holds promise as a future fuel for the shipping industry, it also presents unique challenges that must be addressed to ensure safe and efficient use. Addressing these challenges will require ongoing research and development efforts, as well as collaboration between industry stakeholders and regulatory bodies.

When hydrogen burns, it produces almost invisible flames that emit very little heat, making it difficult to detect fires during the day. Moreover, given hydrogen's increased tendency to leak in comparison to other gaseous fuels, it is imperative to focus on leak prevention and effective leak detection as essential safety measures. If hydrogen escapes in a confined space, it could generate a substantial flammable mixture. The rapid flame speed of hydrogen mixtures creates a risk of fire escalating to explosions in confined spaces. Therefore, it is best to conduct hydrogenation and storage outdoors or in a properly ventilated room. Li et al. [57] employed Fluent software to conduct numerical computations aimed at modeling hydrogen leakage

Table 1 Comparison of CH_2, LH_2, and MH_2 [61]

Type	Storage pressure (bar)	T (K)	H_2 content (%wt)	Energy density (kWh/L)	Cost range (EUR/kg)	Remarks
CH_2	200–1000	293	100	0.3–0.8	0.9–8.4	A 10% total energy loss occurs in the compression process. With the limited energy density, it is only suitable for short-distance transportation.
LH_2	1–12	20	100	2.3	2.4–9.9	The energy loss in the liquefaction process can be as high as 40%. Availability is limited. Recommended for medium and long-distance transportation. Control of boiling is required.
MH_2	20–150	260–245	>8	35–40	0.5–8	Heat management is required. It can be combined with heat recovery for PEMFC.

and diffusion within fuel cell containers. They examined the distribution of hydrogen concentration in various compartments under varying ventilation scenarios, offering insights to support the formulation of high-pressure gaseous H_2 fuel cell vessels.

The main applications of hydrogen in ships include internal combustion engines and fuel cells. One of the primary obstacles in implementing H_2 as a marine fuel is the storage of itself. In order to tackle this issue, a range of storage technologies have been created, which are mainly classified into physical-based and material-based solutions. At present, the most commonly used hydrogen storage methods include compressed hydrogen storage (CH_2), liquid hydrogen storage (LH_2), and metal hydride hydrogen storage (MH_2), as shown in Table 1. The volumetric energy density of LH_2 is 8.49 GJ/m^3 (70.8 kg/m^3 at 1 bar, −253 °C), and CH_2 is 4.82 GJ/m^3 (40.2 kg/m^3 at 700 bar). Compared with gas storage, liquid hydrogen storage can produce higher mass fractions [68], but there are difficulties in handling and long-term storage due to evaporation issues. In addition, a significant quantity of energy is needed for the refrigeration process, which makes up around 30% of the total energy content of hydrogen. An alternative approach to storing hydrogen is through the use of metal hydride. This method involves the reversible reaction of hydrogen with the parent hydride, resulting in the formation of metals or intermetallic alloys [58]. Metal hydride exhibits excellent characteristics, especially at equilibrium pressures and ambient temperatures below 40 bar, with a high bulk hydrogen storage density comparable to LH_2 [59]. However, the main drawbacks of metal hydride storage are its high cost and low weight density, as MH_2 tanks have a high weight. Hence, they are primarily appropriate for stationary purposes and certain modes of transportation, including submarines and compact vessels [60]. Research is ongoing to improve the properties and reduce the cost of metal hydride storage systems.

The longevity and security of ship hydrogen storage tanks play a vital role in ensuring the dependability and effectiveness of ship hydrogen energy systems, along with the safety of ship navigation operations. The heat produced by the discharge of hydrogen from the tanks, as well as hydrogen embrittlement resulting from the unique properties of hydrogen gas, can impact the performance of hydrogen storage tanks. Li et al. [62] conducted a comprehensive review of the research methodologies employed to investigate the fatigue life of connection nodes between marine hydrogen storage tanks and ships under external forces. This aspect holds significant importance in the design and functioning of ship hydrogen energy systems. To ensure the secure and dependable operation of ship hydrogen storage tanks, it is imperative to undertake additional research and testing.

4.2 Marine Hydrogen Internal Combustion Engines

Marine hydrogen engines are prone to abnormal combustion due to hydrogen's low calorific value (120 mJ/kg, diesel is 42.55 MJ/kg), extremely low ignition energy (0.02 MJ, diesel is 0.24 MJ), and high combustion rate (200 cm/s, diesel is 45 cm/s). The larger cylinder diameter, higher average effective pressure, and lower rotational speed of marine engines further exacerbate abnormal combustion. As a result, research and development of marine hydrogen engines have only just begun [63]. On March 7, 2024, Mitsui Shipbuilding Co., Ltd. announced that it had successfully retrofitted a 4S50ME-T engine (7 MW, rated speed 117 rpm, MEP = 2.10 MPa) to shift one cylinder from a ME-GI engine design using LNG fuel to a hydrogen operating mode. This marks the initial achievement of effectively integrating hydrogen combustion into a sizable two-stroke marine engine measuring 50 centimeters in diameter [64].

A lifecycle assessment published in *Science of The Total Environment* found that both hydrogen fuel cells and internal combustion engines have great potential to promote energy transformation, with hydrogen internal combustion engines having better environmental benefits [65]. The object of life cycle assessment for hydrogen fuel vessels is mostly hydrogen fuel cells [66]. It is generally believed that the application of hydrogen fuel cells on ships has more advantages than internal combustion engines. However, it has not been confirmed in actual ship energy systems whether H_2 internal combustion engines or H_2 Fuel cells are more advantageous alternatives.

4.3 Application of Compressed Hydrogen on Ships

To store hydrogen in the form of compressed gas is the simplest and most straightforward approach. This method facilitates the processing and storage of hydrogen at room temperature. After compression, the density of hydrogen reaches 23.32 kg/m^3

at 350 bar and increases to 40 kg/m^3 when the pressure reaches 700 bar. At present, there are five types of storage tanks that can store H_2 under conditions of pressurization. Type IV storage tanks are favored for use in fuel cell-powered vehicles because they can provide up to 5.7% weight capacity. Nevertheless, the expense of storing high-capacity hydrogen is tenfold greater under high-pressure conditions compared to low-pressure conditions, primarily because of the materials needed. Under increased pressure, the hydrogen storage capacity is only 0.81 kWh/L, and the maximum hydrogen density is only 26 g/L. Hence, the storage of compressed hydrogen at significantly higher pressure levels compensates for around 10% of the energy content of hydrogen [67].

CH_2 commonly utilizes storage tanks ranging from 350 to 700 bar to enhance density through hydrogen compression. While the rise in pressure contributes to density enhancement, excessive pressure poses challenges for refueling operations [68]. Moreover, as the pressure increases, compression requires more energy. At present, the primary emphasis of fuel cell ship projects revolves around incorporating PEMFC technology onto vessels of varying sizes, particularly small and medium-sized ones [61]. The most favored method for hydrogen storage among these initiatives is CH_2. The constrained energy density of CH_2 on board hinders its usability and adaptability, especially for ships with high power demands or extensive navigation requirements. This presents obstacles to maintaining a satisfactory transportation range for such vessels. Nonetheless, CH_2 is foreseen to have a notable influence on fuel storage for small to medium-sized ships involved in daily operations and with convenient access to refueling infrastructure.

For example, China's first newly built hydrogen fuel vessel, San Xia Qing Zhou 1, uses CH_2 to store hydrogen on board [69]. The storage system comprises 32 hydrogen cylinders, each with a pressure of 35 MPa, securely fastened using metal frames. These cylinders have a total capacity of 240 kg. FRS Windcat Offshore Logistics [70] launched the first hydrogen-powered Crew Transfer Vessel to the German offshore market in January 2024, heralding the beginning of a new era of Nordic maritime support. Its storage tank has a capacity of 27 steel cylinders, which can hold 207 kg of hydrogen. In addition, the dual fuel unit of the MAN engine has been modified by CMB with a hydrogen injection system, which extracts hydrogen gas from the compressed hydrogen storage tank.

For CH_2 refueling, the CH_2 refueling station mainly consists of a compressor, chiller unit, and pressure regulating valve. A primary obstacle in the technical realm is the extended duration of the refueling procedure. For example, Melideo et al. [71] conducted a case study on the use of hydrogen on short-haul roll-on/roll-off vessels. Compressed hydrogen is stored in a reinforced container at 350 bar. When the daily hydrogen demand for a roll-on/roll-off ship is 1329 kg and the refueling speed is 3.6 kg/min, the total refueling time exceeds 6 h. An additional choice is to consider the use of interchangeable steel cylinders, often containerized, that can be unloaded as necessary. It is important to recognize, however, that this method may only be suitable for small and medium-sized ships and may not be viable for larger vessels due to the prolonged port call time.

4.4 Application of Liquid Hydrogen on Ships

Liquefaction is an effective method for enhancing hydrogen storage density [72]. With a saturation density of 70 kg/m^3 at 1 bar, liquid hydrogen boasts an impressive storage density. When compared to CH_2, the density of LH_2 experiences a significant, indicating that, even when considering evaporation losses and transportation expenses, the long-term transportation of liquid hydrogen is consistently more advantageous. Researchers have also highlighted the benefits of using liquid hydrogen as the main fuel for long-distance travel over hydrogen based on compressed gas, although its efficiency remains a contentious issue with substantial variation [67]. An example under consideration is the Kattamaram passenger ship: a catamaran with a length of approximately 35 m, capable of accommodating 200 passengers, and a cruising speed of 5.4 knots. Usually, the propulsion system of this type of ship is an internal combustion engine fueled with MGO, with a total power of approximately 360 kW. Rivarolo et al. [73] analyzed that liquid hydrogen is more suitable for application on this ship, while the CH_2 and MH_2 methods require higher volume and weight, respectively. Liquid hydrogen presents a more viable option in terms of weight and volume when compared to metal hydrides and compressed hydrogen, both of which necessitate higher volume and weight. For more information on the application of liquid hydrogen on ships, please refer to reference [74].

The liquefaction process of liquid hydrogen requires a significant amount of energy, estimated to be around 33% of its lower calorific value, due to its extremely low boiling point of −253 °C. Furthermore, high-performance insulation is essential for liquid hydrogen tanks aboard ships to minimize heat entry and gas evaporation. However, the cold energy of liquid hydrogen can be recycled, as demonstrated by Lee et al.'s [75] innovative PEMFC system that incorporates an organic Rankine cycle direct expansion cycle (ORC-DEC), resulting in a 6% increase in total output power compared to a standalone PEMFC system.

Liquid hydrogen's low boiling point (20 K or −253 °C) allows it to serve as a coolant for superconductors, potentially leading to lighter, quieter, more compact, and more efficient propulsion systems for ships. According to a study by Mojarrad et al. [76] in 2023, although the initial investment cost of CH_2 is low, LH_2 provides a more efficient energy solution for long-range navigation and can further improve energy efficiency and reduce operating costs when superconducting propulsion systems are implemented. This study analyzed the economy and energy consumption of a ferry capable of accommodating 300 people with different propulsion systems and pointed out that although the CH_2 has the lowest total cost, its energy consumption is the highest (33,000 kWh), and hydrogen consumption is the highest (1970 kg/day), while the LH_2 has an energy consumption of 28,000 kWh and hydrogen consumption of 1667 kg/day. In the case of superconducting propulsion systems, energy consumption is further reduced to 25,000 kWh, and hydrogen consumption is reduced to 1490 kg/day. Nevertheless, in order for LH_2 to compete effectively with CH_2, efforts must be made to reduce the associated costs of liquid hydrogen.

LH_2 tanks are traditionally positioned on the uppermost deck of vessels as a precautionary measure against potential safety concerns arising from the considerable flammability range of hydrogen. The installation of liquid hydrogen storage tanks on the top deck of ships is based on safety considerations for several reasons [77]: (i) Dispersion during leaks: Hydrogen is lighter than air, so it will quickly rise during leaks. Installing on the top deck can more easily release hydrogen into the atmosphere, reducing the risk of hydrogen accumulation inside the ship. (ii) Stay away from living areas: Placing the LH_2 tank on the top deck can separate it from the ship's living and operating areas, thereby reducing the direct risk to personnel in the event of a leak. (iii) Safety system integration: The top deck is usually equipped with safety systems such as hydrogen detectors and emergency release devices to monitor hydrogen concentration and safely discharge hydrogen into the atmosphere when necessary. (iv) Fire protection and control: On the top deck, any hydrogen-related fire is more easily detected and controlled while reducing the risk of fire spreading to other parts. (v) Structure and stability: When designing ships, the impact of hydrogen storage on ship stability and structural integrity will also be considered. The position of the top deck helps to maintain a lower center of gravity, thereby maintaining the stability of the ship.

Establishing an efficient LH_2 refueling system is vital for the widespread adoption of liquid hydrogen as a marine fuel. Pratt et al.'s research [78] provides valuable insights into LH_2 refueling technology, particularly in the detailed analysis of refueling infrastructure design in the SF-BREEZE ferry project. In the study, the author explored two methods of LH_2 transfer: one using pressure-generating units (PBUs) and the other using low-temperature pumps. PBU system comprises an evaporator that crucially vaporizes a small quantity of liquid hydrogen obtained from the fuel tank at the refueling station. Subsequently, it redirects the vapor to the air cushion situated atop the tank in order to elevate the internal pressure. This discrepancy in pressure levels between the fuel tank at the refueling station and the filled fuel tank facilitates the transfer of liquid hydrogen. This method exhibits a notable advantage in terms of cost, as the PBU system is more economically viable than low-temperature pumps. Furthermore, the vaporized hydrogen produced can be utilized for the purpose of cleaning the refueling pipeline and cooling it prior to refueling. Another advantageous feature of the PBU technology is its capacity to prevent the initial transfer of liquid hydrogen from a static intermediate tank to an onboard tank, thus mitigating the loss of hydrogen due to evaporation. In contrast, low-temperature pumps provide an efficient hydrogen compression method, which is more energy-efficient than compressing gaseous hydrogen. Using a low-temperature pump for liquid hydrogen filling is not only faster but also suitable for filling multiple containers simultaneously, especially suitable for fixed hydrogen storage scenarios. To prevent cavitation, low-temperature pumps are usually placed below storage tanks or underwater. When selecting the liquid hydrogen refueling system, it is necessary to comprehensively consider cost, efficiency, and practical application scenarios to determine the optimal refueling plan.

Despite the challenges associated with using liquid hydrogen as a fuel, its potential is becoming increasingly clear through ongoing research. For instance,

Jung et al. [79] developed a liquid hydrogen fuel hybrid ship propulsion system (HSPS) that integrates the LH_2 fuel gas supply system (FGSS), PEMFC, and battery system. They conducted dynamic characteristic analysis on FGSS, and the study showed that under rapidly changing hydrogen mass flow demand, the transient response time of the system can reach 30 s, and the supplied gas temperature is maintained in the range of 11–41 °C. Additionally, a deep reinforcement learning-based energy management strategy (DRL-EMS) for HSPS was proposed, further highlighting the potential of liquid hydrogen in maritime applications [80].

4.5 Summary

The challenges associated with storing compressed hydrogen are twofold. Firstly, the use of gas cylinders leads to a relatively low weight and volume density for the entire system, necessitating the majority of the deck and hold space on board the ship to be dedicated to storing steel cylinders. Secondly, slow refueling times present another issue. A significant quantity of hydrogen, approximately 100 tons, would be necessary to power a sizable maritime vessel. The process of refueling such a vessel would be time-consuming, taking several weeks or necessitating the use of numerous refueling nozzles. An alternative concept for fuel replenishment involves using hydrogen tanks installed in standard 20 or 40-foot containers, which are then loaded onto the ship, known as a box fuel system. While this may be a feasible option for smaller vessels like ferries, large vessels would face extended docking times due to the time required to load and unload all the containers.

The primary challenge associated with liquid hydrogen processes is the low temperature required for storage. The storage tank material necessitates the use of specialized insulation materials to maintain a minimal heat flux density entering the tank. Before liquid hydrogen can be adopted as a global fuel, the development of liquid hydrogen ports remains a significant obstacle.

5 Conclusion

Ammonia, as a potential clean marine fuel, offers several advantages over hydrogen, including better safety, lower cost, simpler storage methods, and higher sustainability. However, it faces challenges in internal combustion engine applications, such as its toxicity, poor combustion performance, and emission issues, notably the increased emissions of NO_x and unburned NH_3, which necessitate a matching after-treatment system. Additionally, there is uncertainty regarding the greenhouse gas emission reduction benefits of N_2O emissions generated by ammonia combustion. Current ammonia internal combustion engines need to be combined with fuels like diesel or natural gas. The high-pressure injection mode of ammonia-diesel dual fuel engines can achieve higher ammonia substitution rates and may offer a cleaner solution. Pre-chamber ignition technology provides a high replacement rate possibility for ammonia-natural gas dual-fuel engines. Although new technologies and understanding of pollutant emission mechanisms indicate progress, more research and development are needed, particularly for large two-stroke marine ammonia fuel internal combustion engines, to promote the carbon-free transformation of large ocean-going vessels. In the realm of fuel cells, SOFCs have demonstrated great potential in navigation, as they can directly utilize the hydrogen produced by ammonia decomposition, avoiding separate decomposition steps and reducing complexity. However, the dynamic response speed of ammonia SOFC is relatively slow, and combining SOFC with PEMFC may increase system complexity and cost. Although alkaline anion exchange membrane ammonia fuel cells have fast start-up speeds and are suitable for mobile devices, there is currently insufficient research on their application in ships.

Methanol, the simplest alcohol, has gained recognition as an alternative fuel for reducing greenhouse gas and pollutant emissions in recent years. Its application in ship fuel has reached a relatively mature stage, especially in methanol-diesel dual-fuel engines. However, challenges persist in implementing methanol in marine engines, including low energy density, potential corrosion issues, and human toxicity. To achieve near-zero emissions, further research and development are necessary to efficiently remove all pollutants from the exhaust and reduce the production cost of green methanol. Despite these challenges, commercial progress in applying methanol as a marine fuel has been made. Leading ship engine manufacturers like MAN Energy Solutions and Anglo Belgian Corporation have developed methanol dual-fuel engines, while China State Shipbuilding Corporation has created a methanol fuel engine. These advancements highlight methanol's cost-effectiveness, superior environmental performance, and technological feasibility for commercial marine applications, making it a strong contender for ship power systems. It is anticipated that methanol, as a marine fuel, will significantly contribute to carbon reduction in the shipbuilding industry over the next 20 years. In conclusion, methanol's potential as a marine fuel is increasingly recognized due to its benefits and ongoing commercial development. Addressing the challenges and pursuing further research will contribute to the successful implementation of methanol in the shipping industry and pave the way for a more sustainable future.

Compressed hydrogen stores hydrogen at room temperature, making it convenient to operate and store with flexible applications. Consequently, compressed hydrogen is currently the preferred hydrogen storage method for small and medium-sized ships, with countries like China and Germany already adopting it. However, compressed hydrogen has the drawback of low energy density, making it difficult to apply to long-distance ships. Although increasing pressure can raise the energy density of hydrogen, higher pressure means higher compression energy. Liquid hydrogen effectively enhances hydrogen storage density and is more advantageous than compressed hydrogen during long-distance voyages. However, its low storage temperature poses a significant challenge for ship applications. Conversely, due to the low temperature, liquid hydrogen can be used as a coolant for superconductors. Researchers have actively explored the application of both compressed hydrogen and liquid hydrogen on ships. Regardless of the chosen hydrogen storage method, establishing an efficient refueling system is crucial. Furthermore, safety issues related to hydrogen cannot be ignored.

In summary, ammonia, methanol, gaseous hydrogen, and liquid hydrogen each present their own advantages and challenges as marine fuels. Ammonia and hydrogen require further technological breakthroughs to overcome existing obstacles, while methanol may be more rapidly applied in ships. To fully realize the potential of these alternative fuels, it is essential to address their respective challenges through continued research, development, and collaboration between industry stakeholders and researchers. This will pave the way for a more sustainable and eco-friendly future in the shipping industry.

Acknowledgments Thanks for the support of the Heilongjiang Province Excellent Youth Science Fund (YQ2023E034). Thanks for the support of Fundamental Research Funds for the Central Universities of China (3072022JC0305).

References

1. 2023 IMO strategy on reduction of GHG emissions from ships. https://www.imo.org/en/OurWork/Environment/Pages/2023-IMO-Strategy-on-Reduction-of-GHG-Emissions-from-Ships.aspx, access online June 9, 2024.
2. Feng, Y., Qu, J., Wu, Y., Zhu, Y., & Jing, H. (2024). Utilizing waste heat from natural gas engine and LNG cold energy to meet heat-electric-cold demands of carbon capture and storage for ship decarbonization: Design, optimization and 4E analysis. *Journal of Cleaner Production, 446*, 141359.
3. Wang, Y., Cao, Q., Liu, L., Wu, Y., Liu, H., Gu, Z., & Zhu, C. (2022). A review of low and zero carbon fuel technologies: Achieving ship carbon reduction targets. *Sustainable Energy Technologies and Assessments, 54*, 102762.
4. Moshiul, A. M., Mohammad, R., Hira, F. A., & Maarop, N. (2022). Alternative marine fuel research advances and future trends: A bibliometric knowledge mapping approach. *Sustainability, 14*(9), 4947.
5. Cullinane, K., & Yang, J. (2022). Evaluating the costs of decarbonizing the shipping industry: A review of the literature. *Journal of Marine Science and Engineering, 10*(7), 946.
6. Dostál, Z., & Ladányi, L. (2018). Demands on energy storage for renewable power sources. *Journal of Energy Storage, 18*, 250–255.
7. Ager, J. W., & Lapkin, A. A. (2018). Chemical storage of renewable energy. *Science, 360*(6390), 707–708.
8. He, M., Sun, Y., & Han, B. (2022). Green carbon science: Efficient carbon resource processing, utilization, and recycling towards carbon neutrality. *Angewandte Chemie, 134*(15), e202112835.
9. Palys, M. J., & Daoutidis, P. (2022). Power-to-X: A review and perspective. *Computers & Chemical Engineering, 165*, 107948.
10. Xing, H., Stuart, C., Spence, S., & Chen, H. (2021). Alternative fuel options for low carbon maritime transportation: Pathways to 2050. *Journal of Cleaner Production, 297*, 126651.
11. Schlapbach, L., & Züttel, A. (2001). *Hydrogen-storage materials for mobile applications. nature, 414*(6861), 353–358.
12. Park, C., Jeong, B., & Zhou, P. (2022). Lifecycle energy solution of the electric propulsion ship with Live-Life cycle assessment for clean maritime economy. *Applied Energy, 328*, 120174.

13. Chai, W. S., Bao, Y., Jin, P., Tang, G., & Zhou, L. (2021). A review on ammonia, ammonia-hydrogen and ammonia-methane fuels. *Renewable and Sustainable Energy Reviews, 147*, 111254.
14. Gill, S. S., Chatha, G. S., Tsolakis, A., Golunski, S. E., & York, A. P. E. (2012). Assessing the effects of partially decarbonising a diesel engine by co-fuelling with dissociated ammonia. *International Journal of Hydrogen Energy, 37*(7), 6074–6083.
15. Wang, N., Huang, S., Zhang, Z., Li, T., Yi, P., Wu, D., & Chen, G. (2021). Laminar burning characteristics of ammonia/hydrogen/air mixtures with laser ignition. *International Journal of Hydrogen Energy, 46*(62), 31879–31893.
16. Lewandowski, M. T., Pasternak, M., Haugsvær, M., & Løvås, T. (2024). Simulations of ammonia spray evaporation, cooling, mixture formation and combustion in a direct injection compression ignition engine. *International Journal of Hydrogen Energy, 52*, 916–935.
17. Scharl, V., & Sattelmayer, T. (2024). Spectroscopic investigation of diesel-piloted ammonia spray combustion. *Fuel, 358*, 130201.
18. Zhou, X., Li, T., Wang, N., Wang, X., Chen, R., & Li, S. (2023). Pilot diesel-ignited ammonia dual fuel low-speed marine engines: A comparative analysis of ammonia premixed and high-pressure spray combustion modes with CFD simulation. *Renewable and Sustainable Energy Reviews, 173*, 113108.
19. Li, S., Li, T., Wang, N., Zhou, X., Chen, R., & Yi, P. (2022). An investigation on near-field and far-field characteristics of superheated ammonia spray. *Fuel, 324*, 124683.
20. Wu, X., Feng, Y., Gao, Y., Xia, C., Zhu, Y., Shreka, M., & Ming, P. (2023). Numerical simulation of lean premixed combustion characteristics and emissions of natural gas-ammonia dual-fuel marine engine with the pre-chamber ignition system. *Fuel, 343*, 127990.
21. Wu, X., Feng, Y., Xu, G., Zhu, Y., Ming, P., & Dai, L. (2023). Numerical investigations on charge motion and combustion of natural gas-enhanced ammonia in marine pre-chamber lean-burn engine with dual-fuel combustion system. *International Journal of Hydrogen Energy, 48*(30), 11476–11492.
22. Liu, L., Wu, Y., Wang, Y., Wu, J., & Fu, S. (2022). Exploration of environmentally friendly marine power technology-ammonia/diesel stratified injection. *Journal of Cleaner Production, 380*, 135014.
23. Ichikawa, Y., Niki, Y., Takasaki, K., Kobayashi, H., & Miyanagi, A. (2022). NH3 combustion using three-layer stratified fuel injection for a large two-stroke marine engine: Experimental verification of the concept. *Applications in Energy and Combustion Science, 10*, 100071.
24. Ichikawa, Y., Niki, Y., Takasaki, K., Kobayashi, H., & Miyanagi, A. (2023). Experimental study of combustion process of NH3 stratified spray using imaging methods for NH3 fueled large two-stroke marine engine. *Applications in Energy and Combustion Science, 13*, 100119.
25. Lu, Z., Ye, J., Gui, Y., Lu, T., Shi, L., An, Y., & Wang, T. (2023). Numerical study of the compression ignition of ammonia in a two-stroke marine engine by using HTCGR strategy. *Energy, 276*, 127578.
26. Zhou, X., Li, T., Chen, R., Wei, Y., Wang, X., Wang, N., et al. (2024). Ammonia marine engine design for enhanced efficiency and reduced greenhouse gas emissions. *Nature Communications, 15*(1), 2110.
27. Xu, X., Wang, Z., Qu, W., Song, M., Fang, Y., & Feng, L. (2024). Optimizations of energy fraction and injection strategy in the ammonia-diesel dual-fuel engine. *Journal of the Energy Institute, 112*, 101455.
28. Wolfram, P., Kyle, P., Zhang, X., Gkantonas, S., & Smith, S. (2022). Using ammonia as a shipping fuel could disturb the nitrogen cycle. *Nature Energy, 7*(12), 1112–1114.
29. Wu, B., Wang, Y., Wang, D., Feng, Y., & Jin, S. (2023). Generation mechanism and emission characteristics of N2O and NO_x in ammonia-diesel dual-fuel engine. *Energy, 284*, 129291.
30. Cairns, E. J., Simons, E. L., & Tevebaugh, A. D. (1968). Ammonia–oxygen fuel cell. *Nature, 217*(5130), 780–781.
31. Zhao, Y., Setzler, B. P., Wang, J., Nash, J., Wang, T., Xu, B., & Yan, Y. (2019). An efficient direct ammonia fuel cell for affordable carbon-neutral transportation. *Joule, 3*(10), 2472–2484.

32. Perčić, M., Vladimir, N., Jovanović, I., & Koričan, M. (2022). Application of fuel cells with zero-carbon fuels in short-sea shipping. *Applied Energy, 309*, 118463.
33. Wu, S., Miao, B., & Chan, S. H. (2022). Feasibility assessment of a container ship applying ammonia cracker-integrated solid oxide fuel cell technology. *International Journal of Hydrogen Energy, 47*(63), 27166–27176.
34. Duong, P. A., Ryu, B., Jung, J., & Kang, H. (2022). Thermal evaluation of a novel integrated system based on solid oxide fuel cells and combined heat and power production using ammonia as fuel. *Applied Sciences, 12*(12), 6287.
35. Duong, P. A., Ryu, B. R., Lee, H., & Kang, H. (2023). Thermodynamic analysis of integrated ammonia fuel cells system for maritime application. *Energy Reports, 10*, 1521–1537.
36. Machaj, K., Kupecki, J., Malecha, Z., Morawski, A. W., Skrzypkiewicz, M., Stanclik, M., & Chorowski, M. (2022). Ammonia as a potential marine fuel: A review. *Energy Strategy Reviews, 44*, 100926.
37. Herbinet, O., Bartocci, P., & Dana, A. G. (2022). On the use of ammonia as a fuel – A perspective. *Fuel Communications, 11*, 100064.
38. European Maritime Safety Agency. (2022). *Potential of ammonia as fuel in shipping*. EMSA. Download from: https://emsa.europa.eu/newsroom/latest-news/item/4833-potential-of-ammonia-as-fuel-in-shipping.html.
39. Ahmed, S., Li, T., Yi, P., & Chen, R. (2023). Environmental impact assessment of green ammonia-powered very large tanker ship for decarbonized future shipping operations. *Renewable and Sustainable Energy Reviews, 188*, 113774.
40. Cardoso, J. S., Silva, V., Rocha, R. C., Hall, M. J., Costa, M., & Eusébio, D. (2021). Ammonia as an energy vector: Current and future prospects for low-carbon fuel applications in internal combustion engines. *Journal of Cleaner Production, 296*, 126562.
41. Zhen, X., & Wang, Y. (2015). An overview of methanol as an internal combustion engine fuel. *Renewable and Sustainable Energy Reviews, 52*, 477–493.
42. Yin, X., Li, W., Duan, H., Duan, Q., Kou, H., Wang, Y., et al. (2023). A comparative study on operating range and combustion characteristics of methanol/diesel dual direct injection engine with different methanol injection timings. *Fuel, 334*, 126646.
43. Cheng, C., Cordtz, R. F., Thomsen, T. B., Førby, N. L., & Schramm, J. (2022). Application of methanol with an ignition improver in a small marine CI engine. *Energy Conversion and Management, 271*, 116311.
44. Sun, W., Jiang, M., Guo, L., Zhang, H., Jia, Z., Qin, Z., et al. (2023). Numerical study of injection strategies for marine methanol/diesel direct dual fuel stratification engine. *Journal of Cleaner Production, 421*, 138505.
45. Malmgren, E., Brynolf, S., Fridell, E., Grahn, M., & Andersson, K. (2021). The environmental performance of a fossil-free ship propulsion system with onboard carbon capture – A life cycle assessment of the HyMethShip concept. *Sustainable Energy & Fuels, 5*(10), 2753–2770.
46. Gong, C., Liu, Z., Su, H., Chen, Y., Li, J., & Liu, F. (2019). Effect of injection strategy on cold start firing, combustion and emissions of a LPG/methanol dual-fuel spark-ignition engine. *Energy, 178*, 126–133.
47. Güdden, A., Pischinger, S., Geiger, J., Heuser, B., & Müther, M. (2021). An experimental study on methanol as a fuel in large bore high speed engine applications – Port fuel injected spark ignited combustion. *Fuel, 303*, 121292.
48. Cesur, I. (2022). Investigation of the effects of water injection into an SI engine running on M15 methanol fuel on engine performance and exhaust emissions. *Energy, 261*, 125203.
49. Shi, J., Zhu, Y., Feng, Y., Yang, J., & Xia, C. (2023). A prompt decarbonization pathway for shipping: Green hydrogen, ammonia, and methanol production and utilization in marine engines. *Atmosphere, 14*(3), 584.
50. Tabibian, S. S., & Sharifzadeh, M. (2023). Statistical and analytical investigation of methanol applications, production technologies, value-chain and economy with a special focus on renewable methanol. *Renewable and Sustainable Energy Reviews, 179*, 113281.

51. https://www.abc-engines.com/en/markets/marine-propulsion/product-solutions/main-propulsion-engines—52, access online March 7, 2024.
52. https://www.manifoldtimes.com/news/cssc-subsidiary-successfully-ignites-methanol-engine-for-the-first-time/, access online March 7, 2024.
53. Chi, X., Chen, F., Mo, T., Li, Y., & Wei, W. (2024). Improve methanol efficiency for direct methanol fuel cell system via investigation and control of optimal operating methanol concentration. *Energy, 290*, 130147.
54. Munjewar, S. S., Thombre, S. B., & Mallick, R. K. (2017). Approaches to overcome the barrier issues of passive direct methanol fuel cell – Review. *Renewable and Sustainable Energy Reviews, 67*, 1087–1104.
55. Leo, T. J., Durango, J. A., & Navarro, E. (2010). Exergy analysis of PEM fuel cells for marine applications. *Energy, 35*(2), 1164–1171.
56. Ogden, J. M. (2002). Hydrogen: The fuel of the future? *Physics Today, 55*(4), 69–75.
57. Li, F., Yuan, Y., Yan, X., Malekian, R., & Li, Z. (2018). A study on a numerical simulation of the leakage and diffusion of hydrogen in a fuel cell ship. *Renewable and Sustainable Energy Reviews, 97*, 177–185.
58. Moradi, R., & Groth, K. M. (2019). Hydrogen storage and delivery: Review of the state of the art technologies and risk and reliability analysis. *International Journal of Hydrogen Energy, 44*(23), 12254–12269.
59. Lototskyy, M. V., Tolj, I., Pickering, L., Sita, C., Barbir, F., & Yartys, V. (2017). The use of metal hydrides in fuel cell applications. *Progress in Natural Science: Materials International, 27*(1), 3–20.
60. Chabane, D., Harel, F., Djerdir, A., Candusso, D., Elkedim, O., & Fenineche, N. (2019). Energetic modeling, simulation and experimental of hydrogen desorption in a hydride tank. *International Journal of Hydrogen Energy, 44*(2), 1034–1046.
61. Dall'Armi, C., Pivetta, D., & Taccani, R. (2023). Hybrid PEM fuel cell power plants fuelled by hydrogen for improving sustainability in shipping: State of the art and review on active projects. *Energies, 16*(4), 2022.
62. Li, J. Q., Xu, H., Wang, J. B., Wang, X. Y., Li, J. C., Kwon, S., et al. (2023). A review of methods to study the fatigue life of nodes connecting marine composite hydrogen storage tanks to ships under the action of external forces. *Journal of Energy Storage, 72*, 108367.
63. Qu, W., Fang, Y., Wang, Z., Sun, H., & Feng, L. (2022). Optimization of injection system for a medium-speed four-stroke spark-ignition marine hydrogen engine. *International Journal of Hydrogen Energy, 47*(44), 19289–19297.
64. World's First Successful Hydrogen Combustion Operation with a Large Marine Engine. https://www.mes.co.jp/english/press/2024/0307_002400.html, access online Mar 8, 2024.
65. Fernández-Ríos, A., Santos, G., Pinedo, J., Santos, E., Ruiz-Salmón, I., Laso, J., et al. (2022). Environmental sustainability of alternative marine propulsion technologies powered by hydrogen-a life cycle assessment approach. *Science of the Total Environment, 820*, 153189.
66. Wang, Z., Zhao, F., Dong, B., Wang, D., Ji, Y., Cai, W., & Han, F. (2023). Life cycle framework construction and quantitative assessment for the hydrogen fuelled ships: A case study. *Ocean Engineering, 281*, 114740.
67. Hoang, A. T., Pandey, A., De Osés, F. J. M., Chen, W. H., Said, Z., Ng, K. H., et al. (2023). Technological solutions for boosting hydrogen role in decarbonization strategies and net-zero goals of world shipping: Challenges and perspectives. *Renewable and Sustainable Energy Reviews, 188*, 113790.
68. Raucci, C., Calleya, J., Suarez De La Fuente, S., & Pawling, R. (2015). *November*. University of Strathclyde.
69. Guan, W., Chen, J., Chen, L., Cao, J., & Fan, H. (2023). Safe design of a hydrogen-powered-ship: CFD simulation on hydrogen leakage in the fuel cell room. *Journal of Marine Science and Engineering, 11*(3), 651.
70. https://www.frs.world/frs-fleet/ships/windcat-55, access online January 12, 2024.

71. Melideo, D., & Desideri, U. (2024). The use of hydrogen as alternative fuel for ship propulsion: A case study of full and partial retrofitting of roll-on/roll-off vessels for short distance routes. *International Journal of Hydrogen Energy, 50*, 1045–1055.
72. Andersson, J., & Grönkvist, S. (2019). Large-scale storage of hydrogen. *International Journal of Hydrogen Energy, 44*(23), 11901–11919.
73. Rivarolo, M., Rattazzi, D., Lamberti, T., & Magistri, L. (2020). Clean energy production by PEM fuel cells on tourist ships: A time-dependent analysis. *International Journal of Hydrogen Energy, 45*(47), 25747–25757.
74. Ustolin, F., Campari, A., & Taccani, R. (2022). An extensive review of liquid hydrogen in transportation with focus on the maritime sector. *Journal of Marine Science and Engineering, 10*(9), 1222.
75. Lee, H., Ryu, B., Anh, D. P., Roh, G., Lee, S., & Kang, H. (2023). Thermodynamic analysis and assessment of novel ORC-DEC integrated PEMFC system for liquid hydrogen fueled ship application. *International Journal of Hydrogen Energy, 48*(8), 3135–3153.
76. Mojarrad, M., Zadeh, M., & Rødseth, K. L. (2023). Techno-economic modeling of zero-emission marine transport with hydrogen fuel and superconducting propulsion system: Case study of a passenger ferry. *International Journal of Hydrogen Energy, 48*(71), 27427–27440.
77. Madsen, R. T., Klebanoff, L. E., Caughlan, S. A. M., Pratt, J. W., Leach, T. S., Appelgate, T. B., Jr., et al. (2020). Feasibility of the Zero-V: A zero-emissions hydrogen fuel-cell coastal research vessel. *International Journal of Hydrogen Energy, 45*(46), 25328–25343.
78. Pratt, J. W. (2017). Feasibility of the SF-BREEZE: A zero-emission hydrogen fuel cell high-speed passenger ferry (No. SAND2017-0692PE). Sandia National Lab.(SNL-CA), .
79. Jung, W., Choi, M., Jeong, J., Lee, J., & Chang, D. (2024). Design and analysis of liquid hydrogen-fueled hybrid ship propulsion system with dynamic simulation. *International Journal of Hydrogen Energy, 50*, 951–967.
80. Jung, W., & Chang, D. (2023). Deep reinforcement learning-based energy management for liquid hydrogen-fueled hybrid electric ship propulsion system. *Journal of Marine Science and Engineering, 11*(10), 2007.

The Role of Green Liquid Hydrogen in Sustainable Transportation: Opportunities and Barriers

Saeid Hassankhani Dolatabadi, Seyedvahid Vakili, and Aykut I. Ölcer

Abbreviations

CapEx	Capital cost
CCS	Carbon Capture and Storage
CH_2	Compressed Hydrogen
CI	Carbon Intensity
CII	Carbon Intensity Indicator
EEDI	Energy Efficiency Design Index
EEXI	Energy Efficiency Existing Ship Index
GHG	Greenhouse Gas
HFC	Hydrogen Fuel Cell
ICE	Internal Combustion Engine
IMO	International Maritime Organization
LH_2	Liquid Hydrogen
LNG	Liquefied Natural Gas
MBM	Market-Based Measure
MDO	Marine Diesel Oil
NFPA	National Fire Protection Association
OpEx	Operational cost

S. Hassankhani Dolatabadi (✉)
Mechanical Engineering Department, University of Victoria, Victoria, BC, Canada
e-mail: saeidh@uvic.ca

S. Vakili
Department of Civil, Environment and Maritime, University of Southampton, Southampton, UK
e-mail: S.Vakili@soton.ac.uk

A. I. Ölcer
Maritime Energy Management, World Maritime University, Malmö, Sweden
e-mail: AIO@wmu.se

C. O. Colpan et al. (eds.), *Fuel Cell and Hydrogen Technologies in Maritime Transportation*, https://doi.org/10.1007/978-3-031-86110-9_3

PEMFC	Proton Exchange Membrane Fuel Cell
PV	Photovoltaic
SEEMP	Ship Energy Efficiency Management Plan
SMR	Steam Methane Reforming
SOFC	Solid Oxide Fuel Cell
TTW	Tank-To-Wake
WTT	Well-To-Tank

1 Introduction

Like other industries, the maritime industry faces a squeezing requirement for sustainable solutions to reduce GHG emissions due to the economic, natural, and social damages caused by climate change and global warming. Fourth, the International Maritime Organization's (IMO) GHG study declared that shipping GHG emissions in the maritime industry increased by more than nine percent from 2012 to 2018. Also, predictions show an eight percent increase in GHG emissions by 2050 caused by population and trade growth [1].

In its efforts to mitigate air emissions from shipping, IMO, as the regulatory body for international shipping, is pursuing an interdisciplinary approach. This approach encompasses operational and technical measures, with ongoing discussions regarding economic measures [2]. Among the technical measures are the Energy Efficiency Design Index (EEDI) and the Energy Efficiency Existing Ship Index (EEXI), while operational approaches include the Ship Energy Efficiency Management Plan (SEEMP) and the Carbon Intensity Indicator (CII). Concurrently, progress is being made on the implementation of economic measures, such as the Market-Based Measure (MBM) [3, 4].

To meet zero-emission shipping, the IMO has revised its initial GHG emissions strategy in 2023. The level of ambition in the revised strategy aims to significantly reduce Carbon Intensity (CI) in shipping by enhancing energy efficiency, targeting a 40% reduction by 2030 compared to 2008 levels. Additionally, the strategy aims for the adoption of zero or near-zero GHG emission technologies, fuels, and energy sources, constituting at least 5% (with a goal of 10%) of the energy used by international shipping by 2030. Milestones include a 20% reduction (with an aspiration for 30%) in GHG emissions by 2030 and a 70% reduction (with a striving target of 80%) by 2040 compared to 2008 levels [5] (see Fig. 1).

To achieve the IMO's goals, various technologies and measures have been proposed. These include utilizing wind-assisted propulsion systems [6, 7], implementing air lubrication systems [8], improving new hull designs [9], utilizing alternative fuels [10], and implementing performance management strategies [11]. Among the various measures aimed at promoting sustainability within the maritime sector, the utilization of alternative fuels has a significant potential to play a pivotal role [12]. Green hydrogen has emerged as a particularly noteworthy

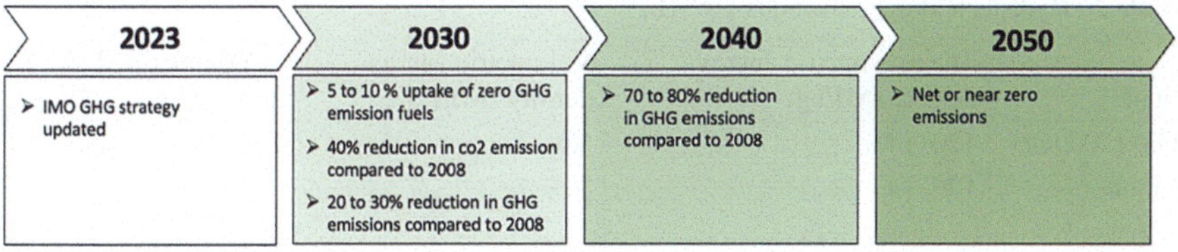

Fig. 1 IMO revised GHG strategy in GHG mitigation

Table 1 Hydrogen production methods [15–17]

Type of hydrogen	Source of energy and hydrogen	Production methods	Current production	Production main emissions	Purity
Brown	Coal	Gasification	96%	CO_2, CO	90%
Gray	Natural gas	Steam reforming		CO_2, CH_4	70–75%
Blue				CO_2 (mostly captured and stored), CH_4	
Pink	Nuclear electricity and water	Electrolysis	<1%	Oxygen	99.99%
Green	Renewable electricity and water	Electrolysis	<4%	Oxygen	99.99%

alternative fuel, garnering widespread attention due to its considerable benefits in emissions reduction and advancements in production, storage, and distribution technologies. This recognition was further underscored at the COP28 agenda, where over 30 countries pledged to endorse a declaration of intent aimed at mutually recognizing certification schemes for renewable and low-carbon hydrogen (green hydrogen) and its derivatives [13].

Considering the above, this book chapter focuses on green LH_2 as a potential source of zero-emission propulsion for vessels. The chapter provides an overview of green LH_2 production, storage, and distribution, followed by an examination of its feasibility for vessel propulsion, both technically and economically. Section 2 covers green LH_2 production and storage, while Sect. 3 elaborates on the feasibility of utilizing LH_2 onboard as a fuel. Section 4 assesses emissions associated with green LH_2 application, followed by an economic evaluation of green LH_2 utilization on vessels in Sect. 5. Section 6 reviews safety considerations, and conclusions are provided in Sect. 7.

2 LH_2 Production and Storage

Hydrogen can be generated through various methods and stored using more than 30 techniques (options), including gases, liquid, and even solid forms [14]. Tables 1 and 2 summarize the most regular production methods and storage techniques of

Table 2 Hydrogen storage methods [16, 18]

State	The gravimetric energy density (MJ/kg)	Volumetric energy density (MJ/lit)	Volumetric density (kg/m3)
Gas (700 bar)	120–143	5.6	41
Liquid (−252.8° c)	120–143	8	70.8

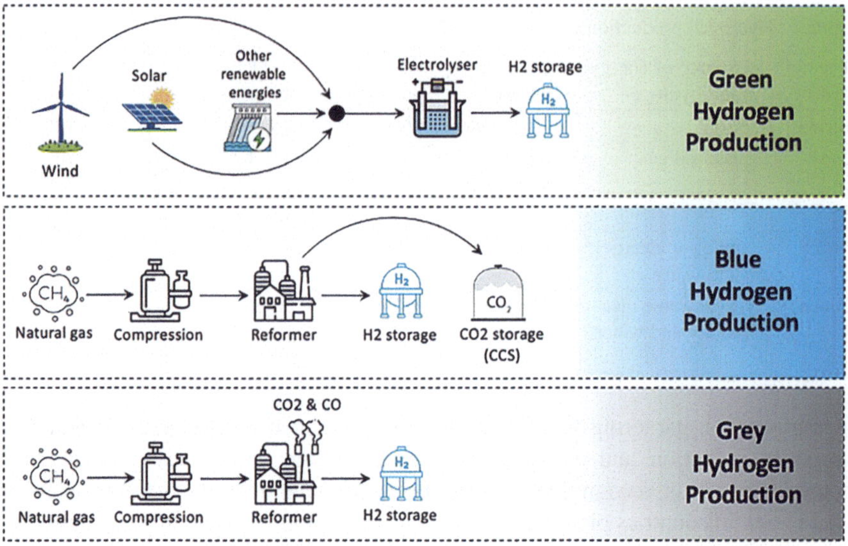

Fig. 2 Hydrogen production methods

hydrogen correspondingly. This chapter focuses on green hydrogen in terms of production and LH_2 regarding storage.

The produced hydrogen by renewable energies is known as green hydrogen (see Fig. 2). Although green hydrogen production costs more than other conventional fossil fuels at the moment, it is estimated the cost of green hydrogen production decline in the coming years due to more innovative measures and technical advancement in this section [19].

LH_2 has a volumetric density of around 70.8 kg/m3, almost double of compressed hydrogen (CH_2) at 700 bar (approximately 42 kg/m3); in other words, storing hydrogen in liquid form provides double the amount of hydrogen in terms of weight [20]. Moreover, LH_2 has lower potential risks, such as lower pressure through the storage compared to CH_2, and has advantages, such as higher purity and more rapid bunkering, leading the industry to consider the higher density benefit of LH_2 storage [21].

Storing LH_2 requires maintaining extremely low temperatures (−252.8 °C) (refer to Fig. 3), which necessitates a high-quality cooling system and results in significant

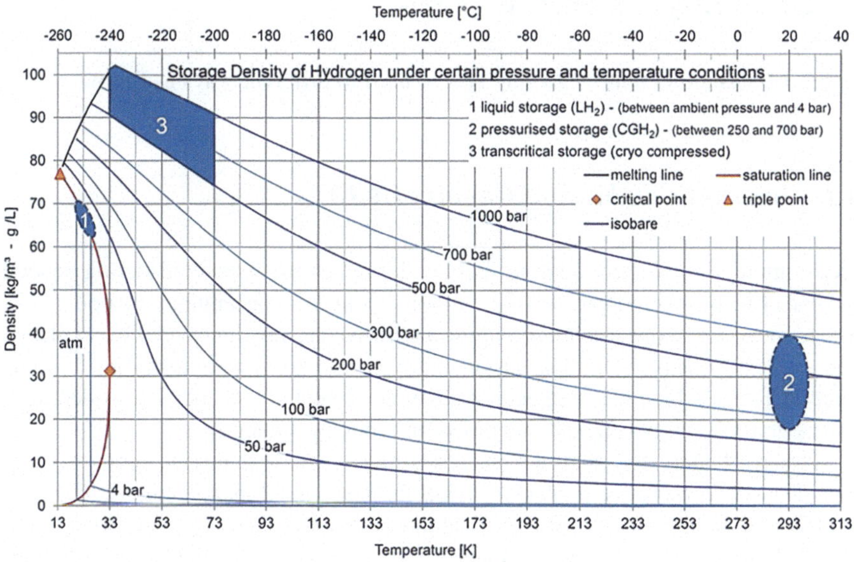

Fig. 3 Hydrogen storage density under certain pressure and temperature [25]

energy consumption [22]. Factors such as the scale of liquefaction, initial hydrogen temperature, and other variables contribute to increased energy consumption during the liquefaction process [23]. Although 3.3 to 3.9 kWh/kg LH_2 is required to liquefy hydrogen from ambient conditions in theory, practical applications show this amount of energy is more remarkable [24]. However, innovative liquefication techniques like magnetic regenerative may reduce the energy demand in the liquefaction process and storage of hydrogen; by current technologies, around 30–40% of stored hydrogen is required to keep hydrogen in liquid [23].

The Boil-off gas of LH_2 through the storage period is one of the most critical challenges that could lead to the loss of hydrogen. Effectively managing boil-off gas and high-quality isolation systems are crucial on vessels to prevent LH_2 loss, whereas both hydrogen production and storage consume energy [20].

According to the DNV alternative fuel information platform, green hydrogen production capacity is less than one million tons of H_2 per year at the moment, and it will be increased to around 12 million tons per year in 2027, based on announced projects around the world. In addition, 21 H_2 production plants are in operation, which would increase to more than 465 plants in the coming years [26].

Despite the challenges inherent in production, transportation, and utilization, research studies and applications are currently underway worldwide to facilitate the use of green LH_2 as a fuel. An illustrative example of this is Kawasaki Heavy Industries' initiative in 2022, which marked the first-ever transportation of LH_2 cargo from Australia to Japan, underscoring the recognized benefits of LH_2 [27].

3 Green LH_2 as a Fuel on Vessels

Green LH_2 as a fuel onboard a ship can be used in different ways as follows:

3.1 Internal Combustion Engines (ICE)

Exploration of alternative marine fuels and engines involves considering several factors. The primary objective is to modify conventional engines to operate on eco-friendly fuels that maintain engine performance, are fuels readily available, and are cost-effective [28]. The sustainability of the maritime industry has become a recent focus, which includes economic, environmental, and social considerations by integrating energy-efficient and emission-reducing strategies.

A range of alternative fuels are being considered for use in marine engines. These include gaseous fuels like methane, ethane, propane, and dimethyl ether [29]. Methanol is also being considered as a potential alternative fuel. It has a liquid form similar to diesel, but its technical maturity is still moderate [30]. Carbon-free fuels like hydrogen and ammonia have gained significant attention from both industry and academia due to their established production methods [18].

One of the main challenges with ammonia combustion is the flame front's slow progression, making it difficult to use in engines. The burning velocity of an ammonia-air mixture is usually between 6 to 8 cm/s, which is notably lower compared to alternative fuels like methane, where burning velocities can reach approximately 35 cm/s [31]. Using methanol as the only fuel in compression-ignition engines presents challenges due to its low reactivity [32]. In 2010, Sayin et al. investigated a four-stroke direct-injection diesel engine that used up to 15% of methanol blends. The study revealed elevated NOx emissions at high loads (around 60%) but reduced smoke and CO_2 emissions due to methanol's low carbon content [33].

The exploration of Hydrogen ICE commenced in 1820 with Cecil [34]. Various research institutions and governmental bodies have initiated research-oriented initiatives, and a subset of these endeavors has demonstrated the efficacy and safety of employing hydrogen in ICE [35].

Tokyo University is one of the first institutes that have been engaged in the development of hydrogen ICE (since 1970), developing applications such as buses and trucks with a hydrogen-fueled engine [36]. One of the notable advantages of employing hydrogen ICE is the lower retooling cost for production for both conventional engines and Wankel engines, where Mazda made strides by creating Wankel engines designed to burn hydrogen [37]. BMW has also been engaged with hydrogen-fueled engines and has tested a car from 2005 to 2007, where the test result was impressive, reaching a speed of 301 km/h [38].

Recently, the development of hydrogen ICE for heavy-duty commercial vehicles has been expanded since the industry has made an effort to reduce emissions to

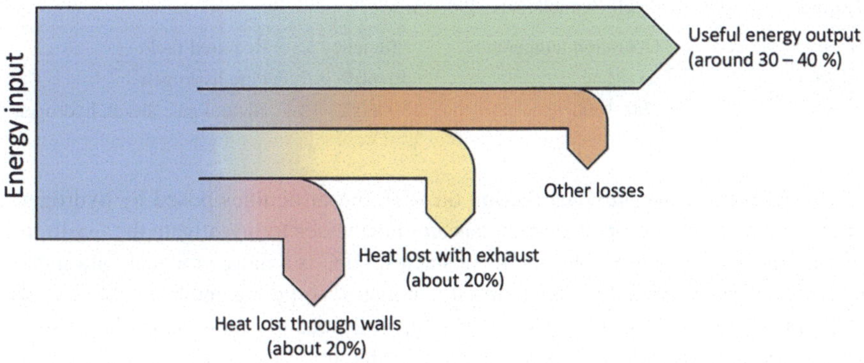

Fig. 4 Energy losses in ICE

minimize the impacts of climate change. The efficiency of a hydrogen 2-stroke ICE can be comparable to that of a traditional 2-stroke ICE, where conventional ICE typically operates at an efficiency of 30 to 40% (Fig. 4). It should be noted like conventional engines, the efficiency of hydrogen ICE is higher at high loads [39].

Using hydrogen alone in compression ignition engines poses a challenge due to its high self-ignition temperature; however, hydrogen finds applications in ICE as a dedicated fuel, in bi-fuel configurations, or in dual-fuel operations due to the elevated stoichiometric air-to-fuel ratio [40], where a mix of liquefied natural gas (LNG) and hydrogen, up to 30%, works well in maritime applications [40].

Burning hydrogen in a 2-stroke ICE is more challenging than a 4-stroke engine, where combustion conditions, ignition, the tiny size of hydrogen, and other technical parameters raise challenges.

3.2 Hydrogen Fuel Cells (HFC)

Unlike traditional batteries that store chemical energy, fuel cells continuously produce electricity as long as fuel and oxygen are supplied. Fuel cells are particularly well-suited for applications requiring continuous power generation, such as vehicles and power plants [41]. Despite notable advancements, the utilization of fuel cells for ships is primarily experimental, with ongoing efforts focused on early-stage design and development.

Fuel cells typically exhibit an energy efficiency range of 40 to 95%, surpassing the efficiency of some other energy generation systems [42]. For instance, a conventional car's ICE (4-stroke engine) operates at about 24–32% efficiency [43], while steam power plants and 2-stroke ICE usually achieve efficiencies in the range of 30–40% [44]. The evaluation of a fuel cell's performance relies on thermodynamic assessments; using hydrogen's lower heating value, the maximum fuel cell efficiency would be 94.5% [45].

Table 3 —Various fuel cell characteristics [50, 52, 53]

Type of fuel cell	Operation temperature	Efficiency	Required fuel
PEMFC	65–85 °c	40–60%	Pure hydrogen
SOFC	800–1000 °c	50–60%	Natural gas, diesel, hydrogen

As outlined in the previous section on ICE, the difficulties posed by hydrogen-based engines have compelled scientists and businesses to investigate the feasibility of utilizing HFC on ships. Despite its obstacles, HFC is a more practical option than hydrogen ICEs, especially considering the industry's ongoing endeavors to electrify ships in the future [46]. However, HFCs typically occupy more space in terms of volumetric density than conventional diesel generators [40].

While different variations of HFC exist, their fundamental operational principles remain consistent [47]. The proton exchange membrane fuel cell (PEMFC) and solid oxide fuel cell (SOFC) are identified as the two most efficient and suitable types of fuel cell within maritime transportation [48–51]. Table 3 summarizes the key characteristics of these two fuel cell types.

The PEMFC presents an intriguing alternative to the SOFC due to several advantages, including reduced noise, greater compactness, enhanced resistance to vibrations (resulting in longer lifespan), availability in the market, and lower pollutant emissions [54]. Nevertheless, the SOFC could be a viable option for larger ships, such as cargo vessels [46].

Many studies have investigated the use of fuel cell technology in marine transportation. For instance, Sapra et al. 2021, investigated the combination of ICE with fuel cells for large ocean-going vessels [55]. Ahn et al. 2018, proposed a hybrid system that utilizes generators and SOFC on ships, where the diesel-electric generator, SOFC, and gas turbine provide a power of 16 MW in the open sea for the vessel [49]. Table 4 presents a summary of some proposed combinations of fuel cell models and applications with other power sources onboard vessels for marine transportation.

Madsen et al. 2020, designed a coastal research vessel called Zero-V that uses HFC as its primary propulsion system. The ship has a beam of 56 feet, a length of 170 feet, a draft of 12 feet, and can travel at a speed of 10 knots. It has two liquid hydrogen tanks with a total capacity of 10,900 kg and a PEMFC that can produce a maximum output of 1800 kW during voyages, covering approximately 2400 nautical miles in 15 days. The authors highlighted that through wave analysis, the vessel significantly diminishes pollutants (hydrocarbons and particulate matter) and GHG emissions by as much as 91.4% compared to a ship of the same size that uses conventional fuels [57].

Additionally, the Hydra was a boat that ran on fuel cells and could carry up to 22 passengers with a capacity of 5kWe. It was operated on the Rhine River near Bonn, Germany, from 1999 to 2001, then used as a ferry boat in Ghent, Belgium, during an electric boat conference in 2000. It received full certification from Germanischer Lloyd for passenger transport [59]. Another noteworthy project is the Zemships, a small passenger ship—the FCS Alsterwasser—manufactured from 2003 to 2013 and unveiled in 2008 in Hamburg. It was one of the first passenger

Table 4 A few applications that use HFC as fuel onboard

Research/ Project	Year	Technologies	Application	Result
[55]	2021	ICE, fuel cell	Ocean-going ship	Efficiency improvement (around 8.3%), emissions reduction (30% NOx reduction, 12% CO_2 reduction)
[49]	2018	ICE, fuel cell, steam turbine	Large ethane carrier	SFOC and gas turbines provide 37.5% of the required power
[56]	2014	ICE, Methanol SOFC	Supply vessel	SOFC provides 20% of the required power at ECO-speed
[57]	2020	HFC	Coastal research vessel	Up to 91.4% emissions reduction
[58]	2023	HFC	Ferry	89% reduction in global warming potential through 30 years lifespan
[59]	2000	HFC	Passenger boat	Could be zero-emissions through operation
[60]	2008	HFC	Passenger boat	Could be zero-emissions through operation
[18]	2023	HFC, wind, solar	Bulk carriers	Zero-emission under specific conditions (vessel size, speed, and route)
[50]	2010	HFC, battery, ICE	Passenger ship	Electrical efficiency of 50–60%, the potential to increase the total efficiency by using a heat recovery system
[61]	2012	HFC, solar, wind, battery, ICE	Passenger boat	75–100% potential reduction of GHG emissions
[62]	2015	HFC	High-speed ferry	Could be zero-emissions through operation

ships powered by fuel cells and could accommodate up to 100 passengers. It used a 96 kW PEMFC and seven lead gel batteries equipped with 12 compressed hydrogen storage tanks at a pressure of 350 bar [60].

In 2010, the MF Vågen was introduced, which featured 12 kW fuel cells and 2–3 kg metal hydride hydrogen storage. The vessel also incorporates 25 kWh lithium batteries and a 10 kW DC motor [50]. In 2012, the Hornblower Hybrid was unveiled, combining a diesel generator, batteries, photovoltaics (PV), wind power, and fuel cells for energy [61]. The SF-BREEZE, a two-motor boat, is equipped with 41 x 120 kW fuel cells and utilizes a type C storage tank with a 1200 kg LH_2 capacity [62]. In 2023, Norway launched the world's first commercial ship powered by hydrogen, the HySeas III, which operates on a route between two Norwegian cities and is powered by a combination of PEM HFC and batteries with a capacity of 600 kW and 768 kWh, respectively [58]. Most of these ships are still in operation today.

In another modeling project, Hassankhani et al. 2023, evaluated the feasibility of a hybrid propulsion system for bulk carriers, combining green hydrogen, wind, and solar energy sources. The research analyzed a variety of ships based on their size, resulting in a total of 4986 ships being assessed. The calculations demonstrated that the proposed hybrid zero-emissions model could be implemented on specific routes and under particular conditions and would be suitable for certain vessel sizes in which the HFCs would serve as the primary power source model [18].

The challenges and difficulties related to using hydrogen ICE in marine applications have led the industry to consider using HFC instead. However, HFC also presents specific challenges, including high costs and the need for extensive infrastructure development on vessels. Integrating fuel cell systems into existing vessel designs is complex, and the limited availability of hydrogen refueling infrastructure in maritime routes and ports further complicates the adoption of this technology. Additionally, fuel cell technology's high initial investment and maintenance costs pose significant financial barriers to its widespread adoption in the maritime industry.

4 Green LH_2 Applications Emissions

Combustion of hydrogen in hydrogen ICE leads to water production, as follows:

$$2\,H_2 + O_2 \rightarrow 2\,H_2\,O$$

Nonetheless, due to the varied composition of air, where nitrogen constitutes the primary component, hydrogen combustion in the air gives rise to the formation of nitrogen oxides, commonly referred to as NOx [63, 64]. This feature parallels the combustion processes observed in other high-temperature fuels such as kerosene, gasoline, diesel, or natural gas [65]. The challenge is intensified by the exceptionally elevated temperatures produced during hydrogen combustion, resulting in the categorization of hydrogen combustion engines as not entirely zero-emission [66]. However, the incorporation of advanced combustion technologies and appropriate engine design holds the potential to mitigate NOx emissions.

Like any ICE, a small amount of oil is necessary for lubrication. Some of this oil may enter the combustion chamber and participate in combustion [67]. As a result, the exhaust gases may contain trace amounts of the combustion products of this oil, usually consisting of very small quantities of SO2 and particulates. However, these levels are several orders of magnitude lower than those observed in the exhaust gases of a gasoline or diesel engine [66]. The significant advantage of hydrogen lies in its lack of CO_2 emissions during combustion, particularly in applications where the priority is to reduce carbon emissions [66].

In applications utilizing hydrogen in fuel cells, the sole emission is water vapor (see Fig. 5). Fuel cells undergo electrochemical processes to convert hydrogen and oxygen into electricity, with water being the primary by-product [68]. This makes HFC an attractive option for environmentally-conscious initiatives, as they offer emission-free energy solutions.

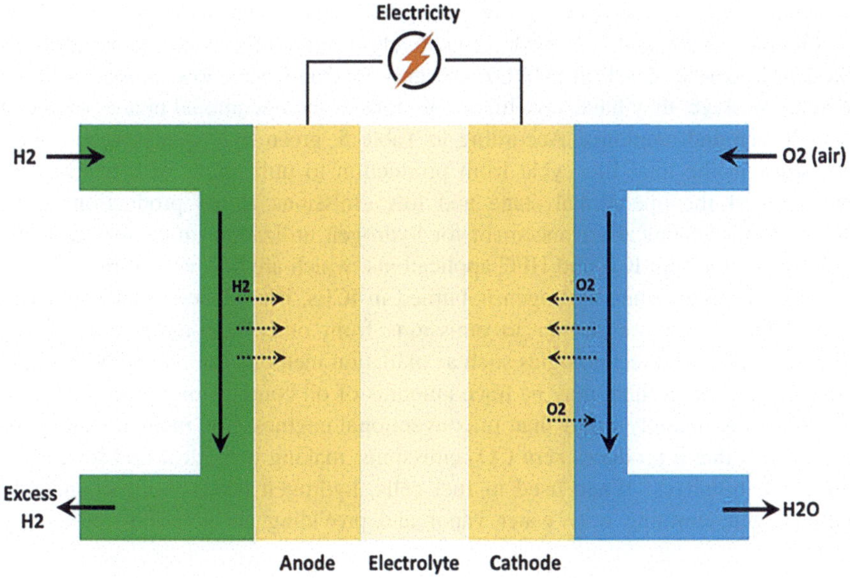

Fig. 5 A typical fuel cell [69]

Table 5 Emissions of different fuels in marine transportation [30, 70, 71]

Fuel	Production phase—WTT				Operation phase—TTW			Technical advancement
	CO_2	N2O	CO_2 eq	NOx	CO_2	NOx	SOx	
	g/MJ				g/kWh			
Fossil-based LNG	26	0.00016	26.03	0.091	412	1.17	0.003	High
MDO	19.6	0.00054	19.7	0.23	524	14.8	0.32	High
Biomass-based methanol	17	0.00022	17.04	0.056	522	3.05	≈ 0	Medium
NG-based methanol	20	0.00029	20.05	0.046	522	3.05	≈ 0	Medium
Brown NH_3	64.8	0.00045	64.89	0.044	0	0.054	≈ 0	Low
Green NH_3	18.5	0.00045	18.59	0.044	0	0.054	≈ 0	Low
Brown H_2	77.9	0.00025–0.0025	77.95–78.4	0.034	0	≈ 0	≈ 0	Low
Green H_2	7.9	0.00041	7.98	0.03	0	≈ 0	≈ 0	Low

The comparative analysis of green LH_2 emissions alongside other fuels is delineated in Table 5 below, considering two distinct phases: well-to-tank (WTT) and tank-to-wake (TTW). As depicted in Table 5, both hydrogen and ammonia exhibit negligible CO_2 and other emissions during the operational stage, and their total emissions depend on the production phase, where the CO_2 eq emission for brown ammonia, brown hydrogen, green ammonia, and green hydrogen is 64.89, 77.95–78.4, 18.59, and 7.98 g/MJ respectively. On the other hand, although fuels like LNG, marine diesel oil (MDO), and grey methanol have low emissions at the production stage, they have very high emissions at the operational phase compared to hydrogen and ammonia. According to Table 5, green hydrogen has the lowest emissions in the total life cycle from production to utilization, with nearly zero emissions at the operational stage and low emissions at the production stage. However, the technical advancement for hydrogen utilization on vessels is in the early stages for both ICE and HFC applications, which are of low maturity.

To summarize, when hydrogen is burned in ICEs, it produces water vapor and some NO_x, which are similar to emissions from other high-temperature fuels. However, advanced technologies such as oxidation methods can reduce NOx emissions [72]. Though there may be trace amounts of oil combustion by-products, the levels are significantly lower than in conventional engines. The main advantage of hydrogen is that it produces zero CO_2 emissions, making it an ideal fuel for carbon reduction initiatives. When used in fuel cells, hydrogen reacts electrochemically with oxygen, emitting only water vapor and providing an emission-free energy source.

5 Initial and Operational Cost of Green LH_2 Propulsion System Onboard

Considering green LH_2 as an alternative fuel for propulsion, two types of expenses need to be taken into account: capital cost (CapEx) and operational cost (OpEx). CapEx refers to the expenses in setting up the infrastructure and technologies necessary to enable the ship to use hydrogen as a propulsion method, while OpEx refers to the ongoing costs associated with operating the vessel, including the expenses of maintenance and hydrogen fuel cost. The main CapEx and OpEx items for LH_2 maritime applications are shown in Fig. 6 below.

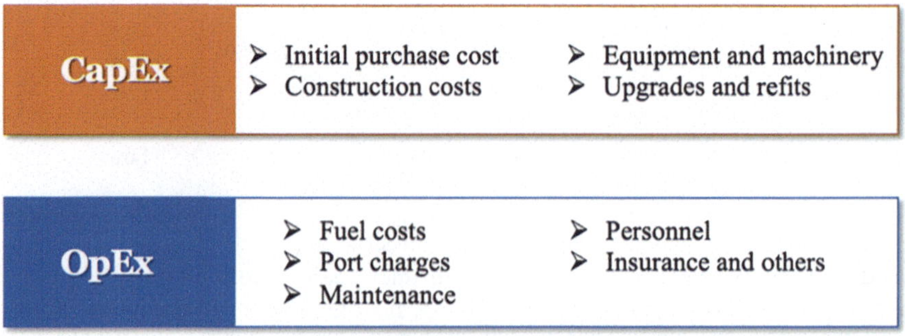

Fig. 6 The main CapEx and OpEx items of maritime applications [73]

Conventional ICEs have an advantage over other types of engines due to their well-established infrastructure, making them more economically viable in terms of the initial investment. Hydrogen engines are emerging as a strong competitor, but they face challenges associated with the ongoing development of infrastructure. On the other hand, HFC systems are generally more expensive initially due to the complex technology and the need for specialized infrastructure, including hydrogen refueling stations [73].

Conventional ICEs have been optimized over time, resulting in high operational efficiency. However, they often require costly maintenance due to their complex combustion systems. In addition, hydrogen engines offer a cleaner and potentially more efficient combustion process, which can lead to reduced maintenance costs compared to conventional engines [74]. Furthermore, HFC systems have fewer moving parts since they operate through electrochemical reactions, resulting in lower maintenance requirements and longer operational lifespan [75]. The overall effectiveness and maintenance costs depend on the specific application and the technology's maturity.

However, fuel costs depend on the vessels' type and size; the costs can reach up to 70% of total OpEx [76, 77]. Therefore, the cost of hydrogen production plays a pivotal role in determining the economic competitiveness of these technologies [78]. Conventional ICEs rely on fossil fuels, with costs susceptible to fluctuations in the global oil markets (see Fig. 7). However, hydrogen engines and fuel cell

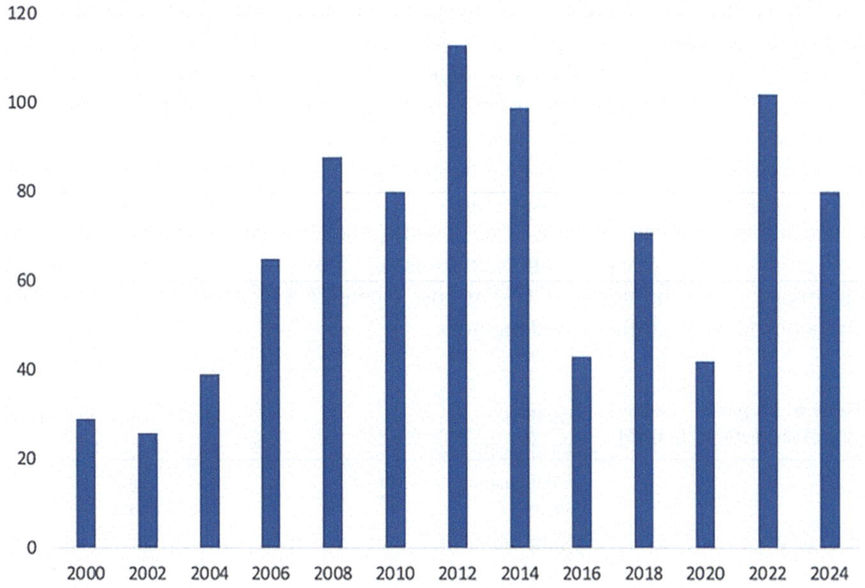

Fig. 7 Fluctuation of average annual Brent crude oil price from 2000 to 2024 (in US dollars per barrel) [79]

systems use hydrogen as a versatile fuel that can be produced through various methods, including electrolysis by renewable energies.

The cost of hydrogen can vary significantly from country to country due to various factors, including production methods, infrastructure development, and government regulations. Countries like Japan and South Korea, which are at the forefront of hydrogen technology adoption, tend to focus on producing green hydrogen, which often results in higher prices due to the costs associated with using renewable energy sources. These regions have invested heavily in building hydrogen infrastructure, which significantly impacts the overall pricing dynamics [80, 81].

In addition, Government policies and incentives play a significant role in determining the prices of hydrogen worldwide. Countries that are committed to sustainable initiatives and carbon neutrality often provide subsidies and incentives to reduce the costs associated with green hydrogen. For instance, in Europe, there are strategic plans to increase the production of hydrogen from renewable sources in line with ambitious climate goals [82]. As the hydrogen market continues to evolve, it is crucial for stakeholders to have a nuanced understanding of the geographical and policy-driven intricacies to navigate the complex terrain of hydrogen pricing [83].

In some regions, hydrogen production primarily depends on fossil fuel processes (grey hydrogen), resulting in lower production costs. For instance, countries like the United States and those in the Middle East use their abundant natural gas reserves to produce hydrogen through steam methane reforming (SMR). Although SMR is a cost-effective method, it raises concerns about carbon emissions, highlighting the need to balance cost efficiency and environmental sustainability when determining hydrogen prices [83].

Table 6 below presents the costs associated with various hydrogen production methods. The cheapest method is coal gasification at $1.34 per kg, but it has the significant drawback of generating high emissions. Conversely, photovoltaic (PV) electrolysis is the cleanest method, producing zero emissions, but it comes at a much higher cost, ranging from $5.78 to $23.27 per kg. Other methods fall in between these extremes in terms of cost, with some methods, like biomass gasification, having a range due to variations in feedstock. Table 6 also highlights the impact of carbon capture and storage (CCS) on cost, where CCS increases costs but reduces emissions from fossil-fuel-based methods.

Table 6 H_2 production cost by various methods [84–86]

Method	Cost ($/ kg H_2)
Coal gasification	1.34
Coal gasification with CCS	1.63
SMR	2.08
SMR with CCS	2.27
Biomass gasification	1.77—2.77
Nuclear electrolysis	3.56–7.0
Wind electrolysis	5.27–9.37
PV electrolysis	5.78–23.27

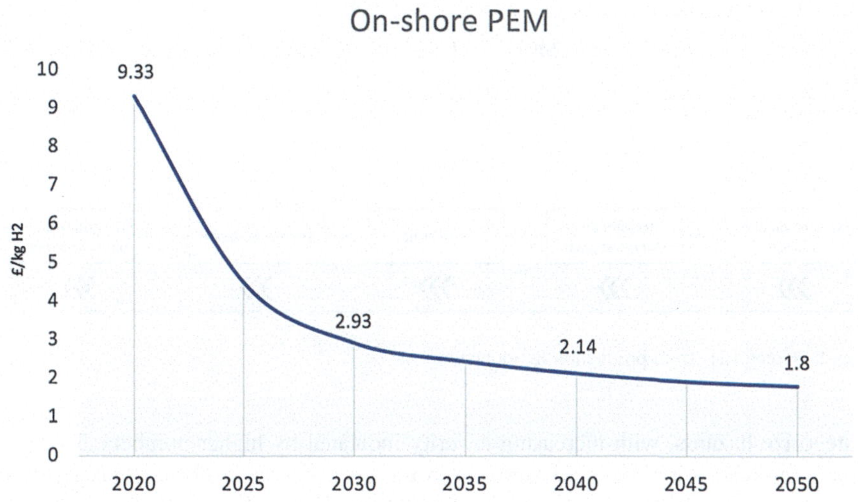

Fig. 8 H_2 production cost using wind power [88]

China is currently making significant investments in green hydrogen infrastructure, focusing on production, storage, and transportation. As per the roadmap presented by China EV100, the projected cost of green hydrogen is expected to be around $6 per kilogram by 2025, $4.5 per kilogram by 2035, and $3 per kilogram by 2050 [87].

The levelized cost of hydrogen for generating hydrogen through PEM electrolysis in the UK offshore and onshore sectors using wind power is illustrated in Fig. 8, showing a promising outlook for hydrogen production costs. Green H_2 is projected to cost around 1.8 €/kg H_2 in 2050, making it a more cost-effective option than LNG [88].

The feasibility of using green LH_2 as a propulsion fuel relies on both CapEx and OpEx. Conventional ICEs have the advantage of having established infrastructure already in place, whereas engines that run on hydrogen and fuel cell systems face challenges related to infrastructure. The economic competitiveness of hydrogen production is affected by fluctuating production costs, which are influenced by production methods and government policies.

6 LH_2 Safety Onboard

To ensure the safe use of hydrogen in global maritime transportation, it's essential to comprehend its inherent hazards fully. Careful consideration must be given to a variety of factors. The National Fire Protection Association (NFPA) in the United States has developed a standardized system known as NFPA 704 for identifying material hazards during emergencies [89]. This system uses a scale of 0 to 4 to

Table 7 NFPA standard regarding the hazard of fuels [89]

	LH_2	HFO/MGO	LNG	Methanol	Ammonia	Biodiesel
Health	0	1	3	1	3	1
Flammability	4	2	4	3	1	1

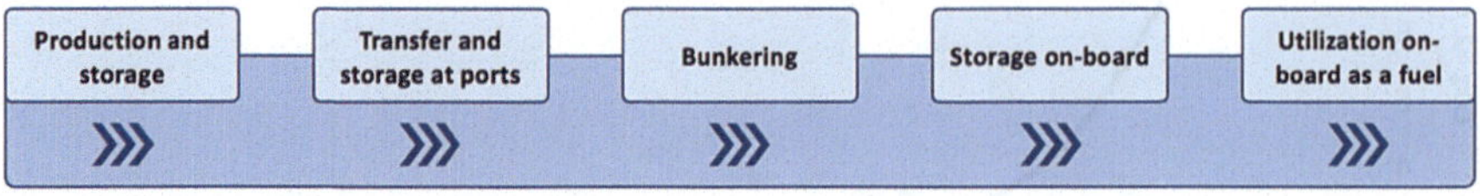

Fig. 9 Green LH_2 from production to application

categorize hazards, with increasing severity indicated by higher numbers. Table 7 briefly compares hydrogen's hazards with those of other typical fuels, taking into account the two NFPA 704 standards of health and flammability risks.

Regarding the use of LH2 as fuel on vessels, specific safety challenges regarding storage, transportation, and utilization need to be addressed. However, even green LH_2 production onshore has its own set of safety considerations that need to be considered, such as the risk of leakage, electrical hazards, and pressure build-up (see Fig. 9). It's essential to remember that utilizing a new alternative fuel comes with its own risks, just like conventional fossil fuels did during their early stages [90].

6.1 LH_2 Storage Risks Onboard

While hydrogen is non-toxic, environmentally benign, and radiation-free, its combustion can pose a significant danger due to its colorless and invisible flame [91]. Hydrogen possesses lower ignition energy (approximately 0.019 MJ) compared to gasoline and methane (0.24 and 0.29 MJ, respectively) [92]. On the other hand, the explosive range of hydrogen in air is significantly broader, ranging from 18% to 58% by volume, whereas it is 1.1% to 3.3% for gasoline and 5.7% to 14% for natural gas [90].

LH_2 storage on a vessel presents unique risks due to cryogenic temperatures, flammability, and potential leaks. However, double-containment systems and advanced monitoring technologies can mitigate these risks [93]. LH_2 needs to be stored at extremely low temperatures, usually around −253 °C. This presents several challenges, particularly concerning cryogenic conditions. Cryogenic storage involves keeping hydrogen in liquid form. However, this method poses risks of embrittlement in materials, transportation, and discharge of LH_2 in cryogenic storage conditions on vessels [93]. To address these risks, strict safety protocols must be observed, and specialized double-walled or double-containment tanks are required to handle the extremely low temperatures involved [94].

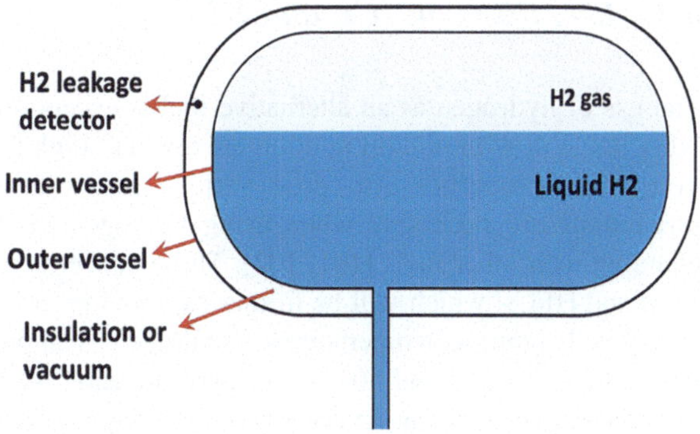

Fig. 10 Typical LH_2 storage tank [93, 95]

A double containment system must be employed to enhance safety and prevent hydrogen leakage into the open air in case of failure or leakage in the primary layer, as shown in Fig. 10 below. This system consists of an extra layer that serves as a backup to the primary layer. The materials used for these containments must be able to withstand cryogenic temperatures without becoming brittle [94].

Furthermore, it is crucial for storage tanks to be well-insulated to reduce boil-off losses due to hydrogen's ultra-low boiling point (about $-253°$ c) and low heat of vaporization. Proper safety precautions must be taken to handle vented hydrogen gas, as it has a wide flammability range with air [96].

Hydrogen is considered safer than other fuels due to its unique properties in case of leakage [97]. Hence, H_2 is 14 times lighter than air and can quickly ascend to the atmosphere at a rate of 20 m/s [90]. Moreover, the likelihood of a hydrogen explosion in open air is very low, owing to its high volatility and diffusion [98]. However, if there is any leakage, hydrogen may cause cold burning and extended duration when stored in liquid form [91].

Another concern regarding hydrogen application is hydrogen embrittlement. H_2 embrittlement occurs when metals come into contact with LH_2, deteriorating their mechanical characteristics, eventually breaking down, and potentially leaking. This process becomes particularly noteworthy when it manifests as cracking, especially in scenarios involving hydrogen pipelines and storage [99]. Embrittlement is influenced by various factors such as metal purity, surface conditions, hydrogen exposure duration, and environmental pressure and temperature. Higher levels of hydrogen purity increase steel's susceptibility to embrittlement [99]. Therefore, materials utilized in LH_2 services are subjected to extreme cryogenic temperatures. Thus, when choosing materials for these applications, it's vital to account for changes in their properties. Key thermal characteristics critical for hydrogen services at cryogenic temperatures include low-temperature thermal contraction and embrittlement [99].

6.2 LH_2 Risks on ICE and HFC

The use of hydrogen as an alternative fuel is exceptional due to its broad flammability range, low minimum ignition energy, and high burning velocity [100]. However, these properties also pose various safety considerations. Specific safety precautions are necessary when using hydrogen in ICEs via direct injection or blending with other fuels [101, 102]. These concerns are shared between hydrogen ICEs and HFCs, which will be further explored in the following.

The self-ignition temperature of hydrogen is approximately 585 °C, while it ranges between 228 and 501 °C for gasoline and 540 °C for LNG [92]. As shown in Table 8, hydrogen has a very low boiling point and a high volatility, causing high diffusion in the air [103]. Additionally, hydrogen's high permeability through various materials, driven by its tiny size, underscores the importance of safety precautions for handling hydrogen. Accurate detection methods for hydrogen are crucial to monitor its concentration, particularly in the event of accidental leaks, given that hydrogen is a tasteless and colorless flammable gas that human senses cannot perceive [104].

A possible consequence of releasing hydrogen from a pressurized system is immediate ignition, leading to a jet fire. Jet fires are dangerous and characterized by rapid velocity, flame lengths, and high temperatures (reaching approximately 1500 °C) [105]. Consequently, incidents involving hydrogen fires may have more severe consequences than those involving conventional fuels. It should be noted that hydrogen jet fires usually occur within ICE cylinders when hydrogen is released through nozzles, triggered by various factors such as sparks from electrical equipment, electrostatic discharges, rapidly closing valves, and similar sources [106].

It's noteworthy that hydrogen fires burn cleanly, emitting water vapor and negligible emissions, making the flame challenging to detect visually. This is attributed to hydrogen's clean-burning nature. However, hydrogen flames' high velocity and rapid heat release pose an escalated risk of fire expansion and exposure to individuals, amplifying the potential threat to human safety [107]. As outlined in Sect. 6, hydrogen has been assigned the highest level of flammability (4) by NFPA [89].

Moreover, after releasing hydrogen into ICE cylinders, there is a possibility of a delay between the release and ignition, which can lead to an explosion. During this period, hydrogen mixes with air, which can cause a vapor cloud explosion or a flash fire [40]. The likelihood of delayed ignition depends on the initial release conditions [108]. Vapor cloud explosions can be hazardous and can be used to determine safety

Table 8 Volatility and boiling points of fuels [103]

Substance	Boiling point	Volatility
Hydrogen	−252.9°c	The most volatile substance
Methane	−161.5°c	Lower than H_2, higher than methanol and ammonia
Ammonia	−33.34°c	Higher than methanol
Methanol	64.7°c	Low volatility

distances from liquid hydrogen vessels. Hence, vapor cloud explosions are more destructive than flash and jet fires [109].

Hydrogen is a promising alternative fuel, but its use requires careful attention to safety risks. Stringent safety protocols are necessary throughout the entire process, from production to utilization in maritime transportation. It is important to note that concerns about the safety of HFCs are mainly related to the possibility of leaks and other faults during storage, as they do not have a combustion chamber like ICEs [110].

7 Conclusions

The maritime industry is at a turning point in its transition to a zero-emission industry. Alternative fuels play a key role in exceeding the transition, and adopting green LH_2 for zero-emission propulsion shows great promise in shaping the industry's future.

Although there have been significant advances in green LH_2 production, storage, and utilization, there are still critical challenges that require collaborative efforts to overcome. Green LH_2 is one of the leading contenders in the race to decarbonize the maritime industry. It has a high gravimetric energy density, which means it can power large cargo ships and long-distance routes for longer periods than battery technologies. Additionally, unlike traditional fuels, LH_2 burns cleanly, producing only water vapor as a by-product. The reduction of harmful emissions, such as sulfur and nitrogen oxides, significantly improves air quality. However, LH_2 has a higher volumetric density compared to conventional fuels, necessitating a larger storage tank onboard.

However, the widespread adoption of green LH_2 faces several obstacles. It is necessary to establish a robust infrastructure for green LH_2 production, storage, and bunkering (refueling) in major ports worldwide. Currently, the production process can be energy-intensive, and the cost of green LH_2 is still higher than that of conventional fuels, which presents a significant barrier. Furthermore, the production and storage of LH_2 require specialized equipment and technologies that result in significant capital expenditures. Despite the development of new production methods powered by renewable energy, such as electrolysis, achieving cost competitiveness with conventional fuels remains a challenge. Moreover, the transportation and bunkering of LH_2 involves complex logistics, vessel retrofitting, and adherence to safety regulations, resulting in operational expenditures. Therefore, optimizing production processes, increasing infrastructure efficiency, and implementing supportive policies such as renewable energy subsidies, carbon pricing mechanisms, and research and development funding to reduce costs and promote the economic viability of green LH_2 as a sustainable fuel option for marine transportation is crucial.

Taking into account the barriers in the production, storage, and use of hydrogen, hydrogen ICEs and HFCs offer viable pathways toward greener propulsion systems.

Despite these barriers, fuel cell technologies, including PEMFCs and SOFCs, offer compelling alternatives for marine propulsion. While HFCs present challenges such as space constraints and infrastructure requirements, ongoing research and development efforts aim to overcome these obstacles and pave the way for widespread adoption in the maritime sector.

As the industry moves towards decarbonization, addressing safety concerns and ensuring regulatory compliance remains essential. In maritime transportation, safe hydrogen use requires a clear understanding of its hazards. LH_2 presents challenges because of its cryogenic temperatures and flammability, which can be mitigated by safety measures like double-containment systems. Hydrogen's properties, such as its lightness and volatility, can impact safety in the event of leakage and embrittlement of metals. Therefore, it is essential to take necessary precautions to ensure the safety of both the environment and people.

It is imperative to invest in infrastructure and research to accelerate the adoption of green LH_2 propulsion systems and facilitate a seamless transition to sustainable maritime transportation. By embracing innovation and fostering global cooperation, the maritime sector can chart a course towards a greener and more resilient future, mitigating the adverse impacts of climate change while promoting economic prosperity.

References

1. Joung, T. H., Kang, S.-G., Lee, J.-K., & Ahn, J. (2020). The IMO initial strategy for reducing Greenhouse Gas (GHG) emissions, and its follow-up actions towards 2050. *Journal of International Maritime Safety, Environmental Affairs, and Shipping, 4*(1), 1–7. https://doi.org/10.1080/25725084.2019.1707938
2. Vakili, S., Ballini, F., Schönborn, A., Christodoulou, A., Dalaklis, D., & Ölçer, A. I. (2023). Assessing the macroeconomic and social impacts of slow steaming in shipping: A literature review on small Island developing states and least developed countries. *Journal of Shipping and Trade, 8*(1), 2.
3. Ölçer, A. I. (2018). Introduction to maritime energy management. In *Trends and challenges in maritime energy management* (pp. 1–12). Springer.
4. Vakili, S., Schönborn, A., & Ölçer, A. I. (2022). Techno-economic feasibility of photovoltaic, wind and hybrid electrification systems for stand-alone and grid-connected shipyard electrification in Italy. *Journal of Cleaner Production, 366*, 132945. https://doi.org/10.1016/j.jclepro.2022.132945
5. Global Maritime Forum. (2023). *The implications of the IMO Revised GHG Strategy for shipping*. Available online: https://cms.globalmaritimeforum.org/wp-content/uploads/2023/11/Insight-brief_The-implications-of-the-IMO-Revised-GHG-Strategy-for-shipping.pdf
6. Nyanya, M. N., Vu, H. B., Schönborn, A., & Ölçer, A. I. (2021). Wind and solar assisted ship propulsion optimisation and its application to a bulk carrier. *Sustainable Energy Technologies and Assessments, 47*(2021). https://doi.org/10.1016/j.seta.2021.101397
7. Podeur, V., Merdrignac, D., Behrel, M., Roncin, K., Fonti, C., Jochum, C., Parlier, Y., & Renaud, P. (2018). Fuel economy assessment tool for auxiliary kite propulsion of merchant ship. *La Houille Blanche, 104*(1), 5–7. https://doi.org/10.1051/lhb/2018001

8. Sindagi, S., Vijayakumar, R., & Saxena, B. K. (2021). Experimental investigation on ship's model in carrying out energy economics of BDR/ALS methodology. *Ships and Offshore Structures, 2021*, 1–10. https://doi.org/10.1080/17445302.2021.1926147
9. Lin, Y., He, J., & Li, K. (2018). Hull form design optimization of twin-skeg fishing vessel for minimum resistance based on surrogate model. *Advances in Engineering Software, 123*(2018), 38–50. https://doi.org/10.1016/j.advengsoft.2018.05.010
10. Christodoulou, A., & Cullinane, K. (2022). Potential alternative fuel pathways for compliance with the 'FuelEU Maritime Initiative'. *Transportation Research Part D: Transport and Environment, 112*(2022), 103492. https://doi.org/10.1016/j.trd.2022.103492
11. Masodzadeh, P. G., Ölçer, A. I., Ballini, F., & Christodoulou, A. (2022). A review on barriers to and solutions for shipping decarbonization: What could be the best policy approach for shipping decarbonization? *Marine Pollution Bulletin, 184*(2022). https://doi.org/10.1016/j.marpolbul.2022.114008
12. Chiong, M. C., Kang, H. S., Shaharuddin, N. M. R., Mat, S., Quen, L. K., Ten, K. H., & Ong, M. C. (2021). Challenges and opportunities of marine propulsion with alternative fuels. *Renewable and Sustainable Energy Reviews, 149*(2021), 111397. https://doi.org/10.1016/j.rser.2021.111397
13. Michaelowa, A. (2022). A vision for international climate finance after 2025. In *Handbook of international climate finance* (pp. 476–486). Edward Elgar Publishing.
14. Patonia, A., & Poudineh, R. (2023). *Hydrogen storage for a net-zero carbon future* (No. 23). OIES Paper: ET.
15. Hermesmann. M., & Müller, T. E. (2022). *Green, turquoise, blue, or grey? Environmentally friendly hydrogen production in transforming energy systems, progress in energy and combustion science* (Vol. 90), 100996, ISSN 0360-1285, https://doi.org/10.1016/j.pecs.2022.100996.
16. Lipman, T. (2011). *An overview of hydrogen production and storage systems with renewable hydrogen case studies*. Conducted under US DOE Grant: DE- FC3608GO18111 A000, Office of Energy Efficiency and Renewable Energy Fuel Cell Technologies Program.
17. Howarth, R. W., & Jacobson, M. Z. (2021). How green is blue hydrogen? *Energy Science & Engineering, 2021*(9), 1676–1687. https://doi.org/10.1002/ese3.956
18. Dolatabadi, S. H., Ölçer, A. I., & Vakili, S. (2023). The application of hybrid energy system (hydrogen fuel cell, wind, and solar) in shipping. *Renewable Energy Focus*. https://doi.org/10.1016/j.ref.2023.05.008. ISSN 1755-0084.
19. Ji, M., & Wang, J. (2021). Review and comparison of various hydrogen production methods based on costs and life cycle impact assessment indicators. *International Journal of Hydrogen Energy, 46*(78), 38612–38635. https://doi.org/10.1016/j.ijhydene.2021.09.142. ISSN 0360-3199.
20. Viswanathan, B. (2016). *Energy sources: Fundamentals of chemical conversion processes and applications*. Elsevier.
21. NASA. (2018). *Innovative liquid hydrogen storage to support space launch system* [Online]. Available from: https://www.nasa.gov/feature/innovative-liquid-hydrogen-storage-to-support-space-launch-system
22. Iordache, I., Schitea, D., Gheorghe, A. V., & Iordache, M. (2014). Hydrogen underground storage in Romania, potential directions of development, stakeholders and general aspects. *International Journal of Hydrogen Energy, 39*(21), 11071–11081.
23. US Department of Energy. (2021). *Hydrogen storage*. Available from: https://www.energy.gov/eere/fuelcells/hydrogen-storage
24. Preuster, P., Alekseev, A., & Wasserscheid, P. (2017). Hydrogen storage technologies for future energy systems. *Annual Review of Chemical and Biomolecular Engineering, 8*, 445–471.
25. ILK Dresden, Moritz Kuhn. (2015). Wasserstoff- und Methan-Versuchsfeld am ILK. (n.d.). Available online: https://www.ilkdresden.de/leistungen/forschung-und-entwicklung/projekt/wasserstoff-und-methan-versuchsfeld-am-ilk

26. DNV alternative fuel information platform (AFI). (2024). Available online: https://www.dnv.com/services/alternative-fuels-insights-afi%2D%2D128171/
27. Recharge. (2022). *Liquid hydrogen as shipping fuel. Pioneering intercontinental H2 carrier gets technical green light* [Online]. Available from: https://www.rechargenews.com/energy-transition/liquid-hydrogen-as-shipping-fuel-pioneering-intercontinental-h2-carrier-gets-technical-green-light/2-1-1211897
28. Karvounis, P., Tsoumpris, C., Boulougouris, E., & Theotokatos, G. (2022). Recent advances in sustainable and safe marine engine operation with alternative fuels. *Frontiers in Mechanical Engineering, 8*, 994942.
29. Kanna, I., Arulprakasajothi, M., & Eliyas, S. (2019). A detailed study of IC engines and a novel discussion with comprehensive view of alternative fuels used in petrol and diesel engines. *International Journal of Ambient Energy, 42*, 1794–1802. https://doi.org/10.1080/01430750.2019.1614994
30. Achuthanandam, K. (2021). LNG's pathways to decarbonization, s.l. DNV-GL.
31. Li, J., Lai, S., Chen, D., Wu, R., Kobayashi, N., Deng, L., & Huang, H. (2021). A review on combustion characteristics of ammonia as a carbon-free fuel. *Frontiers in Energy Research, 9*, 760356.
32. Herdzik, J. (2021). Decarbonization of marine fuels—The future of shipping. *Energies, 14*(14), 4311.
33. Sayin, C., Ozsezen, A. N., & Canakci, M. (2010). The influence of operating parameters on the performance and emissions of a DI diesel engine using methanol-blended-diesel fuel. *Fuel, 89*(7), 1407–1414.
34. Verhelst, S., Wallner, T., & Sierens, R. (2014). Hydrogen-Fueled internal combustion engines. *Hand book of Hydrogen Energy, 35*, 821–902. https://doi.org/10.1201/b17226
35. Karunamurthy, K., & Shinde, B. J. (2021). Recent progress in hydrogen fueled internal combustion engine (H2ICE)—A comprehensive outlook. In *Materials Today: Proceedings*. https://doi.org/10.1016/j.matpr.2021.10.378, ISSN 2214-7853.
36. Ogden, J. M. (2002). Hydrogen: The fuel of the future? *Physics Today, 55*(4), 69–75.
37. Ozcanli, M., Bas, O., Akar, M. A., Yildizhan, S., & Serin, H. (2018). Recent studies on hydrogen usage in Wankel SI engine. *International Journal of Hydrogen Energy, 43*(38), 18037–18045.
38. Wallner, T., Lohse-Busch, H., Gurski, S., Duoba, M., Thiel, W., Martin, D., & Korn, T. (2008). Fuel economy and emissions evaluation of BMW Hydrogen 7 Mono-Fuel demonstration vehicles. *International Journal of Hydrogen Energy, 33*(24), 7607–7618.
39. DOE, U. (2010). *Multi-Year Research, Development and Demonstration Plan: Planned Program Activities for 2005-2015* (pp. 1–34). US Department of Energy, Office of Energy Efficiency and Renewable Energy, Hydrogen, Fuel Cells and Infrastructure Technologies Program (HFCIT).
40. Abohamzeh, E., Salehi, F., Sheikholeslami, M., Abbassi, R., & Khan, F. (2021). Review of hydrogen safety during storage, transmission, and applications processes. *Journal of Loss Prevention in the Process Industries, 72*, 104569. https://doi.org/10.1016/j.jlp.2021.104569. ISSN 0950-4230.
41. Singla, M. K., Nijhawan, P., & Oberoi, A. S. (2021). Hydrogen fuel and fuel cell technology for cleaner future: A review. *Environmental Science and Pollution Research, 28*, 15607–15626.
42. Steele, B. C. (2001). Material science and engineering: The enabling technology for the commercialization of fuel cell systems. *Journal of Materials Science, 36*, 1053–1068.
43. Lohse-Busch, H., Duoba, M., Rask, E., Stutenberg, K., Gowri, V., Slezak, L., & Anderson, D. (2013). *Ambient temperature (20 F, 72 F and 95 F) impact on fuel and energy consumption for several conventional vehicles, hybrid and plug-in hybrid electric vehicles and battery electric vehicle* (No. 2013-01-1462).
44. Joshi, A. (2020). Review of vehicle engine efficiency and emissions. *SAE International Journal of Advances and Current Practices in Mobility, 2*(2020-01-0352), 2479–2507.

45. O'hayre, R., Cha, S. W., Colella, W., & Prinz, F. B. (2016). *Fuel cell fundamentals*. John Wiley & Sons.
46. Ma, S., Lin, M., Lin, T.-E., Lan, T., Liao, X., & Mar, E. F. (2021). Fuel cell-battery hybrid systems for mobility and off-grid applications: A review. *Renewable and Sustainable Energy Reviews, 135*, 110119. https://doi.org/10.1016/j.rser.2020.110119
47. Qian, J.-y., Wu, J.-y., Gao, Z.-x., Wu, A., & Jin, Z.-j. (2019). Hydrogen decompression analysis by multi-stage Tesla valves for hydrogen fuel cell. *International Journal of Hydrogen Energy, 44*(26), 13666–13674. https://doi.org/10.1016/j.ijhydene.2019.03.235. ISSN 0360-3199.
48. De-Troya, J. J., Álvarez, C., Fernández-Garrido, C., & Carral, L. (2016). Analysing the possibilities of using fuel cells in ships. *International Journal of Hydrogen Energy, 41*(4), 2853–2866.
49. Ahn, J., Park, S. H., Noh, Y., Choi, B. I., Ryu, J., Chang, D., & Brendstrup, K. L. M. (2018). Performance and availability of a marine generator-solid oxide fuel cell-gas turbine hybrid system in a very large ethane carrier. *Journal of Power Sources, 399*, 199–206.
50. Tronstad, T., Åstrand, H. H., Haugom, G. P., & Langfeldt, L. (2017). *Study on the use of fuel cells in shipping*. European Maritime Safety Agency.
51. Bassam, A. M., Phillips, A. B., Turnock, S. R., & Wilson, P. A. (2017). Development of a multi- scheme energy management strategy for a hybrid fuel cell driven passenger ship. *International Journal of Hydrogen Energy, 42*(1), 623–635.
52. Baschuk, J. J., & Li, X. (2001). Carbon monoxide poisoning of proton exchange membrane fuel cells. *International Journal of Energy Research, 25*(8), 695e713. https://doi.org/10.1002/er.713
53. Dai, W., Wang, H., Yuan, X. Z., Martin, J. J., Yang, D., Qiao, J., & Ma, J. (2009). A review on water balance in the membrane electrode assembly of proton exchange membrane fuel cells. *International Journal of Hydrogen Energy, 34*(23), 9461–9478.
54. Rivarolo, M., Rattazzi, D., Lamberti, T., & Magistri, L. (2020). Clean energy production by PEM fuel cells on tourist ships: A time-dependent analysis. *International Journal of Hydrogen Energy, 45*(47), 25747–25757. https://doi.org/10.1016/j.ijhydene.2019.12.086, ISSN 0360-3199.
55. Sapra, H., Stam, J., Reurings, J., van Biert, L., van Sluijs, W., & de Vos, P. (2021). Integration of solid oxide fuel cell and internal combustion engine for maritime applications. *Applied Energy, 281*, 115854. https://doi.org/10.1016/j.apenergy.2020.115854
56. Díaz-de-Baldasano, M. C., Mateos, F. J., Núñez-Rivas, L. R., & Leo, T. J. (2014). Conceptual design of offshore platform supply vessel based on hybrid diesel generator-fuel cell power plant. *Applied Energy, 116*, 91–100.
57. Madsen, R. T., Klebanoff, L. E., Caughlan, S. A. M., Pratt, J. W., Leach, T. S., Appelgate, T. B., Kelety, S. Z., Wintervoll, H.-C., Haugom, G. P., Teo, A. T. Y., & Ghosh, S. (2020). Feasibility of the Zero-V: A zero-emissions hydrogen fuel-cell coastal research vessel. *International Journal of Hydrogen Energy, 45*(46), 25328–25343. https://doi.org/10.1016/j.ijhydene.2020.06.019. ISSN 0360-3199.
58. Gomez Trillos, J. C., Wilken, D., Brand, U., & Vogt, T. (2019). HySeas III: The World's first sea-going hydrogen-powered ferry – A look at its technical aspects, market perspectives and environmental impacts. *Nutzung Regenerativer Energiequellen und Wasserstofftechnik, 2019*, 57–80.
59. Weaver, G., & Barrett, S. (2003). Marine applications of fuel cell technology. *Fuel Cells Bulletin, 2003*(1), 11–12. https://doi.org/10.1016/S1464-2859(03)00134-2. ISSN 1464-2859.
60. Bassam, A. M., Phillips, A. B., Turnock, S. R., & Wilson, P. A. (2016). An improved energy management strategy for a hybrid fuel cell/battery passenger vessel. *International Journal of Hydrogen Energy, 41*(47), 22453–22464.
61. Tuswan, T., Misbahudin, S., Junianto, S., Yudo, H., Santosa, A. W. B., Trimulyono, A., ... & Chrismianto, D. (2022). Current research outlook on solar-assisted new energy ships:

Representative applications and fuel & GHG emission benefits. In *IOP conference series: Earth and environmental science* (Vol. 1081, No. 1, 012011). IOP Publishing.
62. Klebanoff, L. E., Pratt, J. W., & LaFleur, C. B. (2017). Comparison of the safety-related physical and combustion properties of liquid hydrogen and liquid natural gas in the context of the SF-BREEZE high-speed fuel-cell ferry. *International Journal of Hydrogen Energy, 42*(1), 757–774.
63. Saravanan, N., & Nagarajan, G. (2008). An experimental investigation of hydrogen-enriched air induction in a diesel engine system. *International Journal of Hydrogen Energy, 33*(6), 1769–1775.
64. White, C. M., Steeper, R. R., & Lutz, A. E. (2006). The hydrogen-fueled internal combustion engine: A technical review. *International Journal of Hydrogen Energy, 31*(10), 1292–1305.
65. Wu, S. R., Chang, W. C., & Chiao, J. (2007). Low NOx heavy fuel oil combustion with high temperature air. *Fuel, 86*(5–6), 820–828.
66. De Boer, P. C. T., McLean, W. J., & Homan, H. S. (1976). Performance and emissions of hydrogen fueled internal combustion engines. *International Journal of Hydrogen Energy, 1*(2), 153–172.
67. Canter, N. (2017). Marine diesel cylinder engine oils: Lubrication challenges impacted by operating conditions and regulations. *Tribology & Lubrication Technology, 73*(9), 10.
68. Hao, H., Mu, Z., Liu, Z., & Zhao, F. (2018). Abating transport GHG emissions by hydrogen fuel cell vehicles: Chances for the developing world. *Frontiers in Energy, 12*, 466–480.
69. Heiser, T. (2019). *Auxiliary inverter solutions for fuel cell vehicle turbo compressor applications, Kebamerica.* Available online: https://www.kebamerica.com/blog/fuel-cell-vehicle-auxiliary-inverter-solutions/
70. Muscat, A., de Olde, E., de Boer, I., & Ripoll-Bosch, R. (2020). The battle for biomass: A systematic review of food-feed-fuel competition. *Global Food Security, 25*, 100330.
71. Hwang, S., Gil, S. J., Lee, G. N., Lee, J. W., Park, H., Jung, K. H., et al. (2020). Life cycle assessment of alternative ship fuels for coastal ferry operating in Republic of Korea. *Journal of Marine Science and Engineering, 8*, 660. https://doi.org/10.3390/jmse8090660
72. Zhu, Z., & Xu, B. (2022). Purification technologies for NOx removal from flue gas: A review. *Separations, 9*(10), 307. https://doi.org/10.3390/separations9100307
73. Horvath, S., Fasihi, M., & Breyer, C. (2018). Techno-economic analysis of a decarbonized shipping sector: Technology suggestions for a fleet in 2030 and 2040. *Energy Conversion and Management, 164*, 230–241.
74. Stępień, Z. (2021). A comprehensive overview of hydrogen-fueled internal combustion engines: Achievements and future challenges. *Energies, 14*(20), 6504.
75. Dhanushkodi, S. R., Mahinpey, N., Srinivasan, A., & Wilson, M. (2008). Life cycle analysis of fuel cell technology. *Journal of Environmental Informatics, 11*(1), 36–44.
76. Chou, M. T., Chou, T. Y., Hsu, Y. R., & Lu, C. P. (2017). Fuel consumption ratio analysis for transiting from various ports and harbours in Asia through the Northern Sea Route. *The Journal of Navigation, 70*(4), 859–869.
77. Lau, Y. Y., Ng, A. K., Fu, X., & Li, K. X. (2013). Evolution and research trends of container shipping. *Maritime Policy & Management, 40*(7), 654–674.
78. Ahmed, S. F., Mofijur, M., Nuzhat, S., Rafa, N., Musharrat, A., Lam, S. S., & Boretti, A. (2022). Sustainable hydrogen production: Technological advancements and economic analysis. *International Journal of Hydrogen Energy, 47*(88), 37227–37255.
79. Statista. (2024). *Average annual Brent crude oil price from 1976 to 2024, survey by EIA, published by en2x.* Available online: https://www.statista.com/statistics/262860/uk-brent-crude-oil-price-changes-since-1976/
80. Chu, K. H., Lim, J., Mang, J. S., & Hwang, M. H. (2022). Evaluation of strategic directions for supply and demand of green hydrogen in South Korea. *International Journal of Hydrogen Energy, 47*(3), 1409–1424.

81. Panchenko, V. A., Daus, Y. V., Kovalev, A. A., Yudaev, I. V., & Litti, Y. V. (2023). Prospects for the production of green hydrogen: Review of countries with high potential. *International Journal of Hydrogen Energy, 48*(12), 4551–4571.
82. Koneczna, R., & Cader, J. (2021). Hydrogen in the strategies of the European Union member states. *Gospodarka Surowcami Mineralnymi, 37*(3), 53–74.
83. Younas, M., Shafique, S., Hafeez, A., Javed, F., & Rehman, F. (2022). An overview of hydrogen production: Current status, potential, and challenges. *Fuel, 316*, 123317.
84. Nikolaidis, P., & Poullikkas, A. (2017). A comparative overview of hydrogen production processes. *Renewable and Sustainable Energy Reviews, 67*, 597–611.
85. El-Emam, R. S., & Özcan, H. (2019). Comprehensive review on the techno-economics of sustainable large-scale clean hydrogen production. *Journal of Cleaner Production, 220*, 593–609.
86. Sinigaglia, T., Lewiski, F., Martins, M. E. S., & Siluk, J. C. M. (2017). Production, storage, fuel stations of hydrogen and its utilization in automotive applications-a review. *International Journal of Hydrogen Energy, 42*(39), 24597–24611.
87. EV100 China. (2020). *The development of hydrogen energy industry in China: A 2020 report*. Beijing.
88. UK Offshore. (2020). *Renewable energy catapult: Offshore wind and hydrogen: Solving the integration challenge*. Available from: https://s3-eu-west-1.amazonaws.com/media.newore.catapult/app/uploads/2020/09/09161500/Solving-the-Integration-Challenge-ORE-Catapultr.pdf
89. NCEC. (2019). *The dangerous goods emergency action code list*. https://the-ncec.com/en/resources/the-dangerous-goods-emergency-action-code-list-2019
90. FCHEA. (2013). *Hydrogen is safe*. Available from: http://www.fchea.org
91. Barbir, F., Braun, J., & Neutzler, J. (1999). Properties of molded graphite bipolar plates for PEM fuel cell stacks. *Journal of New Materials for Electrochemical Systems, 2*.
92. Zini, G., & Tartarini, P. (2011). *Solar hydrogen energy systems: Science and technology for the hydrogen economy*. Springer.
93. Züttel, A. (2004). Hydrogen storage methods. *Naturwissenschaften, 91*, 157–172.
94. Stetson, N. T., McWhorter, S., & Ahn, C. C. (2016). Introduction to hydrogen storage. In *Compendium of hydrogen energy* (pp. 3–25). Woodhead Publishing.
95. Fehrm, B. (2020). *The challenges of hydrogen. Part 5. The hydrogen tank, Leeham News*. Available online: https://leehamnews.com/2020/08/21/bjorns-corner-the-challenges-of-hydrogen-part-5-the-hydrogen-tank/
96. Ratnakar, R., Gupta, N., Kun, Z., van Casimir, D., Fesmire, J., Birol, D., & Balakotaiah, V. (2021). Hydrogen supply chain and challenges in large-scale LH2 storage and transportation. *International Journal of Hydrogen Energy, 46*(47), 24149–24168. https://doi.org/10.1016/j.ijhydene.2021.05.025, ISSN 0360-3199.
97. CEC. (2005). *Failure modes and effects analysis for hydrogen fueling options*. CEC-600-2005-0012004, California Energy Commission: California.
98. Cipriani, G., Di Dio, V., Genduso, F., La, C. D., Liga, R., & Miceli, R. (2014). Perspective on hydrogen energy carrier and its automotive applications. *International Journal of Hydrogen Energy, 39*(16), 8482–8494.
99. Barthelemy, H. (2007). *Compatibility of metallic materials with hydrogen review of the present knowledge*. Presented at International Conference on Hydrogen Safety, San Sebastian, Spain, 17 September 2007, https://www.h2knowledgecentre.com/content/conference277
100. Cashdollar, K. L., Zlochower, I. A., Green, G. M., Thomas, R. A., & Hertzberg, M. (2000). Flammability of methane, propane, and hydrogen gases. *Journal of Loss Prevention in the Process Industries, 13*(3), 327–340.
101. Molnarne, M., & Schroeder, V. (2019). Hazardous properties of hydrogen and hydrogen containing fuel gases. *Process Safety and Environmental Protection, 130*, 1–5.
102. Royle, M., & Willoughby, D. (2011). The safety of the future hydrogen economy. *Process Safety and Environmental Protection, 89*(6), 452–462.

103. Joakim, A., & Stefan, G. (2019). Large-scale storage of hydrogen. *International Journal of Hydrogen Energy, 44*(23), 11901–11919. https://doi.org/10.1016/j.ijhydene.2019.03.063. ISSN 0360-3199.
104. Dutta, S. (2014). A review on production, storage of hydrogen and its utilization as an energy resource. *Journal of Industrial and Engineering Chemistry, 20*(4), 1148–1156.
105. Virtue, B., Mohammadpour, J., Salehi, F., & Abbassi, R. (2023). Safety assessment of hydrogen jet fire scenarios within semi-confined spaces. *Fire, 6*(1), 29. https://doi.org/10.3390/fire6010029
106. Liu, W., Markert, F., Giuliani, L., & Hostikka, S. (2024). Analyzing the gas temperature of a hydrogen jet fire in a compartment with the Fire Dynamics Simulator. *International Journal of Hydrogen Energy, 53*, 1097–1106. https://doi.org/10.1016/j.ijhydene.2023.11.306
107. Najjar, Y. S. (2013). Hydrogen safety: The road toward green technology. *International Journal of Hydrogen Energy, 38*(25), 10716–10728.
108. Zhang, S., Ma, H., Huang, X., & Peng, S. (2020). Numerical simulation on methane-hydrogen explosion in gas compartment in utility tunnel. *Process Safety and Environmental Protection, 140*, 100–110.
109. Li, Z., Pan, X., Meng, X., & Ma, J. (2012). Study on the harm effects of releases from liquid hydrogen tank by consequence modeling. *International Journal of Hydrogen Energy, 37*(22), 17624–17629.
110. Foorginezhad, S., Mohseni-Dargah, M., Falahati, Z., Abbassi, R., Razmjou, A., & Asadnia, M. (2021). Sensing advancement towards safety assessment of hydrogen fuel cell vehicles. *Journal of Power Sources, 489*, 229450. https://doi.org/10.1016/j.jpowsour.2021.229450

Advanced Technologies and Recent Applications on Fuel Cell Powered Marine Transport

Hüseyin Turan Arat, Yunus Onur Yildiz, and Alper Burgaç

This book chapter is dedicated to Prof. Dr Turhan Nejat Veziroğlu (Dr V.), one of the pioneers in introducing hydrogen energy all over the world, who dearly departed on September 5, 2024.

1 Introduction

Humanity has been using transportation technologies for centuries. The transportation sector is constantly evolving. Trying many different ways to get from one point to another, humankind has been on water since ancient times. A large part of the Earth is covered with water. To move on its surface and cross rivers and oceans, people had to produce some "facilitating" devices.

Maritime history has been studied for years. It has been the subject of encyclopedias; maritime associations and journals are dedicated to the field (i.e., International Journal of Maritime History). The systematic history of marine transportation has taken place for one million years in human history, according to local and accessible sources and discoveries. Having a long history from the first people to ancient times, from the Sumerians and Egyptians to the discovery of the continents,

H. T. Arat (✉)
Faculty of Engineering and Architecture, Department of Mechanical Engineering, Sinop University, Sinop, Turkey
e-mail: htarat@sinop.edu.tr

Y. O. Yildiz
Faculty of Maritime Studies, Department of Naval Architecture and Marine Engineering, Bursa Technical University, Bursa, Turkey
e-mail: onur.yildiz@btu.edu.tr

A. Burgaç
Faculty of Engineering, Department of Energy Systems Engineering, Adana Alpaslan Türkeş Science and Technology University, Adana, Turkey
e-mail: aburgac@atu.edu.tr

C. O. Colpan et al. (eds.), *Fuel Cell and Hydrogen Technologies in Maritime Transportation*, https://doi.org/10.1007/978-3-031-86110-9_4

it has evolved with the development of technology. Such factors as the growing need for transportation, travel, and fishing caused the development of navigation and, thus, new geographical discoveries. As a result, navigation has positively affected human life in every way.

One of the most important aspects of navigation is to benefit from the buoyancy of water and the space it takes up. In other words, the deeper the water and the more space it takes up, the bigger and heavier ships can get. This correlation has undoubtedly raised maritime transportation to the top of the transportation sector. As stated in many sources [1–4], nearly 80% of the world's global trade is carried out by maritime transport. After the Industrial Revolution and the acquisition of kinetic energy utilizing an engine, marine vehicles also began to benefit from this kind of energy-using propellers. Steam engines were introduced in the maritime sector in the early 1900s. In fact, the system was simple: the vessel's kinetic energy was not produced by the paddles and oars used by humans but by using a combination of machinery and equipment, creating pressure as a result of combustion and thus revealing kinetic energy.

Although technological progress has reduced the size of engines and, consequently, fuel consumption in automobile transport, it is a physical fact that this could not be possible for other means of transport (trains, ships, and airplanes). Therefore, their energy needs are inevitably high, and in the case of large vessels, they are met by using internal combustion engines.

In this section, several ways in which large vehicles can have a propulsion capability together with a more environmentally friendly, more efficient, and sustainable energy source have been compiled by today's technological progress. The following points are to be analyzed: what these systems are, why we need them, their advantages and disadvantages, the obstacles they face, and what can be done with the latest sources in the future.

The necessity and demand for energy are increasing day by day. The fact that primary energy sources are limited and produce greenhouse gas emissions (GHG) and pollutants due to chemical reactions is something that the whole world is paying attention to, trying to prevent the latter. With the increasing population, the quantity of consumed energy is also increasing, and this (if precautions are insufficient) causes our Earth to degrade faster.

The diagram in Fig. 1 presents these facts in numbers. This figure depicts the total CO_2 emissions, the required energy demand, and the world population in the last decade. The 2024 data are created from the relevant references as the future expectation of the first 6 months' data [6, 8, 12]. It can be quickly concluded that the increasing population causes the energy demand to increase, which in turn causes the rise in emissions.

In the last three decades, many products/studies have been put forward regarding secondary and renewable energy production. In the previous 30 years, the prototypes of wind, solar, geothermal, wave, and biomass energy studies, which date back a century, worked efficiently, meeting increasing energy demand in a greener and more environmentally friendly way.

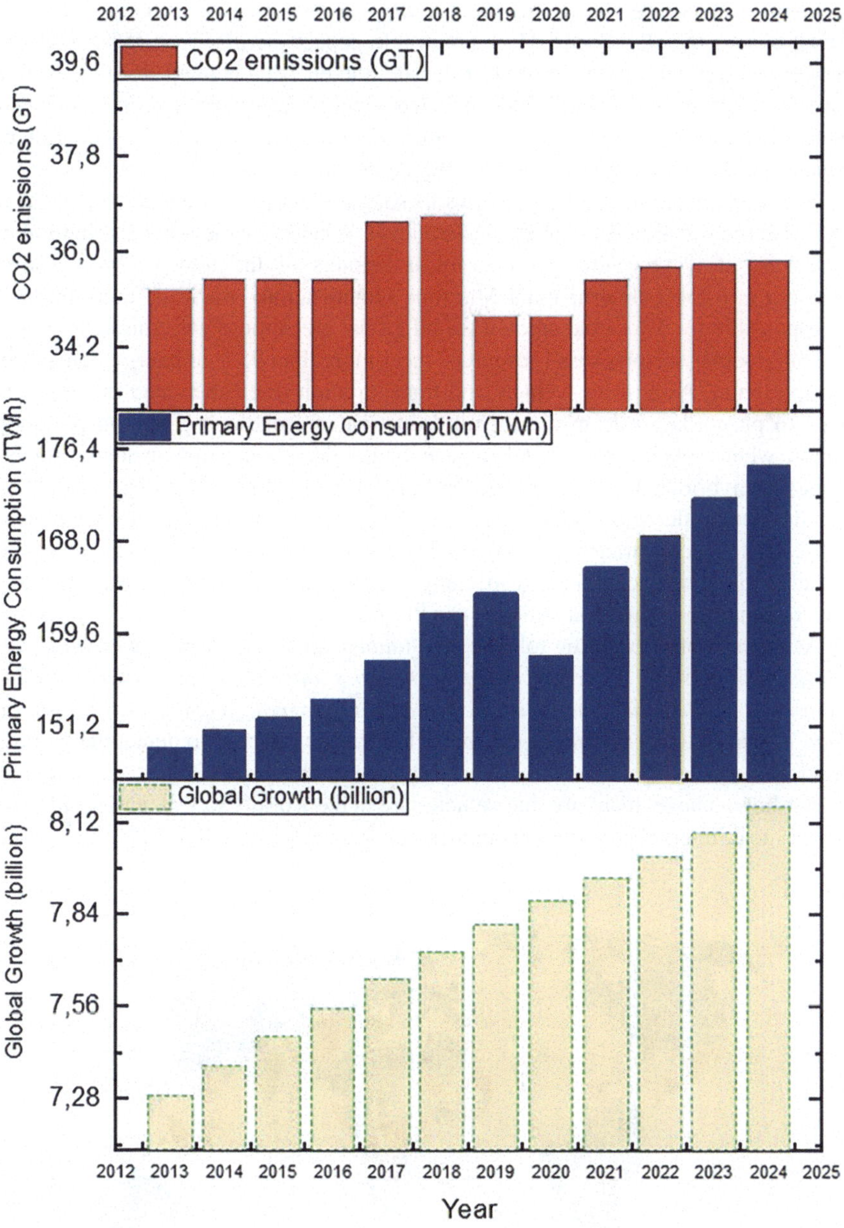

Fig. 1 The global human population, energy consumption, and CO_2 emissions (*only by fossil fuels*) [5–12]

Fossil fuel power plants still meet a large portion of global energy demand. According to the IEA report [13], renewable capacity will cover 35% of global power generation in 2025. In the last decade, the energy obtained from renewable sources (hydropower, solar, wind, etc.) has significantly increased. This increase undoubtedly took place due to the reduction of global warming and the reduction of emissions following the goals of the Paris Agreement.

The determination of all signatories to reduce emissions is admirable, but it is clear that the situation is not at the desired level. When we look at the distribution of emissions, it is seen that 15–17% of it belongs to the transportation sector. According to IMO (International Maritime Organization), maritime transportation is responsible for 3% of the amount (of all global greenhouse gas emissions) [14].

Unlike other sectors, fossil resources meet more than 95% of energy demand in transportation. As a result, vehicle and transportation researchers and practitioners need to play a key role in the urgent reduction of emissions. The transportation sector, where much action is taken to enhance decarbonization in land vehicles alongside hybridization and electric propulsion, has come a long way in the last decade. The production of electric, hybrid, fuel cell vehicles has increased exponentially. This "electric transportation era" has advantages and disadvantages and has already arrived in land vehicles (especially passenger cars). Therefore, ways to make this transition to air and sea vehicles need to be sought.

As an example, according to [15], "*In January 2023, the world's merchant fleet consisted of 105 500 vessels of at least 100 gross tons (gt), of which 56 500 ships were over 1000 gt. and the world fleet reached a carrying capacity of 2.3 billion dead weight tons (dwt) in the same year*". Figure 2 is reserved to depict the world's buildings, ownership, registration, and recycling of ships in 2022.

As shown above, many marine vehicles are in the world's waters, and fossil fuels power most. Internal combustion engines, turbines, nuclear, LNG, LPG, multi-fuel,

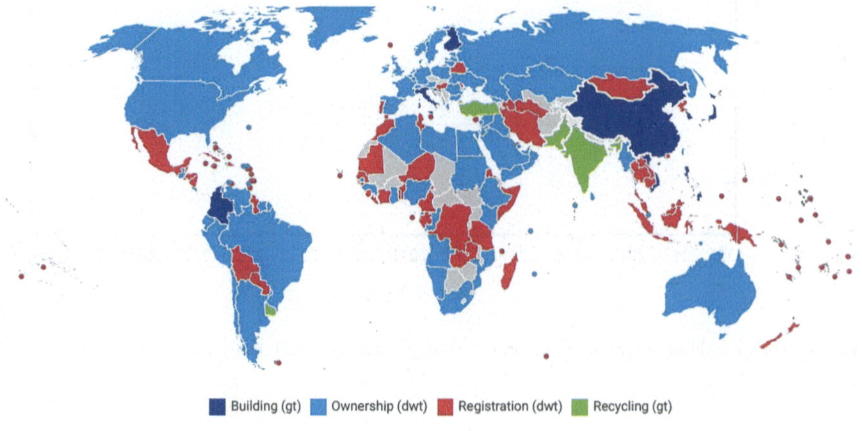

Fig. 2 Building, ownership, registration, and recycling of ships, 2022 [faithfully to the original from 15]

air-independent (Stirling), etc., are widely used in marine applications. As for submarines and boats, they have started installing electrical motors recently. Another fuel option is hydrogen. It is carbon-free and has a higher heating value, making it an excellent choice.

This section presents a new range of systems, and the details will follow in the next sections. This chapter is dedicated to fuel cell marine transportation. Before going into technical detail, here is a fact about how the subject's popularity and importance have increased. Enter "fuel cell marine" into the Scopus search machine. You can see that the number of studies on the topic increases exponentially every year (e.g., *while there were 50 studies in 2013, there were 153 in 2023, Accessed on September 11, 2024;* https://www.scopus.com/). As a result of deep research on the subject, the authors have extracted analyses from 60 studies in the listed references for readers who want to reach the core information on the subject published over the last 2 years (2023–2024).

2 Fuel Cells

Simply put, the process occurring in fuel cells is the reverse of electrolysis. Its essence lies in the combination of fuel and oxygen between plates and the release of electricity. In the first chemistry lessons, students often conduct electricity through water to separate hydrogen from oxygen atoms. Here, the process in question is precisely the opposite.

A more technical explanation can be found in most books and previous authors' works [16, 17]; "A fuel cell is an electrochemical device which converts the chemical energy of the fuel to electrical energy and heat." Different fuel cell types are named according to their operating temperatures and the components they are made of. Several kinds of FCs are compared in Table 1 based on Arat's [16, 17] previous studies.

3 Marine Transport with Fuel Cells

Engineering is a field that produces solutions to people's needs and aims to facilitate their lives and work. Devices that require varied expertise and combine knowledge from different spheres have solved people's "problems" and optimized their energy/cost/material/workforce. A fuel cell is one of the devices used to meet their energy requirements. In general, they have been used in propulsion systems in the maritime sector for the last 25 years. Significant examples of the development and certain studies on FC marine vehicles are given in Table 2.

In the transportation sector, there are certain requirements for the sources of energy used in the vehicle: it should be highly efficient, it should be usable in the vehicle for long periods of time, it should have a clean and environmentally friendly

Table 1 FC technologies and components view [16, 17] (collected in one table, faithfully to the original)

Fuel cell	Polymer electrolyte membrane (PEM)	Alkaline (AFC)	Phosphoric acid (PAFC)	Solid oxide (SOFC)	Molten carbonate (MCFC)
Common electrolyte	Polymeric ion exchange membrane	Immobilized alkaline salt solution	Immobilized liquid phosphoric acid	Ceramic	Immobilized liquid molten carbonate
Apps	Backup power, transportation, special vehicles	Military, space, transportation	Distributed generation	Auxiliary power, distributed generation, electric utility	Distributed generation, electric utility
Efficiency	40–60%	60%	40%	50–60%	50%
Components view/structure and operation temperatures	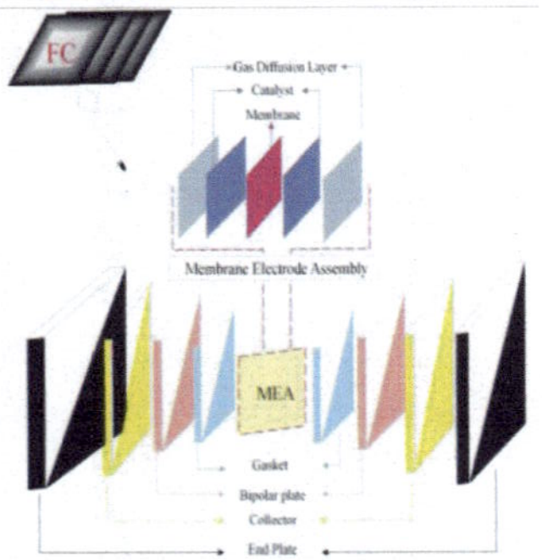		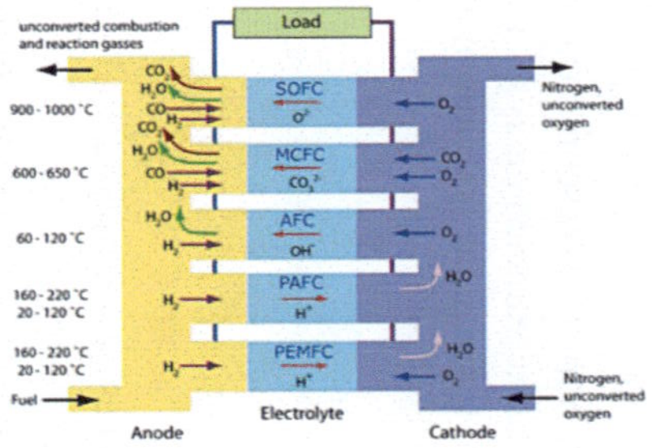		

Table 2 Examples of the development and certain studies on FC marine vehicles [1, 18]

Prototype	Project name/partners	Year	FC type (fuel)	FC capacity (kW)	Explanatory
	RiverCell—Elektra (TU Berlin, BEHALA, DNVGL, etc.)	2015–2016	HTPEM (H_2)	3(pieces) × 100	*Feasibility study *Elektra has a hybrid system: – 3(pieces) × 100 kw FC – 2(pieces) × 200 kW electric motor. – 6(pieces) × 125 kg hydrogen tank (500 bar) – Rechargeable batteries with a total 2050 kWh nominal capacity
	ZemShip - Alsterwasser (proton motors, GL, Alster Touristik GmbH, Linde group Etc.)	2006–2013	PEMFC (H_2)	2 (pieces) × 48	*The 100 kW PEMFC system was developed and tested on a small passenger ship. *ZemShip has a hybrid system: – 2(pieces) × 48 kW FC – 7 (pieces) lead gel battery packs with a total capacity of 560 V and 234 kWh – 12 (pieces) storage tanks with 50 kg of hydrogen at 350 bar pressure – 1(pieces) 100 kW electric motor
	Nemo H_2 (Rederij lovers etc.)	2012-present	PEMFC (H_2)	60–70	* Nemo H_2 has a hybrid system: – 30–50 kW battery system – 60–70 kW FC – 6 (pieces) storage tanks with 24 kg of hydrogen at 350 bar pressure – 30-50 kW battery
	Hornblower hybrid (Hornblower)	2012-present	PEMFC (H_2)	32	*Hybrid ferry with a diesel generator, batteries, PV, wind, and fuel cell

(continued)

Table 2 (continued)

Prototype	Project name/partners	Year	FC type (fuel)	FC capacity (kW)	Explanatory
	Hydrogenesis (Bristol boat trips etc.)	2012-present	PEMFC (H_2)	12	* 12-passenger ferry * Small passenger ship which operates in Bristol
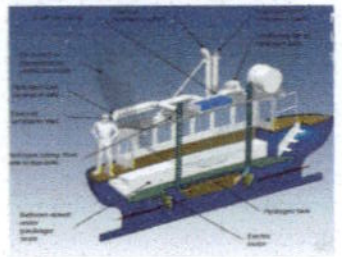	MF Vågen (CMR Prototech, ARENA-project)	2010	HTPEM (H_2)	12	* MF Vågen have: – 12 kw FC – Metal hydride H_2 storage (2–3 kg) – 25 kWh Li-ion batteries – 10 kW DC motor
	Class 212A/214 submarines (CMR Prototech, ARENA-project, ThyssenKrupp marine systems, Siemens)	2003—Present	PEMFC (H_2)	306, 30–50 per module (212A), 120 per module (214)	*Hybrid propulsion using a fuel cell and a diesel engine
	SF-BREEZE (Sandia National lab., red and white Fleet)	2015—Present	PEMFC (liquid H_2)	41(pieces) × 120	*SF-BREEZE have: – 41(pieces) × 120 kw FC – Two motors with 2.5 MW – Single type C (pressurized vessel) storage tank with 1200 kg of LH_2
	MS forester (SchIBZ)	2009–2018	SOFC	500 kW	Diesel fueled 500 kW SOFC

	Aranda	2017–2022	PEMFC	2x82.5	165 kW, 85 kg H_2
	ZULU	2019-present	PEMFC	2x200	2 × 200 PEM and batteries, 350 kh H_2
	FPS Waal	2019-present	PEMFC	6x200	6 × 200 kW Ballard, FCwave, 470 kg H_2
	ZEUS	2018–2022	PEMFC	71	71 kW FC and hybridization with 150 kW batteries
	Sea change	2018–2022	PEMFC	3X120	3 × 120 kW all FC-based first commercial passenger ship

(continued)

Table 2 (continued)

Prototype	Project name/partners	Year	FC type (fuel)	FC capacity (kW)	Explanatory
	Energy observer	2017-present	PEMFC	22	Hybrid. Solar, wind energy. Hydrogeneration from seawater, electrolyzer, and 106 kW Li battery. 63 kg H_2
	Hydrogenia	2019–2021	PEMFC	100	10 m, 162 kW at 3200 rpm, speed: 10 knots, battery: 92 kW
	Hynova 40	2021-present	PEMFC	80	80 kW FC, 22.5 kg H_2, cruising speed:15 knots
	M.F. Hydra	2023-present	PEMFC	2 × 200	Ballard 2x200 kW PEMFC, LH_2; 2X440 kW generator, 80 m^3 H_2 tank capacity

structure, it should have low amortization and maintenance costs, and the refilling systems must be fast and accessible.

It is expected that fuel cells will need to have certain features when being used as energy devices. The reason for classifying the types mentioned in the previous section according to their operating temperatures is that thus they can be classified according to their area of use. Two are generally used in the maritime sector. Examples of these fuel cells can be seen in Table 2.

Ref. [3] drew particular attention to the efficiency rate of internal combustion engines (ICEs) and FCs with 40% and 50–80%, respectively. They touched upon the current development status and policy direction of countries worldwide related to FCs in sea vehicles. They tabulated the typical hydrogen-fueled vehicles used worldwide, as well. They underlined the challenges the use of global hydrogen energy vessels faces.

The other study by Ref. [4] describes the prospects of technical standards of H_2-powered vehicles for maritime zero-emission carbon transformation. They stated that vessels less than 35 m long are much more feasible for hydrogen-driven operation because of their hydrogen storage. They compared the annual average cost of a diesel engine vessel with that of a hydrogen-powered vessel and concluded that the construction cost of a hydrogen-powered vessel is 1.4 times as expensive as the traditional one. A comparison table is given for land FC vehicles and marine FC vessels.

Ref. [14] prepared a detailed review of FC in marine applications. They touched upon the importance of the zero-emission shipping industry, which must be directly linked with water and renewable energy sources. They listed examples of types of FCs used in marine applications. Moreover, they portrayed the status of the systems and the need to bridge the gap between contemporary systems and hybrid or pure FC propulsion systems in maritime vehicles.

In Ref. [19], a crucial difference between conventional fuels and hydrogen with FC is explored, and the gaps in regulation are highlighted.

The report prepared for "Nautilus" by [20] described Phase 1 of Nautilus activities in detail. A hydrogen FC hybrid propulsion system produced auxiliary power for the vessel via lithium batteries.

The attempts to use green energy in ship systems are discussed and well analyzed by [21]. Ship types, decarbonization, requirements for non-emission ships, state of the art in marine transport powered by hydrogen and FCs, types of thrusters and configuration of power systems in marine FC vessels were described.

Ref. [22] conducted a review of energy efficiency and environmental impact of FC ship and pointed out the importance of seawater properties. Also, some regulatory energy-related indices were presented.

Ref. [23] explains why batteries cannot be chosen on ships requiring high power and energy sustainability. In another study by Ref. [24], the author/s modeled a volume requirement for hydrogen fuel used in deep-sea shipping. They concluded that 105.6 MW PEMFC and 6.9 MWh batteries are necessary for a case study vehicle.

A global patent review of FC technologies in marine transport is presented by Ref. [25]. Hydrogen FC, alcohols (methanol, ethanol), ammonia, and gaseous ones

(LNG and LPG) patents were evaluated over the last 5 years, and it was discovered that 2627 patents were granted for hydrogen and FCs. Their number increases yearly, and the results show that the share of FC-related technologies was 27.4%.

A case study of retrofitting short distances on vessels was carried out by Ref. [26]. They stressed the importance of proper hydrogen infrastructures and the sustainability of hydrogen-powered auxiliary systems in Italy.

Ref. [27] analyzed degradation of a PEMFC vessel. The study aimed to generate a useful tool for hybrid FC. They simulated a 5100-h operation of a ferry. They found that fuel consumption increased by 14.6% and concluded that it is crucial to size a fuel tank capacity.

Ref. [28] gives a favorable review of hydrogen, ammonia, and methanol fuels for maritime usage. The review contains strong arguments in favor of FCs and against batteries.

In a thesis study carried out in Italy [18], a ferry with a 110 kW capacity was examined in two different locations. It also mentions many of the latest studies related to FC-powered marine transport.

Ref. [29] estimated the mass of hydrogen necessary for FCs on crew transfer vessels in Germany. They stated that the PEM and SOFC can be feasible for the next generations. They referred to the hybrid system as preferable to single usage of FC because of its dynamics with instantaneous stops.

Well-to-wake analyses were conducted by Ref. [30]. The study presents a different vision, which suggests that alternative and renewable energy sources would provide a greener perspective in using FCs.

The sustainability of the required fuel is crucial for large vessels. Therefore, hydrogen can be transported by natural gas pipelines for household use. Ref. [31] studied the risk of hydrogen transportation for these pipelines and described it in detail. The challenges of FC propulsion on marine transport were described in detail [32]. The Hydra ferry with 200 kW Ballards FCs was inspected for traveling in China by Ref. [33]. The requirements for FC-powered systems in marine and offshore applications were reported by Ref. [34]. Figure 3 shows the complexity

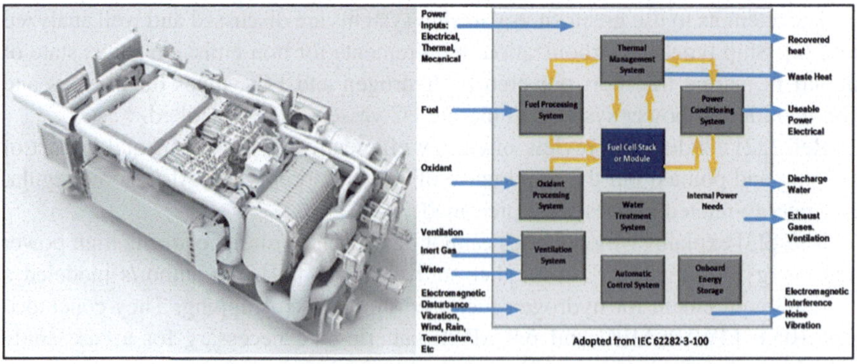

Fig. 3 Marinized FC systems and FC power systems [34, 35] (collected on one figure, faithfully to original)

Table 3 Advantages/benefits and disadvantages/barriers of FC propulsion in marine transport

FC	Advantages	Benefits	Disadvantages	Barriers
	High efficiency	Modularity	Low output power	Hydrogen storage for PEM/slow dynamic response of SOFC or AFC
	Near zero emission	Greener applications with renewable sources	Cost	Hydrogen transportation
	Low noise	No mechanical vibration	Expensive platinum catalyst for PEM	Needed much power, particularly in bunker and vessels
	Low maintenance cost	Less chance of failure	Necessity pure H_2	Regulations
	Increasing popularity	Significant technological know-how	Lots of system devices	Gaps of standardization
	Hybridization potential	Chemical positive opportunities of H_2	Complexity of control systems	Safety problems

and, thus, the importance of marine FC systems. The "Marinized FC system" figure is formed by putting two figures together; the original figures can be found in Refs. 34 and 35 (*PowerCell Sweden AB*). As it can be clearly seen, the FC doesn't look like a stack. It consists of several parts: pumps, control units, separators, cooling facility, power auxiliary, etc.

In the HOPE Project by Ref. [36], Nordic shipping equipped with Hydrogen FC was studied. 10-ton hydrogen storage capacity was tested for 150 nautical miles. The estimated total annual costs of six different types of fuel were analyzed. They generated a system model that can be used on vessels like the Hydra. The prospective studies were prepared by Refs. 37, 38 for Brazil and inland shipping in Germany. Also, Ref. [39] studied the usage of FCs on deep-sea vessels. They mapped the FCs for marine transport and gave their comparative analysis, in which they chose to use LTPEM (low-temperature PEM), HTPEM (high-temperature PEM), SOFC, and AFC. They concluded that FC-powered vessels can be preferred as environmentally friendly ones. Three propulsion systems were examined for PEM at up to 30–40 knots by [40]: waterjets, propellers, and thrusters. Merits and demerits of battery/CH_2/LH_2 were compared with each other.

Table 3 illustrates FC propulsions' effect on marine transport regarding their advantages/benefits and disadvantages/barriers.

4 Regulations, Policies, and LCA

In this section, critical analysis will be given as to economic parameters, life cycle assessments, and regulations required to spread the use of fuel cells in marine technology, to make important breakthroughs that will lead to the development of

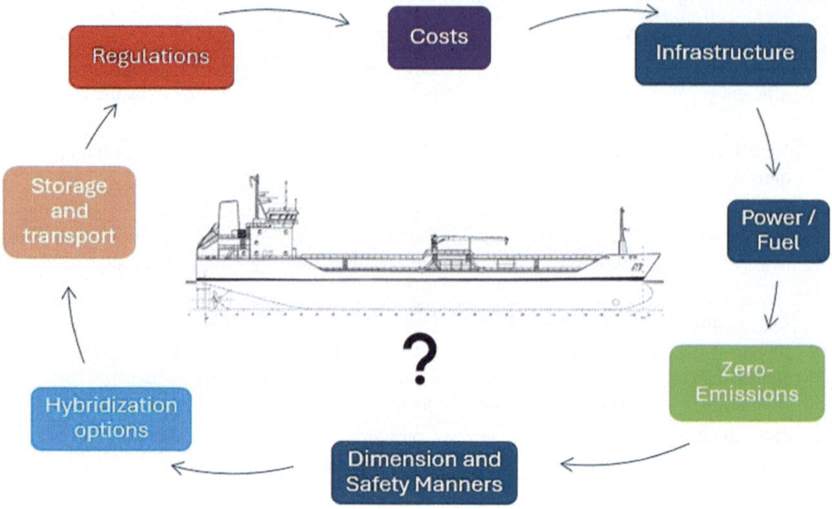

Fig. 4 Questionnaires of ship owners and manufacturers

this technology, and to pave the way to its regular and sustainable use. The importance of this section is illustrated in Fig. 4, which represents a ship manufacturer's or an owner's way of thinking.

Explaining the factors in Fig. 4 and providing information on these systems under this heading will enhance a more detailed understanding of the subject. In a business or engineering product, most product development aspects are listed above; they are also valid for fuel cell ship transportation. Let's look into them one by one, starting from "Costs" and going clockwise.

4.1 Cost

As is known to all, high-value businesses/factories/applications are expensive. Roughly speaking, a ship manufactured in the maritime sector, which takes 85% of the world's trade transportation, is almost equal in financial value to a medium-sized factory (large cargo ships production costs around $80–100 M). Therefore, a good feasibility analysis is necessary before starting the construction of a ship. To produce the energy that will move ships, there must be "an engine department" and "a propulsion," which constitute a vast part of the total cost. The study by Ref. [41] gives an excellent example of a cost strategy.

4.2 *Infrastructure*

When a problem is encountered in ship transportation, robust, ongoing, fast decision-making mechanisms and intelligent solutions are crucial. To solve fuel, personnel, weather/ship conditions, cargo status, and extraordinary instant problems, there must be a solid infrastructure on ships or in ports. Safety and security rules must be strictly followed, especially on ships where hydrogen systems are installed. Insufficient infrastructure causes unnecessary costs, time, and labor.

4.3 *Power/Fuel*

On deep sea vessels and most manufactured cargo ships, transportation is expressed in gross tonnage. An object of this weight requires tremendous power to move. Heavy engines are required to produce thousands of horsepower or M.W. power. On the other hand, two important parameters can affect power production: the speed required from the ship and the environmental conditions. Maritime vehicles differ from land vehicles in their power/weight optimization. For example, the FC stacks of a fuel cell marine vehicle should be placed in a way suitable for the wobbling and stabilization caused by the undulation. The choice of fuel and systems is also very important. The analysis of which fuel cell to choose in the power system is crucial.

4.4 *Zero-Emission*

The shipping industry, which alone produces 3% of the world's emissions, must minimize them within the framework of the decisions and regulations. As all the studies referred to in the reference part, IMO (International Maritime Organization) aims for zero emissions in 2050. Solutions such as ships equipped with alternative energy generators, navigating with engines powered by electricity from renewable energy sources, additional systems used to save energy, and using hybrid engines running on low-emission fuel need to be implemented soon.

4.5 *Dimension and Safety*

Ships developed for safe transportation/travel cannot be manufactured outside standards. Laws and regulations should be reviewed regularly alongside the development of technology. In addition, design parameters should be determined so as not to affect fuel cell applications' operating efficiency. Since the size of ships is directly proportional to the increase in power required, running energy analyses, and examining safety parameters well in naval architecture and marine engineering is essential.

4.6 Hybridization Options

Diesel engines or turbines are not environmentally friendly, batteries take up space and have a short lifespan, long charge/discharge time span, and wind, sea, and wave energy as renewable energy sources are difficult to use on ships. All known types of propulsion have advantages and disadvantages (including FCs). To minimize the latter, hybridization methodology can be an essential transition step, turning each other's demerits into merits.

4.7 Storage and Transportation

Storage and transportation that should be mentioned here concern hydrogen. Hydrogen storage is essential in PEM and SOFC and is generally preferred in ship applications. Large hydrogen tanks are required in PEMs using pure hydrogen, whereas the prime concern for SOFCs operating on high-temperature platforms is hydrogen and temperature balancing and communication devices. If liquid hydrogen is used or hydrogen is fed to a diesel engine, the devices required for phase change and special storage areas will be necessary. Implementers often encounter such cases as critical points in design parameters. In addition, the transportation of hydrogen in ports and security and explosion precautions are also considered essential parameters.

4.8 Regulations

Regulations and standards are essential for engineering applications. A system must be produced, controlled, and maintained according to specific rules to operate efficiently and continuously. After the pandemic and the Paris Agreement, after the increase in the use of hydrogen, which is one of the most important factors for decarbonization and climate change, rapid paperwork on hydrogen has begun in the entire industry and transportation sector, as there were few standards and regulatory documents regarding its production, distribution, transportation, and use. A systematic improvement of standards for using hydrogen or fuel cells in navigation has occurred for a decade now. However, this requires the people involved to understand the subject thoroughly.

In July 2023, IMO revised its GHG emission strategy and aimed to reach net-zero emissions by 2050 [42]. To achieve this goal, the decisions taken in the past play a role. The history of IMO, as well as its activities and policies, is examined in detail by [43–45]. reported that some concerns about fuel efficiency and regulations are included in the "*Energy Efficiency Design Index*" (EEDI), the "*Ship Energy Efficiency Management Plan*" (SEEMP), and the "*International Convention for the Prevention of Pollution from Ships*" (MARPOL) [46]. has added some new indicators, such as EEXI ("*Energy Efficiency Existing Ship Index*") and "*carbon intensity indicator*" (CII). In the [47] report, the regulations of marine transport and risk assessments are given in detail.

In conclusion, the importance of the studies conducted on Life Cycle Assessment (LCA) must be mentioned. In its simplest terms, LCA is a feasibility analysis method that includes all the methods used to evaluate the environmental impact of products from beginning to end; raw materials, production, distribution, usage, waste disposal, and recycling are examined in this regard.

Readers can find the latest studies about LCA [31, 39, 41, 42, 45, 46, 48–55, and] in the reference list.

5 Conclusion and Future Recommendations

In the next 20 years, with the planet's population increasing, humanity will face a rise in demand for energy. To leave a livable world for future generations, everyone involved is working to reduce emissions, keep the Earth and atmosphere temperatures at regular values, minimize climate change, and use renewable and alternative energy sources effectively.

As one of the most effective sectors, maritime is becoming more technological and greener with the increasing number of ships, transactions, and commercial volume. When evaluated within the scope of GHG emissions, the transformation of the maritime sector, which accounts for 3% of the world's emissions, is essential.

This chapter emphasizes the importance of hydrogen and fuel cells as an efficient energy carrier that facilitates decarbonization in the maritime sector.

The main purpose of this study is to provide readers with simply explained, comprehensive, and up-to-date information on fuel cell maritime transportation, on which many studies have been conducted in the last 20 years.

Therefore, current reviews, reports, state policies, and case studies published in 2023–2024 were looked into. Of the 189 found documents, 51 were analyzed to compile the most recent and valuable information.

Basic definitions of fuel cells, which types are used in ship transportation, what their advantages and disadvantages are that will determine their future in the maritime sector, current examples of fuel cell marine vehicles, the status of supported projects, regulations, policies, and details of LCA analyses are summarized and presented in the chapter.

Generally, PEM, SOFC, MCFC, and AFC types are used in maritime applications. The most frequently used ones are PEM and SOFC. The benefits of fuel cells include their being more efficient than batteries, having better unit energy density, causing less noise pollution, and being environmentally friendly. Costs, storage difficulties, power/weight ratio, huge energy requirement, currently developing standards and regulations, and expensive sustainability are among the barriers FCs face when used in navigation.

Authors list the following future suggestions, taking into account the information above:

1. Being environmentally friendly, FCs will likely become much more popular due to the ongoing technological progress and decrease in unit costs. Also, they will soon take an essential share in the maritime sector.
2. In the transition period, hybridizing energy sources can be an important bridge/step in unit cost and energy efficiency.
3. Ships are transportation devices that require vast amounts of power. Significant energy sources must meet these requirements. Although the efficiency of fuel cells is higher than that of internal combustion engines (55–50/35–40%), a lot of space for fuel cell rooms and hydrogen storage systems will be necessary. Space optimization should be done for fuel cells and systems.
4. Regulations, policies, and rules should indicate how to use hydrogen in marine vehicles. The transportation and storage of hydrogen in ships and ports must be safe.
5. For running projects, fuel cell implementers must consider that sea vehicles have dynamics that are different from those of land or air vehicles.
6. Support from companies/countries with experience in the relevant technology to those without can enhance forming an FC community in the world maritime industry.
7. There should be studies on using battery-powered vehicles under 40 m in length, fuel-cell systems for vehicles between 60 and 100 m, and hybrid systems for larger vehicles. They can even be grouped according to energy/efficiency/power/cost optimization prospects.
8. Hydrogen production involving the use of seawater or solar energy in marine vehicles should be supported.
9. It is recommended that an FC in marine vehicles unit (FCMVU) be established under IMO's umbrella, which will also provide regular training/practice to manufacturers and ship owners.

We respect all readers; we hope they will share their ideas with us.

Acknowledgments The authors express their gratitude to the authors of the articles and reports listed in the reference section. Also, the authors would like to thank *Volha Yashchanka-Yıldız* for her contribution in proofreading this chapter.

References

1. Sürer, M. G., & Arat, H. T. (2022). Advancements and current technologies on hydrogen fuel cell applications for marine vehicles. *International Journal of Hydrogen Energy, 47*(45), 19865–19875. https://doi.org/10.1016/j.ijhydene.2021.12.251
2. Sürer, M. G., & Arat, H. T. (2024). Energy and exergy analysis of hydrogen production from seawater using waste heat recovery system on a model ship. *International Journal of Exergy, 44*(1), 53. https://doi.org/10.1504/IJEX.2024.138741
3. Li, J. C., Xu, H., Zhou, K., & Li, J. Q. (2024). A review on the research progress and application of compressed hydrogen in the marine hydrogen fuel cell power system. *Heliyon, 10*, e25304.
4. Wang, Z., Li, M., Zhao, F., Ji, Y., & Han, F. (2024). Status and prospects in technical standards of hydrogen-powered ships for advancing maritime zero-carbon transformation. *International Journal of Hydrogen Energy, 62*, 925–946.
5. Friedlingstein, et al. (2020). *The global carbon budget 2020*. Earth System Science Data.
6. Retrieved September 15, 2024, from https://globalcarbonatlas.org/
7. Liu, Z., Deng, Z., Davis, S. J., et al. (2024). Global carbon emissions in 2023. *Nature Reviews Earth and Environment, 5*, 253–254. https://doi.org/10.1038/s43017-024-00532-2
8. Retrieved September 15, 2024, from https://carbonmonitor.org/
9. U.S. Energy Information Administration. (2023). *Energy Institute-Statistical Review of World Energy (2024)—with major processing by Our World in Data*
10. Ritchie, H., Rodés-Guirao, L., Mathieu, E., Gerber, M., Ortiz-Ospina, E., Hasell, J., & Roser, M. (2023). *Population growth*. OurWorldInData.org. https://ourworldindata.org/population-growth [Online Resource].
11. Ritchie, H., Rosado, P., & Roser, M. (2023). *Energy*. OurWorldInData.org. 'https://ourworldindata.org/energy' [Online Resource].
12. Retrieved September 15, 2024, from https://www.worldometers.info/world-population/
13. IEA. (2023). *Electricity market report 2023*. IEA. https://www.iea.org/reports/electricity-market-report-2023, Licence: CC BY 4.0.
14. Fu, Z., Lu, L., Zhang, C., Xu, Q., Zhang, X., Gao, Z., & Li, J. (2023). Fuel cell and hydrogen in maritime application: A review on aspects of technology, cost and regulations. *Sustainable Energy Technologies and Assessments, 57*, 103181.
15. UNCTAD. (2023). *Handbook of statistics*. UNCTAD. ISBN: 978–92–1-002951-3, Retrieved September 15, 2024, from https://unctad.org/system/files/official-document/tdstat48_en.pdf
16. Tanç, B., Arat, H. T., Baltacıoğlu, E., & Aydın, K. (2019). Overview of the next quarter century vision of hydrogen fuel cell electric vehicles. *International Journal of Hydrogen Energy, 44*(20), 10120–10128.
17. Arat, H. T., & Çeper, B. A. (2024). Investigation of instantaneous temperature variations in hydrogen PEM fuel cell stack under cold climate conditions. *International Journal of Hydrogen Energy*. https://doi.org/10.1016/j.ijhydene.2024.07.294
18. Brusasco, A. (2024). *High-fidelity modelling of a fuel cell system for shipping industry* (Doctoral dissertation, Politecnico di Torino).
19. Inal, O. B. (2024). Decarbonization of shipping: Hydrogen and fuel cells legislation in the maritime industry. *Brodogradnja: An International Journal of Naval Architecture and Ocean Engineering for Research and Development, 75*(2), 1–13.
20. Pal, N., Boudreau, B., Monroe, N., Vaughn, E., Zaag, N., Sookhoo, R., & Klebanoff, L. E. (2024). Project nautilus: Introducing a hydrogen fuel cell system as a retrofit for a hybrid electric vessel. *International Journal of Hydrogen Energy, 53*, 1457–1476.
21. Wang, Z., Dong, B., Yin, J., Li, M., Ji, Y., & Han, F. (2023). Towards a marine green power system architecture: Integrating hydrogen and ammonia as zero-carbon fuels for sustainable shipping. *International Journal of Hydrogen Energy, 50*, 1069.
22. Wang, Z., Dong, B., Wang, Y., Li, M., Liu, H., & Han, F. (2024). Analysis and evaluation of fuel cell technologies for sustainable ship power: Energy efficiency and environmental impact. *Energy Conversion and Management: X, 21*, 100482.

23. Van Rheenen, E. S., Padding, J. T., Slootweg, J. C., & Visser, K. (2024). Hydrogen carriers for zero-emission ship propulsion using PEM fuel cells: An evaluation. *Journal of Marine Engineering & Technology, 23*(3), 166–183. https://doi.org/10.1080/20464177.2023.2282691
24. McKinlay, C. J., Turnock, S. R., Hudson, D. A., & Manias, P. (2024). Hydrogen as a deep sea shipping fuel: Modelling the volume requirements. *International Journal of Hydrogen Energy, 69*, 863–873.
25. Sun, M., Tong, T., Jiang, M., & Zhu, J. X. (2024). Innovation trends and evolutionary paths of green fuel technologies in maritime field: A global patent review. *International Journal of Hydrogen Energy, 71*, 528–540.
26. Melideo, D., & Desideri, U. (2024). The use of hydrogen as alternative fuel for ship propulsion: A case study of full and partial retrofitting of roll-on/roll-off vessels for short distance routes. *International Journal of Hydrogen Energy, 50*, 1045–1055.
27. Saponaro, G., Stefanizzi, M., Torresi, M., & Camporeale, S. M. (2024). Analysis of the degradation of a proton exchange membrane fuel cell for propulsion of a coastal vessel. *International Journal of Hydrogen Energy, 61*, 803–819.
28. Osman, A. I., Nasr, M., Lichtfouse, E., Farghali, M., & Rooney, D. W. (2024). Hydrogen, ammonia and methanol for marine transportation. *Environmental Chemistry Letters, 22*, 1–8.
29. Cramer, H., Fitz, A., Niemi, A., Skobiej, B., & Sill Torres, F. (2023). Estimating hydrogen usage of a crew transport vessel fleet for offshore windfarm maintenance. *WMU Journal of Maritime Affairs, 22*(4), 509–525.
30. Ramsay, W., Fridell, E., & Michan, M. (2023). Maritime energy transition: future fuels and future emissions. *Journal of Marine Science and Application, 22*(4), 681–692.
31. Amer, M. O., Hoseyni, S. M., & Cordiner, J. (2024). Fuelling the future with safe hydrogen transportation through natural gas pipelines: A quantitative risk assessment approach. *Transactions of the Indian National Academy of Engineering, 9*, 1–19.
32. Van Hoecke, L., Laffineur, L., Campe, R., Perreault, P., Verbruggen, S. W., & Lenaerts, S. (2021). Challenges in the use of hydrogen for maritime applications. *Energy & Environmental Science, 14*(2), 815–843.
33. Antolini, A. (2024). Maritime issues—Utilisation of hydrogen as fuel: Maritime application of hydrogen propulsion: Progress and challenges of hydrogen fuel propulsion for different ship type projects. *JTTRI, 1*, 1–14.
34. American Bureau of Shipping. (2023). *Requirements for fuel cell power systems for marine and offshore applications*. ABS.
35. Oravasaari, T. (2024). *Fuel cell auxiliary systems in a cruise ship*. Bachelor's Thesis, Turku University of Applied Sciences.
36. Hansson, J., Davíðsdóttir, B., Fridell, E., Jivén, K., Yum, K. K., Latapí, M., Lundström, H., Parsmo, R., Stenersen, D., Wimby, P., & Burgren, J. (2023). *HOPE—hydrogen fuel cells solutions in Nordic shipping project summary: A Nordic maritime transport and energy research Programme project*. DiVA. ISBN: 978–91–7883-515-7.
37. Carvalho, F., Osipova, L., & Zhou, Y. (2023). *Life-cycle greenhouse gas emissions of hydrogen as a marine fuel and cost of producing green hydrogen in Brazil*. ICCT working paper 2023-11.
38. MariGreen Project Report. (2018). Perspectives for the use of hydrogen as fuel in inland shipping, feasibility study, Retrieved September 15, 2024, from https://www.dst-org.de/wp-content/uploads/2018/11/Hydrogen-Feasibility-Study-MariGreen.pdf
39. Guadagna, G. (2024). *Mærsk Mc-Kinney Møller Center for Zero Carbon Shipping (MMMCZCS)—Fuel cell technologies and applications for deep-sea shipping*. Retrieved September 15, 2024, from https://cms.zerocarbonshipping.com/media/uploads/documents/Fuel-Cell-Technologies-and-Applications-for-Deep-Sea-Shipping.pdf
40. Doornebos, P., Francis, M., le Poole, J., & Kana, A. A. (2023). Design and feasibility of a 30-to 40-knot emission-free ferry. *International Shipbuilding Progress, 70*(2), 81–114.
41. Ganjian, M., Farahabadi, H. B., Alirezapouri, M. A., & Firuzjaei, M. R. (2024). Optimal design strategy for fuel cell-based hybrid power system of all-electric ships. *International Journal of Hydrogen Energy, 50*, 1558–1571.

42. IMO, KCM. (2024). Analysis on interaction of CII with GFS: Calibrating IMO energy efficiency and fuels targets. Retrieved September 15' 2024, from https://cleanshipping.org/wp-content/uploads/2024/03/202401_CII_GFS_EE_targets_for_IMO-3.pdf
43. Latapí, M., Davíðsdóttir, B., Cook, D., Jóhannsdóttir, L., Radoszynski, A. M., & Karlsson, K. (2024). Hydrogen fuel cells in shipping: A policy case study of Denmark, Norway, and Sweden. *Marine Policy, 163*, 106109.
44. Tadros, M., Ventura, M., & Guedes Soares, C. (2023). Review of the IMO initiatives for ship energy efficiency and their implications. *Journal of Marine Science and Application, 22*(4), 662–680.
45. Iqbal, R., Liu, Y., Zeng, Y., Zhang, Q., & Zeeshan, M. (2024). Comparative study based on techno-economics analysis of different shipboard microgrid systems comprising PV/wind/fuel cell/battery/diesel generator with two battery technologies: A step toward green maritime transportation. *Renewable Energy, 221*, 119670.
46. Marichal, G. N., Ávila Prats, D., Conesa, A., Rodríguez, J. A., & Iglesias, G. (2024). The future of energy in ships and harbors. *TransNav: International Journal on Marine Navigation and Safety of Sea Transportation, 18*(1), 45.
47. European Maritime Safety Agency. (2023). *Potential of hydrogen as fuel for shipping*. EMSA.
48. Sánchez, P. J. B., Márquez, F. P. G., Papaelias, M., & Lee, D. (2024). Life cycle cost analysis of an autonomous underwater vehicle that employs hydrogen fuel cell. *Ocean Engineering, 299*, 117300.
49. Guven, D., & Kayalica, M. O. (2024). Environmental and economic assessment of hydrogen-powered ferries for inland transportation. *Ocean Engineering, 301*, 117556.
50. Mojarrad, M., Thorne, R. J., & Rødseth, K. L. (2024). Technical and cost analysis of zero-emission high-speed ferries: Retrofitting from diesel to green hydrogen. *Heliyon, 10*(6), e27479.
51. Dong, D. T., Schönborn, A., Christodoulou, A., Ölcer, A. I., & González-Celis, J. (2024). Life cycle assessment of ammonia/hydrogen-driven marine propulsion. *Proceedings of the Institution of Mechanical Engineers, Part M: Journal of Engineering for the Maritime Environment, 238*(3), 531–542.
52. Trillos, J. C. G., Wilken, D., Brand, U., & Vogt, T. (2021). Life cycle assessment of a hydrogen and fuel cell ropax ferry prototype. *Progress in Life Cycle Assessment, 2019*, 5–23.
53. Wang, H., Aung, M. Z., Xu, X., & Boulougouris, E. (2023). Life cycle analysis of hydrogen powered marine vessels—Case ship comparison study with conventional power system. *Sustainability, 15*(17), 12946.
54. RHC Report (2023). The regulation of hydrogen fuel propulsion in maritime vessels. Retrieved September 15, 2024, from https://assets.publishing.service.gov.uk/media/64b13b4f48826b00103a9e4c/the_regulation_of_hydrogen_fuel_propulsion_in_maritime_vessels.pdf
55. Matthé, J., Jain, P., & Pierre, J. (2022). *Hydrogen in maritime: Opportunities and challenges*. WSP Report. Retrieved on September 15, 2024, from https://www.wsp.com/en-gl/insights/hydrogen-in-maritime-opportunities-and-challenges

Fuel Cells for Autonomous Underwater and Surface Marine Vehicles with Energy Storage System

Youcef Belkhier, Florent Becker, and Jean-Frédéric Charpentier

1 Introduction

Autonomous and unmanned underwater vehicles (AUVs and UUVs) are innovative underwater vehicles designed for autonomous underwater operations. These vehicles are equipped with sophisticated technology that allows them to navigate, collect data, and perform tasks without a direct human onboard intervention. AUVs and UUVs have found diverse applications across various fields, owing to their ability to access underwater environments that may be challenging or hazardous for human intervention. Key features of AUVs include advanced navigation systems, sensors, and communication capabilities, enabling them to operate independently for extended periods. These vehicles can be pre-programmed with specific mission parameters or operate in real-time, adapting to changing conditions in their environment. Applications of AUVs/UUVs span a wide range of applications such as oceanographic research, underwater exploration, and naval defense [1, 2].

Despite their widespread use in naval propulsion, traditional energy/propulsion technologies are not fully suitable for AUVs and UUVs. If traditional fuels are used, extraction, transportation, and utilization of such resources can have a negative impact on the ocean environment and are associated with operational constraints. Indeed, the use of such solutions leads to performance limitations, resulting in higher energy consumption and operational expenses [3]. The sophisticated maintenance and operation of mechanical components, such as internal combustion engines and propellers, pose issues, necessitating frequent maintenance operations and incurring downtime that may have an influence on mission schedules and cost-effectiveness.

Y. Belkhier · F. Becker · J.-F. Charpentier (✉)
Institut de Recherche et d'Etudes Navales-French Naval Laboratory (EA 3634, IRENav), French Naval Academy, Brest, France
e-mail: youcef.belkhier@ecole-navale.fr; florent.becker@ecole-navale.fr; jean-frederic.charpentier@ecole-navale.fr

C. O. Colpan et al. (eds.), *Fuel Cell and Hydrogen Technologies in Maritime Transportation*, https://doi.org/10.1007/978-3-031-86110-9_5

Furthermore, internal combustion engine vibrations and noise can disturb marine organisms and harm the fragile underwater environments where underwater radiated noises are also a key challenge for scientific and defense missions where UUVs and AUVs are very useful. Given these difficulties, it is imperative to investigate and implement environmentally conscious and sustainable energy/propulsion options, especially those designed for AUVs [4].

While fuel cells (FC) offer several advantages for powering AUVs, relying solely on them presents significant inconveniences. Fuel cells often require a warm-up period before delivering optimal power, delaying AUV deployment. They may also struggle to respond quickly to sudden spikes in power demand, making it difficult to handle transient loads efficiently. Furthermore, fuel cells are complex systems with numerous components, such as fuel reformers, membranes, and catalysts, all requiring regular maintenance. The underwater environment, with its high pressure, salinity, and potential for biofouling, exacerbates these maintenance challenges. Additionally, the infrastructure for storing and refueling hydrogen or other fuels needed for fuel cells is still in development and can be cumbersome to manage, especially in remote or deep-sea missions [5, 6]. Therefore, integrating energy storage systems (ESS) with fuel cells is essential to mitigate these issues, providing a reliable and responsive power source for AUVs and UUVs. To improve efficiency and lessen their impact on the environment, hybrid underwater vehicles, which combine fuel cells and storage systems, have become a viable option for these vehicles in recent years. Underwater vehicles equipped with these hybrid power systems have a lot to offer, especially when emissions limitations and discretion are highly required. The fundamental strength of a hybrid energy system is its capacity to integrate several energy sources, maximizing power generation and energy chain efficiency. Fuel cells produce no adverse air quality effects and very little greenhouse gas emissions by using an electrical process to convert hydrogen or other clean fuels into electricity [7, 8]. This process produces a constant, reliable, and continuous power source. ESS, on the other hand, are perfect for balancing the supply and demand for energy because they can store extra energy produced by the fuel cells at times of low demand or high efficiency. ESS also plays a crucial role in delivering high power outputs during periods of high demand, like abrupt maneuvers or accelerations. Several ESSs can be used. Combining batteries and supercapacitors (SCs) allows to efficiently manage power and energy in the system. Batteries allow for storing power and optimizing the global behavior and the sizing of the hybrid system (energy management), but their lifetime can be limited if the number of deep charging/discharging cycles is too high. Supercapacitors, thanks to their quick charge and discharge times and high cycling capabilities [9, 10] allow the managing of high dynamics transients to maintain the other sources in their optimal ranges and to minimize the number and depth of charging/discharging cycles of the batteries which are directly related to battery system aging. SCs' quick energy delivery and storing capabilities enhance the overall performance of the hybrid energy source and storage system as well, guaranteeing a steady power supply under a range of operating circumstances. Hybrid power systems have potential, but they are not without difficulties. There are weight and space limitations associated with the assembly of different energy storage technologies onboard. The hybrid underwater vehicle's performance may be impacted by the fuel cells, batteries, and supercapacitors' added weight and volumes. Furthermore, advanced energy and power management systems are necessary to guarantee the smooth synchronization and ideal distribution of electricity across fuel cells, batteries, and supercapacitors. To ensure optimum efficiency and reduce waste, these systems must constantly assess consumption and availability from various sources and distribute power effectively. Such sophisticated energy management system design, implementation, and upkeep can raise total costs and complexity [11].

A small number of studies are centered on fuel cell-based AUV devices, while others investigated energy management strategy (EMS) for multi-energy undersea systems. Dynamic programming coupled with a fuzzy logic control system is adopted in Ref. [12]. Using simulation studies, Ref. [13] investigated particle swarm optimization techniques and dynamic programming to maximize the distribution of power among the powertrain and the engine in all-electric propulsion marine systems. They were successful in confirming the algorithms' usefulness for ship energy management. Ref. [14] used a low-pass filter to improve power signal frequency division and implemented a rule-based technique to manage the supercapacitor and battery energy. Still, this approach depends on the knowledge of the engineers, thus more improvement in dynamic reaction efficiency is required. In Ref. [15], a sizing and optimizing strategy to enhance the fuel cell efficiency of an AUV is proposed. Maximum efficiency range identification of a hybrid AUV is proposed in Ref. [16] for EMS optimization. Ref. [17] presents an EMS that optimizes using particle swarm optimization techniques. Fuel cell degradation is included as an objective function in the Q network method, which relies on deep learning and was created in Ref. [18] to improve fuel consumption equilibrium as well as energy storage system optimization.

The work presented in this chapter combines practical energy consumption data with high-fidelity simulation power models to give an in-depth assessment of EMS for highlighting three power configurations of AUV electric propulsion using FC as the primary energy source (fuel cell only, fuel cell associated with battery and fuel cell associated with a combination of batteries and supercapacitors). Energy efficiency, electrical power accuracy, and fuel consumption are the performance factors considered in this assessment. Since AUV drive cycles are not standardized, the study uses a power profile using field trial data gathered by sampling AUVs described in the literature. The Machine State Control (MSC) is evaluated based on the different power configurations. The use of a highly variable power profile from actual field trial data ensures a realistic representation of AUV operating conditions, illustrating the relevance and applicability of the study's conclusions.

2 The Hybrid AUV Description

In the current study, power configurations of an AUV using a system architecture roughly similar to that described in Ref. [4] are considered, with the inclusion of supercapacitors for load adjustment. In the hybrid energy system of the studied AUV, batteries are employed to effectively balance the power generated by the FC, while SCs are utilized for load correction. In the general studied configuration described in Fig. 1, FC, batteries, and SC are interconnected to the DC bus via DC/DC converters to a common DC bus. At the same time, the three-phase AC propulsion motor of the AUV is interfaced with the DC bus through a classical DC/AC motor drive (rectifier and PWM VSI). Although alternative configurations are feasible, the studied configuration closely resembles the one described in Ref. [4] to minimize alterations to the energy/propulsion structure (mass, volume, weight, etc.) of the case study under consideration. In our study, a simulation tool to determine the behavior of the multi-source hybrid system has been elaborated under the MATLAB/Simulink® environment. Subsequent subsections will present the subsystem models used to build this simulation tool.

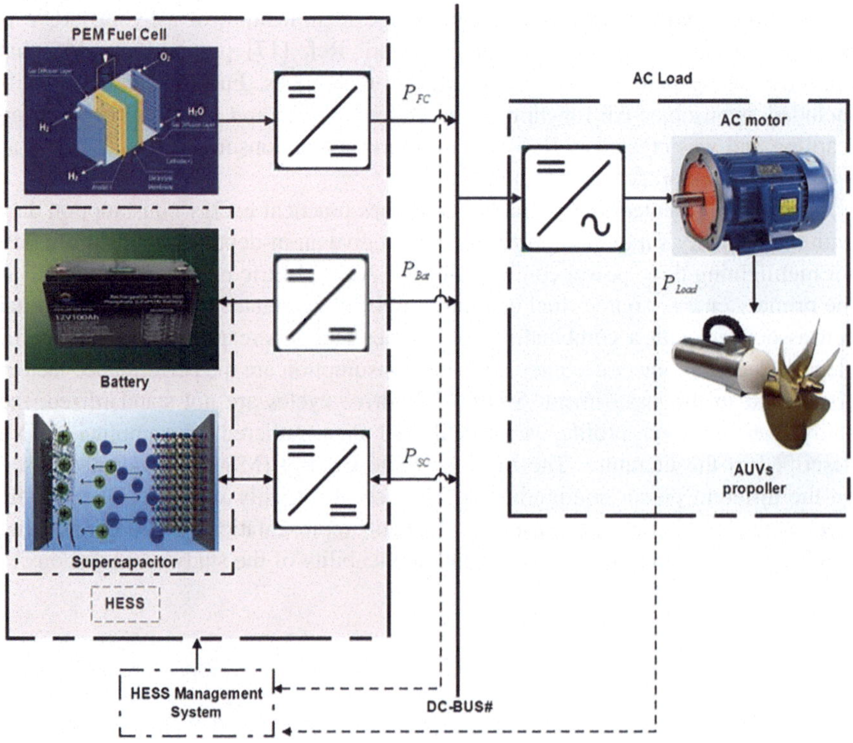

Fig. 1 Schematic of the studied AUV

2.1 Mathematical Modeling of PEM Fuel Cells

The mathematical model of a PEM FC illustrated in Fig. 2 can be quite complex, involving a set of differential equations that describe various physical and chemical processes occurring within the cell. In this study, a simplified representation of the mathematical model typically used for PEM FC is implemented in the simulation tool [15, 20]:

$$V_{\text{Nernst}} = -\frac{\Delta G}{2F} - \frac{RT}{2F}\text{lnln}\left(\frac{p_{H_2O}p_{\text{ref}}^{0.5}}{p_{H2}p_{O_2}^{0.5}}\right) \tag{1}$$

$$V_{\text{cell}} = V_{\text{Nernst}} - A_{\text{cell}}\ln\left(\frac{i_{\text{cell}}}{i_0}\right) - i_{\text{cell}}R_{\text{incell}} - B_{\text{cell}}\ln\left(1 - \frac{i_{\text{cell}}}{i_L}\right) \tag{2}$$

where $p_{\text{ref}}^{0.5}$ is the pressure, A_{cell} and B_{cell} are the FC's coefficients, R_{incell} is the inherent resistance, i_L is the FC's maximum current, R is the gas constant, T is the temperature, F is Faraday's constant, V_{cell} represents the stack's cell voltage, p_{H2} and p_{O2} are the partial pressures of oxygen and hydrogen, respectively, i_{cell} is FC's intrinsic exchange current, ΔG is the Gibbs free energy, V_{Nernst} is the cell thermodynamic potential, and i_0 represents the cell exchange current. The used FC parameters are given in Table 1.

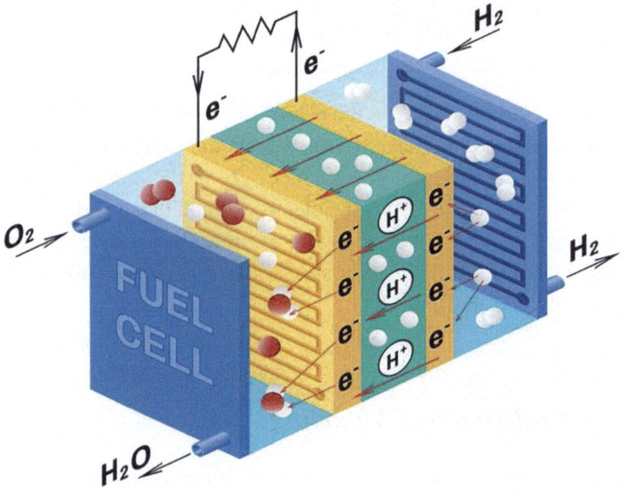

Fig. 2 PEM fuel cell basic representation

Table 1 Fuel cell parameters

Parameters	Value
Fuel cell model	PEMFC-1.26 kW-24 V_{dc}
Rated power	2 kW
Minimum power	0.378 kW
Number of cells	42
Nominal stack efficiency	46%
Operating temperature	55 °C
Nominal composition [H_2, O_2, H_2O]	[99.95, 21, 1]
Nominal supply pressure [fuel (bar), air (bar)]	[1.5, 1]
Nominal voltage	23 V
Nominal current	52.24 A

2.2 Mathematical Modeling of the Supercapacitor

The presented model uses the equations commonly used in the mathematical modeling of supercapacitors, which can be expressed as follows [14, 21]:

$$V_{SC} = \frac{N_p}{C_T} Q_C - R_{SC} I_{SC} \tag{3}$$

$$C_T = \frac{N_p}{N_s} C \tag{4}$$

$$C = \left[\frac{d}{N_e E E_e A_i} + \frac{2 N_e R T}{F Q_c \, sinh\left(\frac{Q_c}{N_e^2 A_i \sqrt{8RTEE_e C}} \right)} \right] \tag{5}$$

where I_{SC} is the output current, N_e is the electrode layers number, R_{SC} is the module resistance, E is the electrolyte permittivity, V_{SC} is the output voltage, E_e is the free space permittivity, C_T is the total capacitance, Q_C is the electric charge, N_p is the parallel-connected cells number, A_i is the electrodes and electrolyte shared area, N_s is the series-connected cells number, d is the Helmholtz layer length, and C is the Molar concentration. Figure 3 shows a supercapacitor's description, and the given parameters are tabulated in Table 2.

2.3 Battery Mathematical Modeling

The mathematical model used for the battery is given by the following equations [15, 22]:

$$V_{\text{bat}} = E_{\text{bat}} - I_{\text{bat}} * R_{\text{bat}} \tag{6}$$

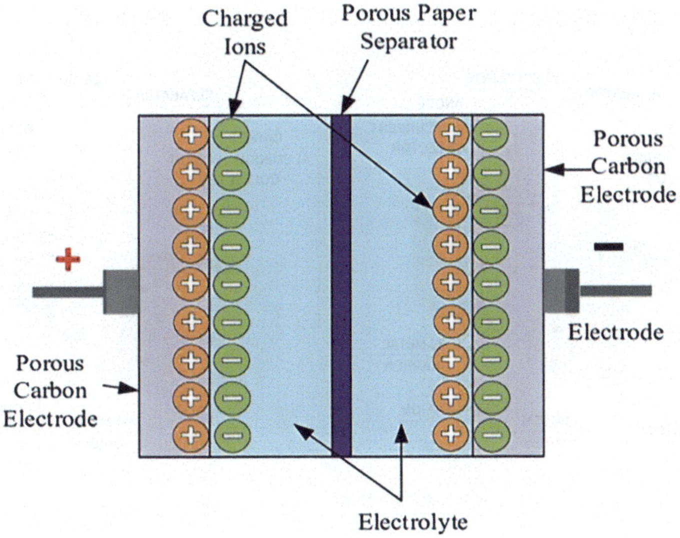

Fig. 3 Supercapacitor description

Table 2 Supercapacitor parameters

Parameters	Value
Operating temperature	25 °C
Equivalent DC series resistance	150e-3 Ω
Initial voltage	270 V
Rated voltage	291.6 V
Rated capacitance	15.6 F
Number of series capacitors	6
Number of parallel capacitors	1

$$I_{bat} = \frac{E_{\text{bat}} - \sqrt{E_{\text{bat}}^2 - 4P_{\text{bat}}R_{\text{int}}}}{2R_{\text{int}}} \tag{7}$$

$$SOC\ (t) = SOC(0) - \int \frac{I_{\text{bat}}}{Q_{\text{bat}}} dt \tag{8}$$

where E_{bat} is the open-circuit voltage, I_{bat} is the battery current, R_{int} is the internal resistance of the battery, P_{bat} is the battery power, Q_{bat} is the capacity, V_{bat} is the electromotive force, and $SOC(0)$ and $SOC(t)$ are the battery's initial state of charge and current state of charge, respectively. Figure 4 gives the battery electrochemical schematic diagram configuration. The studied battery parameters are presented in Table 3.

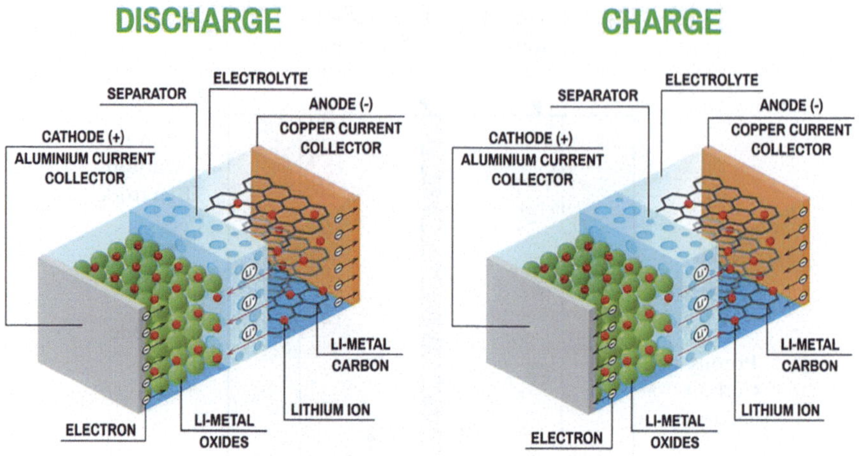

Fig. 4 Battery electrochemical schematic diagram

Table 3 Battery parameters

Parameters	Value
Type	Lithium-ion
Battery response time	20 s
Nominal voltage	48 V
Initial SOC	65%
Rated capacity	52 ah

2.4 Power Converters Modeling

Through DC/DC converters, the load's DC/AC converter is connected to the FC, batteries (BAT), and SCs, enabling voltage conversion and accurate control of the DC-bus voltage and FC/BAT/SC power sharing. While the battery and supercapacitor systems combine a buck-boost, the FC converter uses a boost configuration. In actuality, the two converters are operated in buck mode for recharging and boost mode for discharging. A DC/AC converter manages the AUV's variable-speed propulsion in accordance with the mission profile. Standard median average methods are used to model every converter, as in Ref. [19].

3 Energy Management Strategy

In this section, we will delve into the application of rule-based machine state control (MSC), which can be found in detail in Refs. [4, 19], to optimize energy utilization for underwater vehicles equipped with hybrid power systems. By systematically testing various power configurations of the AUV, the work aims to demonstrate the distinct advantages of a hybrid setup over traditional single-source power systems. Rule-based control involves using predefined rules and algorithms to manage energy distribution effectively based on real-time data and computational methods. This strategy focuses on maximizing FC efficiency by operating it within its optimal range, minimizing hydrogen consumption by leveraging battery storage during low power demand periods, and extending the operational life of the system by distributing the load appropriately between the fuel cell, the battery, and the supercapacitors. We will examine different configurations, including fuel cell-only, FC/battery-only, and FC/battery/SC hybrid systems, to understand their impact on performance. The analysis will highlight how hybrid systems, managed by rule-based control, can achieve extended missions, enhance operational adaptability, and greater sustainability in challenging underwater environments. The tested EMS aims to minimize operating expenses, increase FC efficiency, reduce hydrogen consumption, and prolong the system's operational lifespan. This investigation will highlight innovative strategies in underwater vehicle technology, paving the way for more efficient and sustainable operations. State transition rules are the backbone of the MSC. These rules govern when and how a machine transitions from one state to another. For instance, a machine might be programmed to switch from an idle state to an off state after a specified period of inactivity, thereby conserving energy that would otherwise be wasted. Similarly, during periods of peak energy demand, machines can be strategically shifted to lower energy states to avoid excessive energy consumption and reduce costs. These transitions must be carefully managed to ensure that energy savings do not negatively impact the operational efficiency or the lifespan of the equipment. The used state rules for this study are close to those presented in Ref. [19].

4 Results and Discussion

In this section, the results obtained from simulations of the power system configurations shown in Fig. 1 are presented, and the performance evaluation of the three power configurations used is estimated. Additionally, we discuss the implications of these results on energy efficiency, sustainability, and overall performance of AUVs equipped with hybrid power systems. By highlighting the advantages and limitations of each power system configuration and EMS, we aim to provide recommendations for the optimal design of AUV power systems in challenging underwater environments.

Figure 5 shows the simulated hybrid AUV simulation model with EMSs under MATLAB/Simulink®. In this chapter, mission data using actual tests conducted with a Hugin 3000 AUV is used in the simulation. Ref. [4] presents these trials. The AUV sea trials on the Swedish west coast, as shown in Fig. 6, provide the data needed in this work for the suggested EMS assessment of efficiency. This two-dimensional trial path is used to create a reference load profile. This load profile is shown in

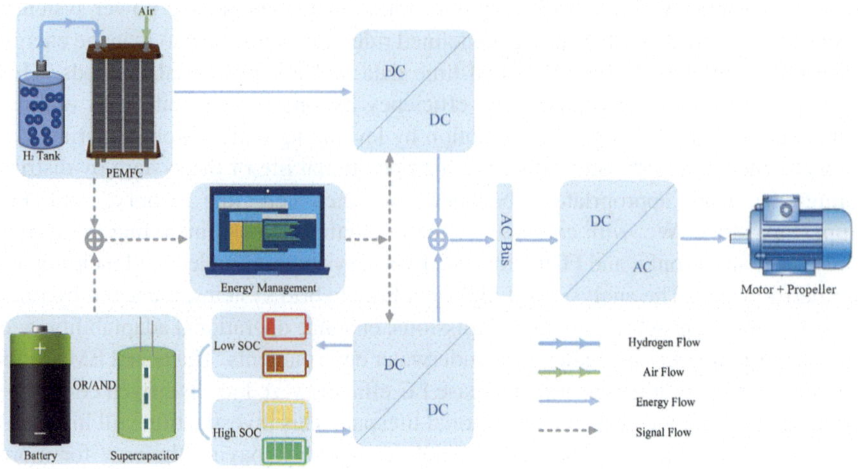

Fig. 5 The simulated hybrid AUV model with EMSs under MATLAB/Simulink

Fig. 7, which shows the current that the propulsion system will need to receive from the 52 V DC bus to follow the trajectory of sea trials (Fig. 6).

The rule-based EMS (MSC) has been used to assess the operation of AUV power setups. Several factors will be compared, including hydrogen consumption, battery stability (which ensures autonomy and minimal deterioration effects), and DC stability (which ensures the capacity to manage various systems).

According to the results of Fig. 8, it is obvious that the use of a fuel cell without ESS will be very detrimental in terms of fuel cell aging and operating points. In this case, the fuel cell undergoes rapid power variations that are highly detrimental to its operation efficiency and lifetime.

The second configuration allows the maintenance of the FC during a large part of the mission profile at over 50% of its rated power and with low power variation. It leads obviously to a more favorable use of the FC source in terms of aging. However, the number of charge/discharge cycles of the battery pack is high, and the charging/discharging cycles can be deep. This frequent use of the battery to compensate for the power demand variation can be highly detrimental in terms of battery aging, especially if the AUV is used in resident or deep-sea situations where battery replacement can be a key problem.

In the third configuration, the FC is maintained between 1 kW and 2 kW with very stable FC operations for a relatively long time. The use of supercapacitors, which allows the management of quick power variations and shaving of the power peaks, allows the minimization of the number of battery charge/discharge cycles and is highly favorable to the battery lifetime enhancement.

In terms of fuel consumption, Fig. 9 presents the hydrogen consumption in the three studied configurations. The third configuration allows for a significant reduction in hydrogen consumption during the mission profile compared to the basic FC-only case and the FC/battery case (first and second case). This result is related to

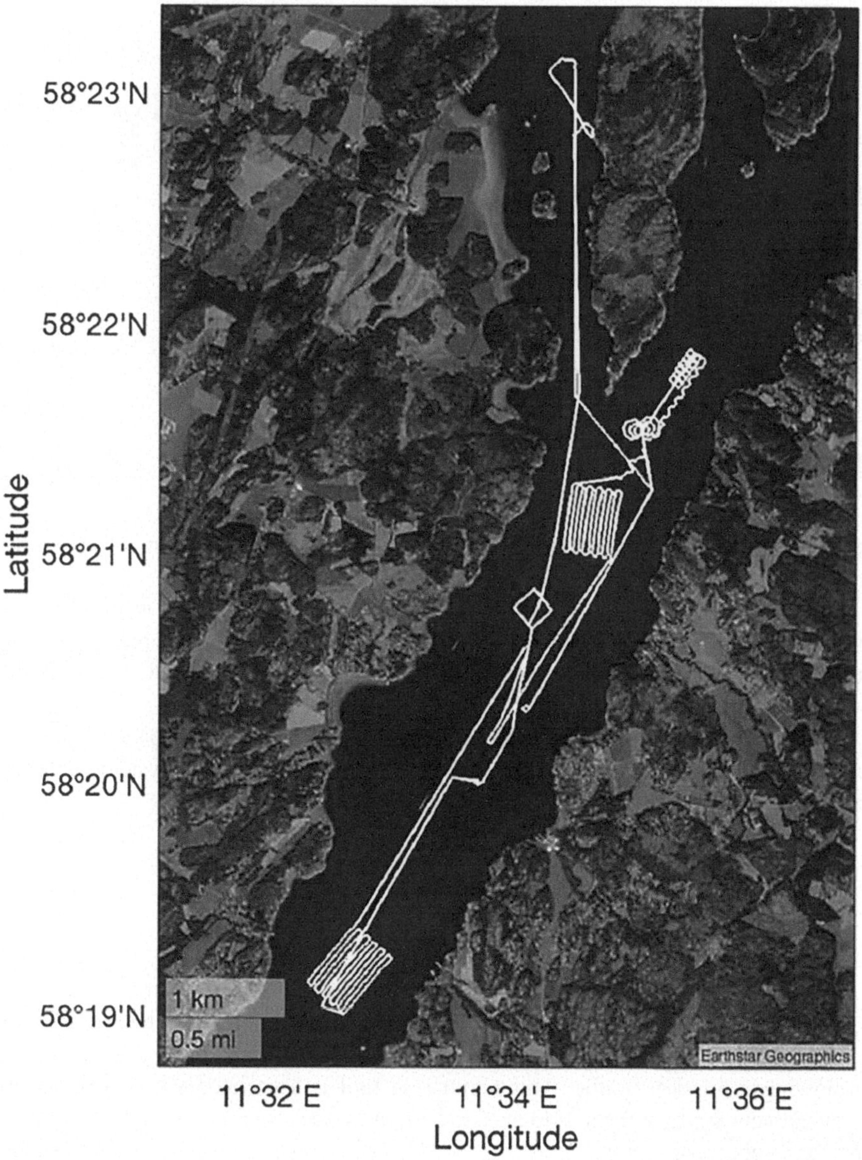

Fig. 6 Path of the field tests conducted in June 2019 at Stora Borno, Sweden, using the Hugin 3000 AUV [4]

better use of the FC and the minimization of energy exchanges between the battery and DC bus. The second solution leads to a higher hydrogen consumption, which can be explained by the high power and energy, exchanged between the battery and

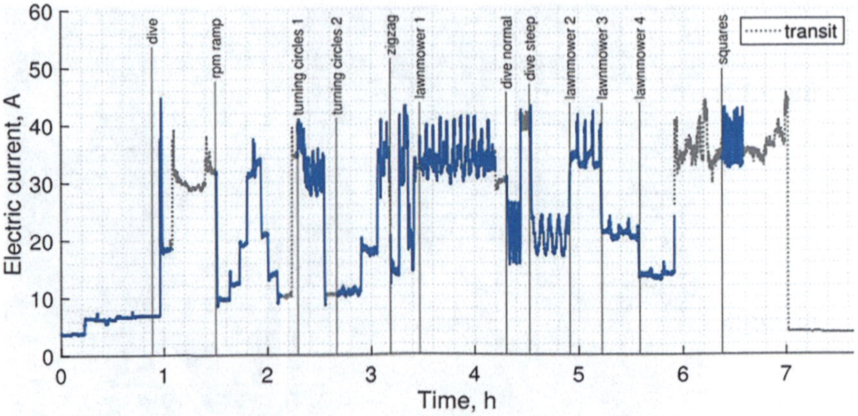

Fig. 7 The reference load profile [4]

the DC bus. These battery/DC bus exchanges are penalized in terms of energetic efficiency by battery and corresponding DC/DC converter losses.

Considering the DC bus voltage variations presented in Fig. 10, it is obvious that the use of the ESS storage allows stabilizing the DC bus voltage in comparison with the FC-only case, where the FC variation limitations lead to a DC bus voltage variation of about 20%.

In a general way, the association of FC with battery and supercapacitor systems for hydrogen-powered AUVs leads to significant improvements in system behavior and lifetime.

5 Conclusion

This chapter first introduces the context of the use of hydrogen and FC for the energy/propulsion system of AUV. A global simulation tool has also been presented to be able to simulate a realistic hybrid FC-based energy system of an AUV following a mission profile. This simulation tool is associated with a rule-based management strategy for hybrid energy systems associating the FC to ESSs (batteries and supercapacitors).

By integrating practical power consumption data with high-fidelity simulation power models and utilizing power profiles derived from field trial data, the study offers valuable insights into the energy efficiency, electrical power accuracy, and fuel consumption of the studied systems. The inclusion of real-world operating conditions enhances the relevance and applicability of the conclusions drawn from the assessment.

Three configurations have been considered for a realistic AUV and realistic mission profile case: firstly, the use of an FC without energy storage systems;

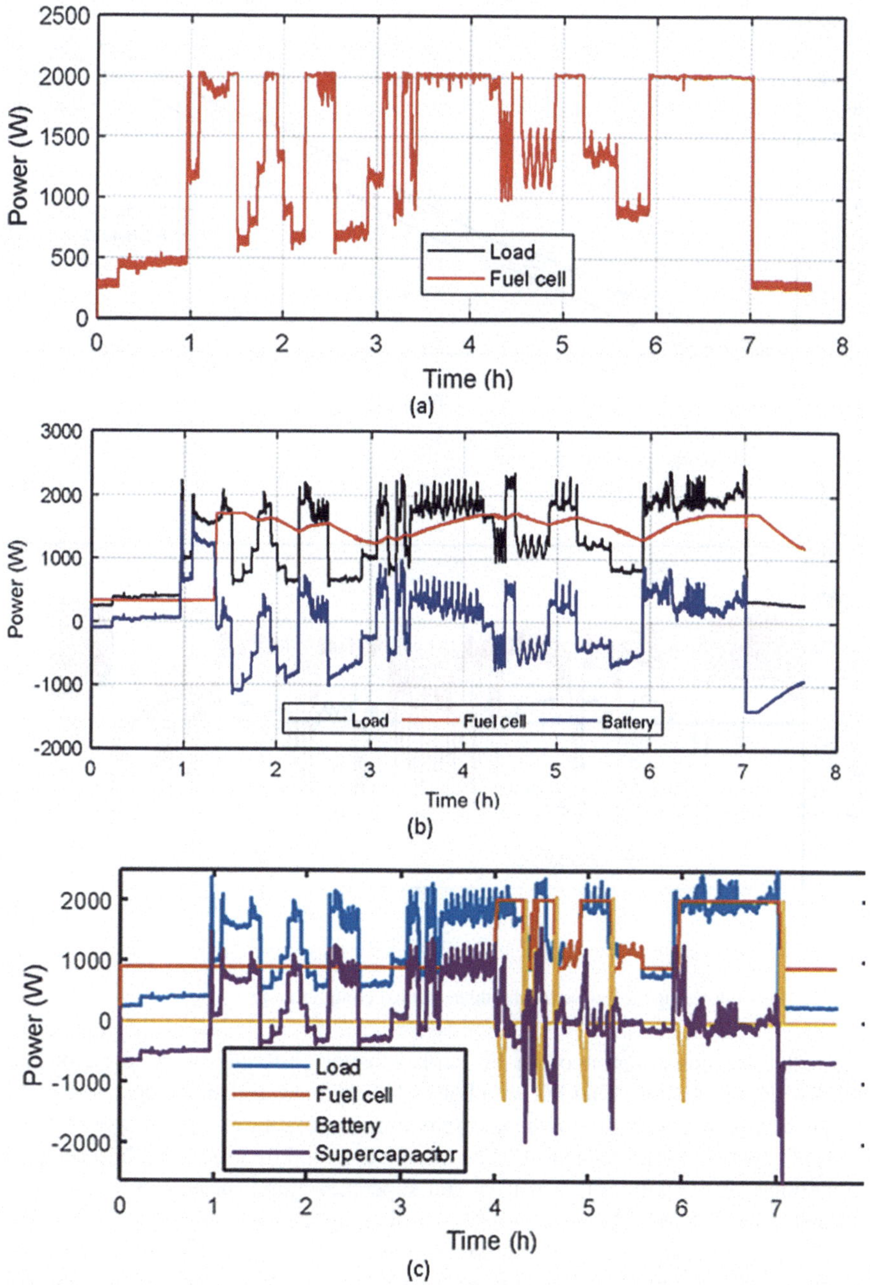

Fig. 8 Power exchanges. (**a**) Fuel cell only, (**b**) fuel cell/battery, (**c**) fuel cell/battery/supercapacitor

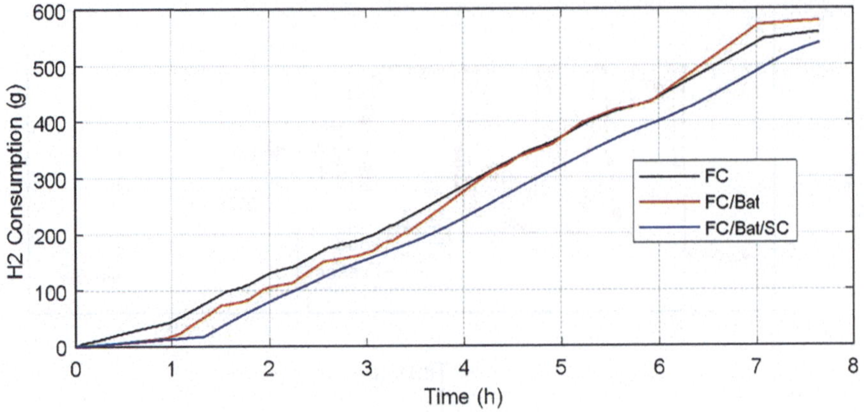

Fig. 9 First configuration H_2 consumption

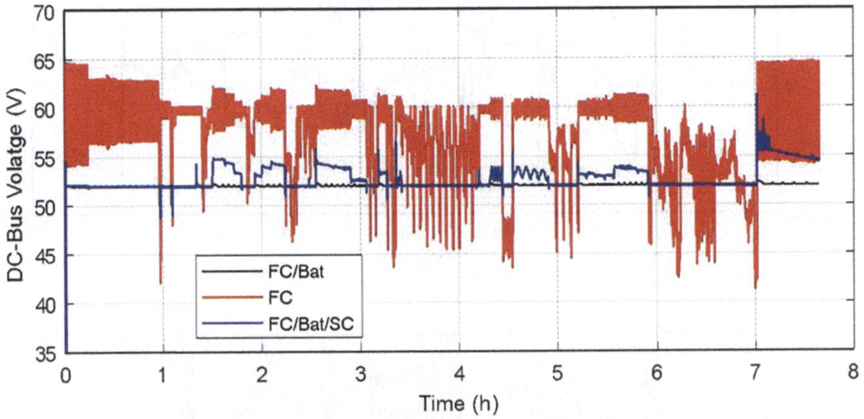

Fig. 10 DC-link voltage response for the three studied configurations

secondly, the combination of an FC with a battery energy storage system; and thirdly, the integration of an FC with both a battery bank and supercapacitors.

Simulation results highlight the necessity of incorporating energy storage systems with FC for hydrogen-powered AUVs. Among the configurations, the third setup—combining an FC with both a battery and supercapacitors—proves to be the most efficient in terms of fuel consumption, system aging, and reliability.

These findings underscore the importance of selecting appropriate FC and energy storage system configurations tailored to specific AUV power requirements to optimize energy consumption and enhance operational performance effectively.

References

1. Yang, Y., Xiao, Y., & Li, T. (2021). A survey of autonomous underwater vehicle formation: Performance, formation control, and communication capability. *IEEE Communications Surveys & Tutorials, 23*(2), 815–841.
2. Wibisono, A., Piran, M. J., Song, H. K., & Lee, B. M. (2023). A survey on unmanned underwater vehicles: Challenges, enabling technologies, and future research directions. *Sensors, 23*(17), 7321.
3. Deutsch, C., Chiche, A., Bhat, S., Lagergren, C., Lindbergh, G., & Kuttenkeuler, J. (2020, September). Energy management strategies for fuel cell-battery hybrid AUVs. In *2020 IEEE/OES autonomous underwater vehicles symposium (AUV)* (pp. 1–6). IEEE.
4. Deutsch, C., Chiche, A., Bhat, S., Lagergren, C., Lindbergh, G., & Kuttenkeuler, J. (2022). Evaluation of energy management strategies for fuel cell/battery-powered underwater vehicles against field trial data. *Energy Conversion and Management: X, 14*, 100193.
5. Sezgin, B., Devrim, Y., Ozturk, T., & Eroglu, I. (2022). Hydrogen energy systems for underwater applications. *International Journal of Hydrogen Energy, 47*(45), 19780–19796.
6. Sürer, M. G., & Arat, H. T. (2022). Advancements and current technologies on hydrogen fuel cell applications for marine vehicles. *International Journal of Hydrogen Energy, 47*(45), 19865–19875.
7. Verma, J., & Kumar, D. (2021). Recent developments in energy storage systems for marine environment. *Materials Advances, 2*(21), 6800–6815.
8. Punov, P., & Gechev, T. (2021). Energy management of a fuel cell hybrid ultra-energy efficient vehicle. *International Journal of Hydrogen Energy, 46*(38), 20291–20302.
9. Zhang, M., Gu, H., & Guo, J. (2022, December). Energy management strategies for hybrid ships and unmanned underwater vehicles. *Journal of Physics: Conference Series, 2395*(1), 012051.
10. Albarghot, M. M. (2020). *Sizing, dynamic modeling and simulation of a solar-wind-hydrogen power system of the MUN explorer autonomous underwater vehicle* (Doctoral dissertation, Memorial University of Newfoundland).
11. Damian, S. E., Wong, L. A., Shareef, H., Ramachandaramurthy, V. K., Chan, C. K., Moh, T. S. Y., & Tiong, M. C. (2022). Review on the challenges of hybrid propulsion system in marine transport system. *Journal of Energy Storage, 56*, 105983.
12. Zhao, Z. H. (2022). Improved fuzzy logic control-based energy management strategy for hybrid power system of FC/PV/battery/SC on tourist ship. *International Journal of Hydrogen Energy, 47*(16), 9719–9734.
13. Kanellos, F. D., Tsekouras, G. J., & Hatziargyriou, N. D. (2014). Optimal demand-side management and power generation scheduling in an all-electric ship. *IEEE Transactions on Sustainable Energy, 5*(4), 1166–1175.
14. Abeywardana, D. B. W., Hredzak, B., Agelidis, V. G., & Demetriades, G. D. (2016). Supercapacitor sizing method for energy-controlled filter-based hybrid energy storage systems. *IEEE Transactions on Power Electronics, 32*(2), 1626–1637.
15. Chiche, A., Lindbergh, G., Stenius, I., & Lagergren, C. (2021). A strategy for sizing and optimizing the energy system on long-range AUVs. *IEEE Journal of Oceanic Engineering, 46*(4), 1132–1143.
16. Wang, T., Li, Q., Wang, X., Qiu, Y., Liu, M., Meng, X., & Chen, W. (2020). An optimized energy management strategy for fuel cell hybrid power system based on maximum efficiency range identification. *Journal of Power Sources, 445*, 227333.
17. Peng, X., Chen, H., & Guan, C. (2023). Energy management optimization of fuel cell hybrid ship based on particle swarm optimization algorithm. *Energies, 16*(3), 1373.
18. Tang, X., Zhou, H., Wang, F., Wang, W., & Lin, X. (2022). Longevity-conscious energy management strategy of fuel cell hybrid electric vehicle based on deep reinforcement learning. *Energy, 238*, 121593.

19. Motapon, S. N., Dessaint, L. A., & Al-Haddad, K. (2013). A comparative study of energy management schemes for a fuel-cell hybrid emergency power system of more-electric aircraft. *IEEE Transactions on Industrial Electronics, 61*(3), 1320–1334.
20. Sahri, Y., Belkhier, Y., Tamalouzt, S., Ullah, N., Shaw, R. N., Chowdhury, M. S., & Techato, K. (2021). Energy management system for hybrid PV/wind/battery/fuel cell in microgrid-based hydrogen and economical hybrid battery/super capacitor energy storage. *Energies, 14*(18), 5722.
21. Gao, H., Wang, Z., Yin, S., Lu, J., Guo, Z., & Ma, W. (2021). Adaptive real-time optimal energy management strategy based on equivalent factors optimization for hybrid fuel cell system. *International Journal of Hydrogen Energy, 46*(5), 4329–4338.
22. Bassam, A. M., Phillips, A. B., Turnock, S. R., & Wilson, P. A. (2016). An improved energy management strategy for a hybrid fuel cell/battery passenger vessel. *International Journal of Hydrogen Energy, 41*(47), 22453–22464.

State-of-Art Energy Management Strategies for Hybrid Fuel Cell Applications for Ships

Andrea Coraddu, Sara Tamburello, Charlotte Löffler, Halit Ege Ceyhun, Lindert van Biert, and Luca Oneto

1 Introduction

The global carbon dioxide (CO_2) emissions of shipping alone were, in 2018, equal to 1,056 million tonnes [1]. This represents a share of about 2.89% of the total global CO_2 emissions. As the emissions caused by ships are expected to be 90–130% of the 2008 greenhouse gas (GHG) emissions in 2050, the International Maritime Organization (IMO) aims to reduce this number to at least 50% [1]. Hydrogen, renewable energy sources like wind and solar, and carbon capture technologies are being extensively researched to help meet global decarbonization goals and achieve net-zero emissions by 2050. Hydrogen, in particular, could contribute to reducing 20% of annual global CO_2 emissions by 2050 through its applications in industries such as steelmaking, transportation, and power generation, complementing renewable energy technologies [2, 3]. The operating profiles of ships have diversified significantly, leading to a notable evolution in power and propulsion architectures, as vessels adapt to varying mission requirements and technological advancements, such as hybrid propulsion systems and dynamic EMS [4]. This adaptation not only enhances efficiency but also ensures that ships are better suited to varied operational demands. As outlined by [1], the development of these sophisticated architectures necessitates precise design and control systems. Properly implemented, these systems can result in substantial fuel savings, thereby making ships both greener and

A. Coraddu (✉) · S. Tamburello · C. Löffler · H. E. Ceyhun · L. van Biert
Delft University of Technology, Faculty of Mechanical Engineering, Delft, The Netherlands
e-mail: a.coraddu@tudelft.nl; s.tamburello-1@tudelft.nl; c.loeffler@tudelft.nl; h.e.ceyhun@tudelft.nl; l.vanbiert@tudelft.nl

L. Oneto
Università Degli Studi di Genova, Department of Informatics, Bioengineering, Robotics and Systems Engineering, Genoa, Italy
e-mail: luca.oneto@unige.it

C. O. Colpan et al. (eds.), *Fuel Cell and Hydrogen Technologies in Maritime Transportation*, https://doi.org/10.1007/978-3-031-86110-9_6

safer, as discussed in [5]. Moreover, recent advancements in maritime propulsion, particularly with the integration of electric and hybrid-electric systems, have opened up new possibilities for control strategies. These systems allow for more precise control over power distribution, enabling optimization based on operational needs, such as load demand and environmental factors. However, further research is required to fully understand and develop these strategies, especially in large-scale and ocean-going vessels, where system complexity and operational requirements pose significant challenges [6, 7]. The evolution of ship design, particularly in response to stricter environmental regulations like the Energy Efficiency Design Index and the Carbon Intensity Indicator, is driving innovation aimed at improving energy efficiency and reducing emissions [8, 9, 10, 11]. Advances in propulsion systems, hull optimization, and the incorporation of low-carbon fuels are examples of how ship design adapts to meet the increasing demands for sustainability and compliance with international maritime standards [12].

However, many existing maritime architectures continue to rely heavily on classical control theories. While these systems often incorporate advanced components and modern architectures, they frequently fail to significantly improve fuel efficiency and emission reductions. Proportional-Integral-Derivative (PID) controllers are recognized as one of the most widely used control mechanisms in industrial applications, including ship propulsion and power management systems. The popularity of PID controllers stems from their simplicity, robustness, and proven reliability in managing dynamic systems, making them particularly suited for applications where real-time stability is critical. They are frequently favored because they require relatively simple tuning compared to more advanced controllers and provide satisfactory performance for most control tasks, including in maritime environments. PID controllers are extensively utilized due to their ability to manage single-input, single-output systems, such as those found in propulsion systems, with high reliability. Despite the rise of more complex, model-based controllers, the PID's balance of practicality and effectiveness ensures its continued dominance in various fields, including marine engineering. Despite their strengths, these classical systems encounter limitations, particularly in dynamic and complex environments where advanced, data-driven control strategies—such as neural networks and fuzzy logic—could provide more flexibility and efficiency. The slow transition to these advanced technologies is exacerbated by the conservative nature of the maritime industry, where safety and regulatory compliance are prioritized, often at the expense of adopting newer control methods [13, 14]. Additionally, as highlighted by Nguyen et al. [15], the integration of traditional control systems into modern architectures increases both system complexity and costs. This mismatch between outdated control strategies and the growing demand for technological advancement creates a critical gap in achieving optimal energy efficiency and environmental sustainability in maritime operations. The systems discussed here are typically classified as hybrid powerplants or hybrid ships, emphasizing the need for better integration between conventional and cutting-edge technologies [6].

Integrating Energy Storage Systems (ESS) is a key point for hybrid ships [16]. From the storage point of view, lead-acid batteries have not been considered for large-scale present and future applications due to limited power and energy density capacity. Lithium-ion batteries (LIBs) are widely used as ESS for hybrid ships due to their high energy density and efficiency. However, producing these batteries is highly energy-intensive, often relying on high-heat processes primarily using fossil fuels, such as coal. As a result, manufacturing these batteries contributes significantly to GHG emissions, with estimates showing that for each ton of lithium extracted, approximately 15 tons of CO_2 are emitted. Furthermore, around 77% of LIBs are produced in China, where coal dominates the energy supply, exacerbating the carbon footprint of their production [17, 18].

To put this into perspective, authors of [19] have reviewed 79 papers on the environmental impact of the production of LIBs, concluding that the average cumulative energy demand and GHG emissions for battery production are 328 kWh and 110 kg CO_2 equivalent per kWh of storage capacity, respectively. And with that, the impact (both environmental and social) on the areas where lithium is mined is not even discussed. As we all strive to solve the global crisis problems and not reallocate them, the lifetime of these batteries should be extended as much as possible. The second focus in hybrid propulsion layouts of ships is using alternative fuels. One of those alternative fuels is hydrogen, which can be used via fuel cells [20]. Integrating component health awareness into the Energy Management System (EMS) plays a critical role in enhancing the overall performance and lifetime of energy systems, particularly in hybrid propulsion systems. Neglecting health adaptation mechanisms can lead to significant inefficiencies. For example, hydrogen consumption can increase dramatically, from around 6.5% to as much as 24%, as demonstrated in studies focused on fuel cell-battery systems, depending on how well the EMS adapts to the evolving health of components during operation [21]. The health of key components such as fuel cells and batteries affects not only the energy efficiency but also the overall lifespan of the system. This is crucial because the limited lifetime of components, such as fuel cells, remains one of the major barriers to the widespread adoption of hybrid propulsion technologies. In such systems, taking health factors into account ensures more consistent performance and reduces wear, thereby extending the lifetime of the components. However, the challenge of effectively integrating health-aware mechanisms into EMS design remains significant. Although research has highlighted the importance of component health monitoring in prolonging system lifespan and improving efficiency, much work is still needed to develop comprehensive EMS models that adapt to real-time health indicators of the system's components. This calls for advancements in both monitoring technologies and algorithmic strategies to ensure that the system can make informed decisions based on the current health of its components, paving the way for a more reliable and efficient transition to hybrid propulsion systems. However, integrating health awareness factors into the EMS design still has a long way to go [21].

Hybrid fuel cell applications for ships refer to integrating fuel cell technology with one or more other energy sources to power a vessel's propulsion and onboard systems. Hybrid fuel cell systems represent a forward-looking solution for the maritime industry. They aim to meet stricter environmental regulations while maintaining operational efficiency and performance, and they are increasingly recognized as a promising solution for reducing GHG emissions and pollutants in the maritime industry. These systems offer significant environmental benefits by generating electricity without harmful emissions, making them a cleaner alternative to traditional diesel-powered propulsion systems. Hybrid fuel cell systems combine fuel cells, typically powered by hydrogen or other alternative fuels, with energy storage devices like batteries and, in some cases, traditional internal combustion engines. These components work together to ensure optimized power delivery, improved fuel efficiency, and reduced emissions. The hybrid system aims to maximize efficiency, reduce emissions, and improve energy management onboard ships. Key Components of hybrid fuel cell systems for ships are

- **Fuel Cells**: to convert chemical energy from fuels directly into electricity through an electrochemical reaction with oxygen or another oxidizing agent, favored for their high efficiency and low emission levels.
- **Energy Storage Systems**: batteries or supercapacitors that store excess energy generated by the fuel cells. This stored energy can be used during peak power demands or when the fuel cell is off.
- **Power Management System**: to control how power is distributed between the fuel cells, energy storage, and the ship's propulsion and electrical systems, ensuring optimal performance and efficiency.
- **Auxiliary Power Units**: include traditional internal combustion engines that can work in conjunction with the fuel cells to provide additional power when needed.

Recent projects demonstrate how fuel cell technology, combined with renewable fuels like hydrogen and methanol, can substantially decarbonize maritime operations, supporting the IMO's goals to reduce emissions by 50% by 2050 [22, 23]. Moreover, fuel cells have intrinsic higher efficiency compared to traditional internal combustion engines [24]. By combining different energy sources, hybrid fuel cell systems increase the ship's energy system flexibility and adaptability to different operating conditions while complying with stringent regulatory frameworks. Furthermore, the silent operation of fuel cells significantly reduces noise pollution, which benefits marine life and crew comfort on board vessels [23]. Among different fuel cell types, polymer electrolyte membrane fuel cells (PEMFCs) powered by hydrogen have been the most applied technologies in projects involving fuel cells on ships [25]. PEMFCs are usually integrated with an ESS, creating a hybrid PEMFC/ESS powertrains configuration that allows zero-emission propulsion.

This chapter targets maritime engineers, policymakers, and researchers interested in the evolution and implementation of EMS within hybrid PEMFC/ESS systems. This chapter gathers and synthesizes the latest developments in EMS for hybrid fuel cell applications in ships, an area characterized by technological innovation and potential environmental impact. For this reason, we begin by recalling the preliminaries related to the fundamentals of hybrid fuel cell systems and EMS, providing an overview of key international research projects that have advanced the application of fuel cell technologies in the maritime sector. The chapter explains the architecture of hybrid PEMFC/ESS powertrains, examining the specific components of the PEMFC system, including its structure, operational principles, and factors influencing performance. Then, we further dissect the ESS, which plays a crucial role in complementing PEMFCs by managing the balance between power generation and energy demands. The combined benefits of hybrid PEMFC/ESS systems, such as improved efficiency, reduced emissions, and fuel flexibility, are weighed against their challenges, including technical complexity, costs, and durability concerns. Then, the focus shifts to EMS, which governs how power is distributed between the PEMFC and ESS. These strategies are broken down into three main approaches: (i) rule-based (RB), (ii) optimization-based (OB), and (iii) learning-based methods (LB). Each type of EMS is explained in detail, starting with RB approaches that use pre-defined rules for power distribution, followed by OB methods, which utilize algorithms to optimize fuel cell efficiency and energy use. The final approach discussed, LB EMS, involves using machine learning techniques to dynamically adapt and optimize the system based on operational data. The actual state-of-the-art EMS provides a more focused discussion on the most current advancements in each of the three EMS categories. RB strategies, though simple, are evaluated for their limitations in flexibility, while OB strategies are praised for their ability to achieve greater energy savings through more complex algorithms. LB strategies, being the most advanced, hold significant promise for future applications due to their adaptability and ability to handle complex maritime operations. A comparative analysis of these strategies highlights their relative strengths and weaknesses in terms of efficiency, scalability, and implementation costs. We will then detail the challenges and future directions. The chapter identifies major obstacles to the adoption of hybrid PEMFC/ESS systems, including technological, regulatory, and economic challenges. It also explores potential future developments, such as advancements in materials, energy management algorithms, and integration with renewable energy sources, which could address these challenges and improve the viability of these systems in maritime applications. Finally, the conclusion summarizes the findings, emphasizing the critical role of hybrid fuel cell systems in achieving sustainable shipping and the need for ongoing innovation to overcome current limitations in both technology and EMS.

This chapter is structured to provide a thorough analysis of hybrid fuel cell systems in maritime applications. It begins with foundational concepts in the preliminaries in Sect. 2, setting the stage for state-of-the-art analysis and discussions. The document then covers EMS in Sect. 3, breaking them down into RB, OB, and LB approaches. Section 4 tackles the ongoing technical, economic, and regulatory challenges while outlining potential areas for future research. The chapter concludes with a summary of key insights and recommendations in Sect. 5.

2 Preliminaries: Fundamentals of Hybrid Fuel Cell Systems for Ships and Energy Management Strategies

In this section, we first revisit the key concepts surrounding hybrid fuel cell systems and EMS, beginning with a review of fundamental principles. This includes an overview of significant international research projects that have propelled the use of fuel cell technologies within the maritime industry. The discussion explores the architecture of hybrid PEMFC/ESS powertrains, focusing on the specific components of the PEMFC system, including its design, operational mechanisms, and performance factors. Following this, we focus on the role of the ESS, which is integral to balancing power generation and energy demand alongside the PEMFC. The combined benefits of hybrid PEMFC/ESS systems, such as enhanced efficiency and reduced emissions, are contrasted with the challenges they face, such as hydrogen storage, cost, and durability issues. From there, we transition to EMS, which dictates how power is distributed between the PEMFC and ESS. The fundamentals of the three primary approaches (RB, OB, LB) are introduced in this section.

2.1 Overview of International Research Projects for Fuel Cell Systems in Maritime Applications

The application of fuel cell technologies, including PEMFCs, high-temperature PEMFCs, solid oxide fuel cells (SOFCs), and molten carbonate fuel cells (MCFCs), into ship energy systems has experienced a significant increase in recent years. Authors [25, 26] reported an analysis of the developed and ongoing projects and real-world applications for the use of fuel cells in shipping. Table 1 gives an overview of past and ongoing projects from 2000 to date, including the installed fuel cell type and power onboard, the logistic fuel, and the type of application. Figures 1 and 2 present the distribution of fuel cell types and logistic fuel types used in maritime applications, respectively. Figure 1 highlights the dominance of PEMFC technology, which accounts for 75% of the fuel cell types used in the analyzed projects. This prevalence is likely due to PEMFC's high degree of maturity, high efficiency, specific power and power density, and other advantageous technical characteristics that will be further discussed in Sect. 2.5. The remaining portion of the chart is shared between SOFC, which contributes 14.3%, and MCFCs, representing 10.7%. Both SOFC and MCFC offer advantages in terms of fuel flexibility and higher operating temperatures, which can improve thermal efficiency, but their deployment is less frequent due to their lower degree of maturity, lower power density, complex cooling requirements, and high costs.

Figure 2 illustrates the distribution of logistic fuels used in these fuel cell systems, with hydrogen being the predominant fuel, powering 81% of the projects. This indicates the increasing global emphasis on zero-emission technologies and the role hydrogen is expected to play in decarbonizing maritime transportation.

Table 1 Overview of research projects for fuel cell systems in maritime applications [25]

Project name	Time	Country	FC power	FC type	Logistic fuel	Application	Ref.
FCSHIP	2002–2004	EU	–	MCFC	Diesel	RoPax vessel	[34]
FellowSHIP	2003–2018	Norway	320 kW	MCFC	LNG	Offshore supply	[35]
MC-WAP	2005–2010	Italy	150 kW	MCFC	Diesel	RoPax	[36]
ZEMSHIP	2007–2014	Germany	96 kW	PEMFC	H_2	Inland vessel	[37]
Nemo H2	2008–	Netherlands	65 kW	PEMFC	H_2	Passenger boat	[38, 39]
PaXell	2009–2016	Germany	60 kW	PEMFC	MeOH	Cruise ship	[40]
SchIBZ	2009–2018	Germany	100 kW	SOFC	Diesel	General cargo ship	[37, 41]
TESEO	2012–2015	Italy	50 kW	PEMFC	H_2	Yacht	[42, 43]
RiverCell	2015–2022	Germany	90 kW	PEMFC	MeOH	Inland vessel	[44]
Sea Change	2016–2022	USA	360 kW	PEMFC	H_2	Passenger ferry	[45, 46]
ELEKTRA	2017–2019	Germany	300 kW	PEMFC	H_2	Canal tug	[24]
Maranda	2017–2022	EU	165 kW	PEMFC	H_2	Arctic research ship	[47]
HFCMARINE	2018–2020	Denmark	200 kW	PEMFC	H_2	Ferry	[48–52]
HYSEAS III	2018–2022	Scotland	600 kW	PEMFC	H_2	RoPax ferry	[53]
TecBIA	2018–2022	Italy	140 kW	PEMFC	H_2	Research vessel	[54, 55]
H2PORTS	2019–2023	Spain	70 kW	PEMFC	H_2	Reach Stacker	[56]
PaXell 2	2019–2022	Germany	N/A	PEMFC	MeOH	Cruise ship	[57]
Hydrogenia	2019–2021	South Korea	100 kW	PEMFC	H_2	Small boat	[45, 58]
MF Hydra	2020–	Norway	400 kW	PEMFC	H_2	RoPax ferry	[59]
ShipFC	2020–2024	Norway	2 MW	SOFC	NH_3	Offshore vessel	[60]
Nautilus	2020–2024	EU	60 kW	SOFC	LNG	Cruise ship	[61]
HyShip	2021–2024	Norway	3 MW	PEMFC	H_2	RoRo	[62]
HIMET	2021–2022	UK	500 kW	PEMFC	H_2	Ferries	[63]
Helenus	2022–2027	EU	500 kW	SOFC	LNG	Cruise ship	[64]

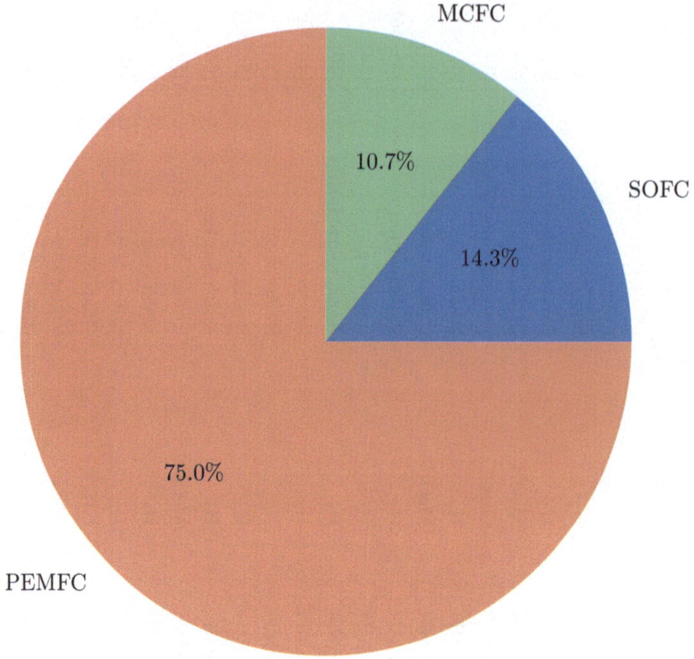

Fig. 1 Distribution of fuel cell types used in maritime applications

Methanol and diesel each account for 9.5%, showcasing their continued relevance in certain applications where hydrogen storage or infrastructure is limited. However, as hydrogen production and distribution infrastructure is expected to improve and higher energy density storage solutions advance, we can anticipate a further shift towards hydrogen as a logistic fuel to power zero-emission fuel cell systems.

As highlighted by the projects' overview, the PEMFC is currently recognized as a technology with a higher potential for decarbonizing the shipping sector, especially in the short and medium-term scenario. The following subsections present the hybrid PEMFC/ESS system architecture, the fundamentals of its main components, and the benefits and challenges related to their applications.

2.2 *Hybrid PEMFC/ESS Powertrain Architecture*

A hybrid powertrain is generally composed of a main power generator coupled with an ESS. In a hybrid PEMFC/ESS system for all-electric ships, the PEMFC is the main power generator, while the most common ESS are batteries and supercapacitors.

In the current state-of-the-art technology, alternating current (AC) grids with fixed frequencies are usually implemented on all-electric ships [26]. However,

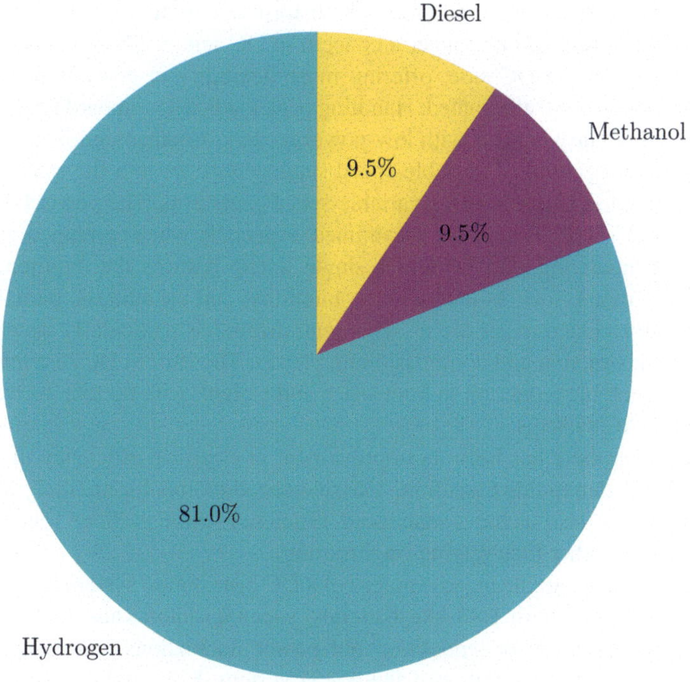

Fig. 2 Breakdown of logistic fuel types used in marine fuel cell systems

when compared to traditional AC distribution systems, the advantages of direct current (DC) distribution have led to its increasing popularity over the past decade [27]. In a DC-distributed system, the voltage from AC generators is first rectified through an AC-to-DC converter before power is transmitted to consumers via a DC bus. Propulsion and thruster loads are then driven by electric motors, which are powered by variable frequency converters (DC to AC). Additionally, other loads, such as hotel services and low-voltage AC distributions, are supplied via fixed-frequency converters and isolation transformers. A key component in this system is the input circuit (IC), a solid-state device used at connection points between different sections of the DC bus. This device isolates faults like short circuits, preventing them from affecting other parts of the DC grid. Power electronic converters are central to the operation of the entire DC-distributed network. DC distribution enables the use of variable-speed generators instead of fixed-speed ones, contributing to significant reductions in fuel consumption and emissions. Furthermore, DC systems offer greater simplicity compared to their AC counterparts, allowing for easier fault prediction and the development of effective protection measures. DC distribution systems present several key benefits, including significant power transmission advantages. For instance, distributing power at 1000 V DC instead of 690 V AC can reduce cable requirements by up to 40%.

Additionally, issues like total harmonic distortion, common in AC systems with frequency converters, are no longer a concern in DC setups. Shore connections are also simplified on the DC side, offering more flexibility in ports as network frequency mismatches are eliminated. Handling peak loads with onboard ESS enhances operational flexibility in ports with low-power feeders. Another significant benefit is the seamless integration of variable-speed shaft generators into the system, which can be managed similarly to other variable-speed generators. The power take-in and power take-off solutions are also streamlined, especially when combined with ESS. Moreover, centralizing all drives in a single lineup reduces the requirements for ambient conditions, such as temperature, humidity, and cleanliness, particularly in critical locations like thruster rooms. This centralization is particularly advantageous during the construction and commissioning phases. The shift to DC distribution has resulted in a notable reduction in both space and weight, with up to a 30% decrease compared to AC systems.

Furthermore, there has been an improvement in electrical efficiency of approximately 0.5–1% due to this transition. Variable-speed motors for auxiliary loads like fans and pumps also achieve enhanced efficiency through flow regulation by adjusting speed rather than relying on throttling.

One of the most significant advantages of a DC distribution system is its ability to seamlessly integrate with ESS-like batteries, supercapacitors, and fuel cells. This integration is essential for smoothing out power fluctuations and enhancing the overall efficiency and reliability of the power system. For instance, the DC distribution system can easily accommodate these energy storage devices to handle load variations and optimize power flow, providing more flexibility in energy management [28]. Additionally, this system minimizes the need for complex synchronization and reactive power issues often encountered in traditional AC distribution systems [29, 30]. These ESS help stabilize load variations from thrusters and other major loads, leading to smoother engine operation [31]. Moreover, installing ESS onboard is expected to generate significant fuel savings [32]. The detailed functionality and integration of ESS will be discussed in the following section [33].

Hybrid powertrains are classified into three main architectures depending on the type of connection between the main power unit and the ESS: series, parallel, and series–parallel [6]. Hybrid PEMFC/ESS systems commonly present series architectures, where all power generators and the corresponding converters are connected to a main electrical grid. The electric power flow generated from both the PEMFC and the ESS is first driven to an electric bus, which flows to the propulsion and the auxiliary power loads of the ship. Hybrid PEMFC-ESS systems usually have a bidirectional electric power flow between the electric bus and the ESS so that the ESS can be charged by the PEMFC when the produced power is higher than the power demand. A schematic example of a hybrid PEMFC/battery powertrain configuration is shown in Fig. 3.

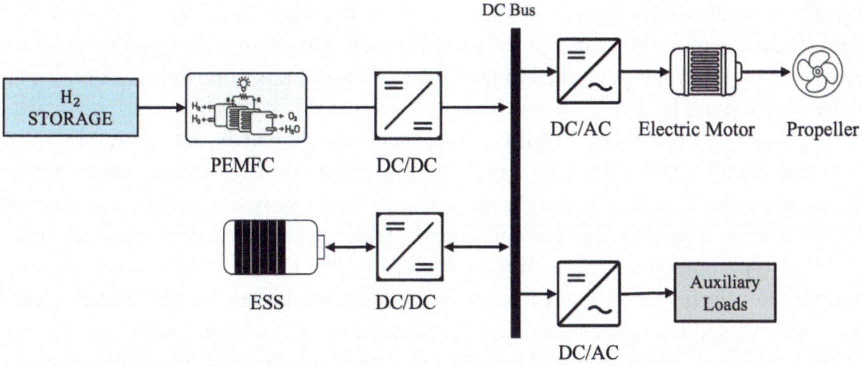

Fig. 3 Schematic of a hybrid PEMFC/ESS powertrain system

2.3 *PEMFC System*

For hybrid PEMFC/ESS systems, the main power source is represented by the PEMFC. A PEMFC generates electricity through an electrochemical reaction between hydrogen and oxygen, producing water and heat as by-products. The core of the PEMFC consists of a proton-conducting electrolyte membrane sandwiched between two electrodes, the anode and the cathode. At the anode, hydrogen is split into protons (H^+) and electrons (e^-) through the following reaction:

$$H_2 \rightarrow 2H^+ + 2e^- \tag{1}$$

Protons travel through the proton-conducting membrane, while electrons flow through an external circuit, generating electrical power. At the cathode, protons, electrons, and oxygen combine to form water:

$$\frac{1}{2}O_2 + 2H^+ + 2e^- \rightarrow H_2O. \tag{2}$$

The overall cell reaction is

$$H_2 + \frac{1}{2}O_2 \rightarrow H_2O. \tag{3}$$

The electrolyte membrane of a low-temperature (LT) PEMFC typically consists of perfluorsulfonic acid (PFSA), which becomes conductive to protons when hydrated [65]. The need for a hydrated membrane requires the operating temperature to remain below the boiling point of water. LT-PEMFCs typically operate above 65 °C to prevent flooding by condensed product water but below 85 °C to prevent dehydration and subsequent degradation of the polymer membrane [66]. Water management is a critical aspect of LT-PEMFC operations. Dehydration of the membrane leads to a reduction in ionic conductivity, while flooding of the electrode inhibits access of the reactants to the reaction sites. The membrane hydration can be sustained by the water generated in the electrochemical reaction, while excess water must be removed [67].

High-temperature (HT) PEMFCs have been developed based on stable solid polymer membranes that achieve high proton conductivity at temperatures above the dew point of water. A polybenzimidazole (PBI) polymer matrix doped with phosphoric acid (H_3PO_4) is most frequently used as the electrolyte material. This PEMFC type operates at temperatures between 120 °C and 180 °C [68]. Despite potentially offering several advantages, such as higher kinetics of the electrochemical reaction, reduced irreversibilities due to the absence of liquid water, and higher tolerance to fuel impurities, HT-PEMFCs are currently at a less mature state compared to their Low-Temperature counterpart. At the actual state of the art, LT-PEMFCs have been the most applied technology in past and ongoing projects for applications in the maritime sector [26].

PEMFC electrodes are composed of carbon support filled with catalyst nanoparticles (typically platinum-based) in an ionomer matrix. The catalyst layer is critical for speeding up the electrochemical reactions, ensuring efficient electricity generation. The electrolyte membrane, anode, and cathode are sandwiched together to form the membrane electrode assembly (MEA). External to the electrodes, the gas diffusion layer (GDL) ensures uniform distribution of the reactant gases and removal of the excess water produced at the cathode. Finally, like most external components, the bipolar plates play a role in fuel and air distribution, water and heat management, separating air and fuel, and conducting electric current. Bipolar plates for PEMFCs can be made of graphite, (coated) metals, or composites [69].

A single PEMFC generates low voltages, ranging from 0.6 to 0.8 V. The actual power output of a single PEMFC depends on the size of the cell and its current density, typically measured in watts per square centimeter (W/cm^2). Single cells can produce power in the range of 10 to 200 W, but this is insufficient for most real-world applications. To achieve higher power levels, multiple fuel cells are connected in series to form stacks, allowing them to produce higher power outputs, from several up to hundreds of kilowatts. Stacks are subsequently combined in modules to even reach megawatts of total power.

Auxiliary equipment is required to run the cell and allow safe operation and power production of a PEMFC system: the balance of plant (BoP). The BoP of a general PEMFC system can be divided into four subsystems: (i) fuel processing and supply system, (ii) air processing and supply system, (iii) cooling system, and (iv) power conditioning, control, and monitoring systems. The fuel processing system consists of all the components required to ensure that hydrogen entering the PEMFC stack has the proper characteristics. This includes pressure reducers and a humidifier to regulate the pressure and humidity, a condensate collector to remove liquid water, and a hydrogen recirculation pump to reuse unreacted hydrogen. The air supply line includes a blower to deliver atmospheric air to the stack, an air filter to remove contaminants that could poison the PEMFC stack, a humidifier to maintain optimal membrane hydration, and a condensate collector [23]. The cooling system

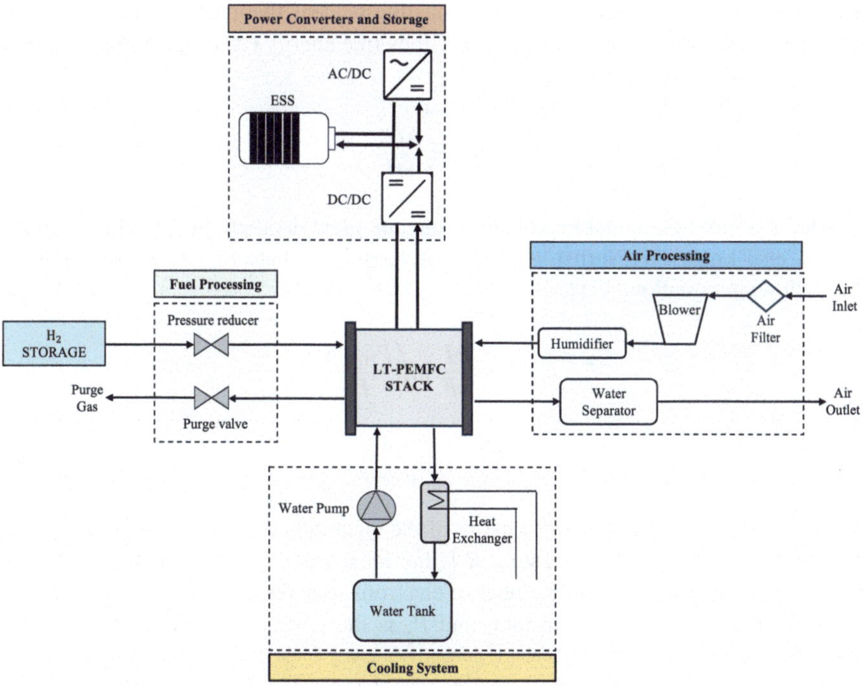

Fig. 4 Schematic example of a PEMFC/ESS system including the BoP: the fuel processing system for the hydrogen flow rate feeding the PEMFC anode and collecting the unreacted fuel, the air processing system to filter and humidify the inlet oxidant air at the cathode and remove the excess water, and a water-based liquid cooling system to control the stack temperature

includes all the components required to remove the excess heat from the PEMFC, maintaining the stack at the selected operating temperature. While several cooling methods exist, liquid cooling is often preferred due to its superior heat transfer capacity and efficiency [70]. The main components of a liquid cooling system include a circulating pump to ensure adequate refrigerant flow, a refrigerant reservoir, a heat exchanger to dissipate heat to the environment or a recovery system, and a deionizer to keep refrigerant conductivity low and avoid short-circuiting. The power conditioning, control, and monitoring system contains all the instrumentation required for system control and data acquisition. This includes safety valves, the control system, and a power inverter/converter. Figure 4 shows a schematic example of a PEMFC system, including the BoP.

2.3.1 PEMFC Performance

The PEMFC load is characterized by the electrochemical nature of its energy conversion process. Differently from traditional engines, fuel cells directly convert the chemical energy of the fuel into electricity, potentially reaching the same ideal efficiency of an ideal heat cycle. The maximum ideal theoretical efficiency (η_{rev}) attained by the cell is determined by the Gibbs free energy (ΔG) and enthalpy (ΔH) of its electrochemical reaction:

$$\eta_{rev} = \frac{\Delta G}{\Delta H}. \tag{4}$$

The maximum attainable voltage from an ideal isothermal PEMFC without losses, also known as Nernst voltage or reversible voltage (E_{rev}), is expressed by the following equations [71]:

$$E_{rev} = E^0 + \frac{RT}{nF} \ln\left(\frac{P_{H_2}\sqrt{P_{O_2}}}{P_{H_2O}}\right), \tag{5}$$

$$E^0 = \frac{-\Delta G^0}{nF}. \tag{6}$$

Where ΔG^0 is the Gibbs free energy of the electrochemical reaction at operation temperature and reference pressure, R is the ideal gas constant, T is the operating temperature (Kelvin), n is the number of electrons transferred in the electrochemical reaction, F is the Faraday constant, and P_x is the partial pressure of the reactant x (bar).

The actual operating voltage declines as the current density increases due to several internal losses:

- Activation losses (V_{act}), originating from the polarization potential needed to drive the electrochemical reaction, dominating the cell performance at low current densities.
- Ohmic losses (V_{ohm}), due to the ionic resistance of the electrolyte and the electronic resistance of the electrodes, dominating the cell performance at intermediate current densities.
- Concentration losses (V_{conc}), caused by mass transport limitation and characteristic of high current densities.

Therefore, the actual operating voltage of the cell is affected by losses as a function of the current density (i) and is described by the following equation:

$$V_{cell}(i) = E_{rev} - V_{act} - V_{ohm} - V_{conc}. \tag{7}$$

Once these losses are accounted for, voltage efficiency may be defined as

$$\eta_{voltage} = \frac{V_{cell}}{V_{rev}}. \tag{8}$$

In addition, the fuel conversion or fuel utilization is usually less than 100% to prevent fuel starvation and purge inert components and contaminants from the anode compartment. The fraction of fuel that is effectively oxidized in a fuel cell is referred

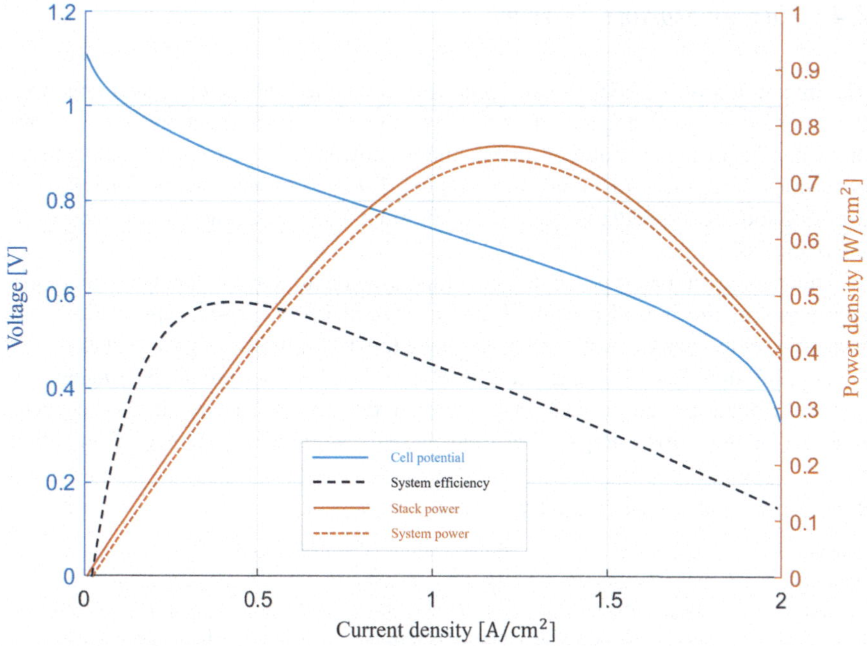

Fig. 5 Simplified example of operational characteristics of a PEMFC system. System efficiency peaks around 50% of the rated power

to as fuel utilization (u_f). The actual PEMFC stack efficiency can be subsequently calculated as

$$\eta_{\text{stack}} = \eta_{\text{rev}} \cdot \eta_{\text{voltage}} \cdot u_f. \tag{9}$$

Finally, the system efficiency is obtained by accounting for the BoP components consumption (P_{BoP}), such as air compressor losses, coolant pumps, blowers, sensors, and control systems. The overall system efficiency, therefore, is

$$\eta_{\text{system}} = \frac{P_{\text{stack}} - P_{\text{BoP}}}{P_{\text{stack}}} \cdot \eta_{\text{stack}}. \tag{10}$$

Figure 5 shows a typical PEMFC system load, power density, and efficiency curves as a function of the current density. PEMFC systems typically have relatively good efficiencies in part load since the electrochemical losses decrease when the operating current reduces. Consequently, the system efficiency usually increases with stack efficiency when the load is reduced. However, the relative contribution of the BoP consumption increases as the load is reduced, eventually nullifying the positive effect of decreasing electrochemical losses. Fuel cell manufacturers often specify minimum load fractions to ensure reliable operation, usually falling in the range from 10% to 30% of rated power [67].

2.4 Energy Storage System

The integration of the ESS is a key point for marine hybrid systems, as it enhances the operational performance of the power systems. ESS functions range from providing backup power and ensuring smooth power delivery to optimizing engine operations for reduced fuel consumption [72]. Table 2 outlines the core functions of the ESS and their specific purposes, highlighting how each function contributes to the overall efficiency of the system.

The choice of the type of ESS is fundamental to ensure the best operating conditions of the hybrid system (Fig. 6). Several ESSs for ship applications are available on the market, with different characteristics in terms of energy and power density, lifetime, cost, efficiency, and safety. Today, batteries are the leading energy storage technology due to their higher energy density, lower cost than other technologies at the same energy level, and already acquired experiences from other

Table 2 Functions and purposes of ESS for marine applications

Name	Description	Purpose
Spinning reserve	The ESS unit stays connected and operational without charging or discharging energy. If generating capacity is lost, it automatically takes over the load for a set period	• Backup power source • Fewer engines needed online • Enhanced efficiency at higher loads • Decreased engine running hours
Enhanced ride through	Like spinning reserve, but localized to a subsystem, such as a thruster or drilling drive	ESS solutions can provide UPS-like functionality for all or specific sections of the power system • New method to achieve higher ERN numbers • Higher power capacity
peak shaving	The unit absorbs load variations, allowing engines to handle only the average system power	• Smooth the power delivered by engines • Offset the necessity to start a new engine • Reduced engine running hours • Enhanced fuel efficiency
Strategic loading	The unit charges and discharges to optimize engine performance, minimizing fuel consumption while accounting for ESS efficiency	• Charging and discharging ESS to optimize the operating point of the diesel generators • Power is assigned at the peak efficiency
Enhanced dynamic performance	The unit smooths abrupt load changes and gradually adjusts engine transitions, automatically incorporating peak shaving when used	• Instant power assistance to support running diesel generators • Enable use of slower power devices, such as LNG/dual fuel engines or fuel cells
Zero-emissions operation	The unit powers the system so that engines can be turned off	• Zero emissions in the harbor • Quiet engine room

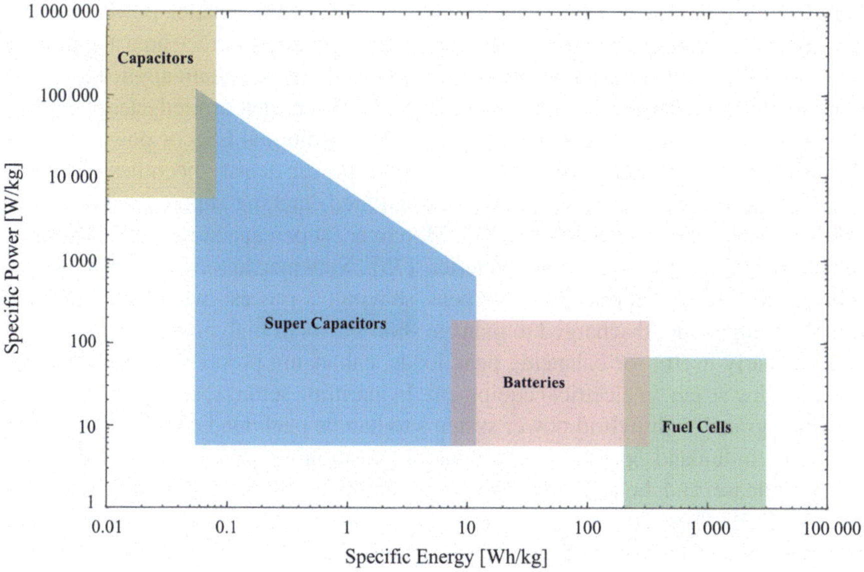

Fig. 6 Functions and purposes of ESS for marine applications

transportation sectors. Structurally, batteries are formed from electrodes, electrolytes, and separators, while the performance depends on the electrode material. LIBs are the most common ESS in shipping applications, thanks to their high energy densities and durability [72]. Furthermore, significant improvements in terms of safety and regulatory aspects of maritime LIBs have been made in recent years, increasing their commercial availability for the sector. The cost of LIBs has also reduced in recent years [72]. The widespread use of LIBs in more recent projects demonstrates these recent advancements and their increase in commercial availability [26].

The operating life of a battery is mainly affected by the depth of discharge (DoD) of its cycles (the percentage of battery capacity that has been discharged, expressed as a percentage of maximum capacity) [73]. Subsequently, their application is more suitable for certain types of ships to increase battery lifetime. In the case of ferries and coastal cruise ships with short voyage times, batteries are suitable ESS because the DoD is relatively low, preventing battery degradation and enhancing its operating lifetime. In the case of tugs and dynamic positioning vessels that require high electric power for a short time, the ESD can be used for a limited time in response to peak operations.

While batteries store energy through electrochemical reactions, capacitors store electricity in the form of electrostatic energy. Ceramic and electrolytic capacitors use a dielectric to store this electrostatic energy, while supercapacitors (or electric double-layer capacitors) are forms of capacitors which utilize a liquid electrolyte to create a Helmholtz layer at the interface of the solid and liquid. In this way, supercapacitors bridge the gap between low-energy, high-power capacitors and high-energy, low-power LIBs, offering higher capacitance, greater power density, and enhanced cycling capability due to their larger electrode surface area and thinner dielectrics [72]. Supercapacitors are particularly suitable for certain applications due to their ability to charge and discharge rapidly. However, their limited energy density is a significant drawback, necessitating the use of additional ESS or power generation solutions for sustained long-term operation. Power density becomes crucial in scenarios where short bursts of power are required, making supercapacitors more effective solutions than batteries [74]. Moreover, supercapacitors generally have longer operating lifetimes than batteries [75]. Supercapacitors are increasingly utilized in maritime applications for peak shaving, a process where they are frequently charged and discharged to manage short bursts of high power demand. This is particularly useful for balancing peak loads, enhancing power system efficiency, and reducing strain on electrical equipment. In maritime settings, supercapacitors are often integrated with hybrid power systems to handle peak load conditions, such as during high-demand operations like dynamic positioning [76]. They provide rapid energy release and help avoid costly grid upgrades by reducing the peak load demand [77]. They can, for example, be used for absorbing loads from the heave compensation of cranes in offshore drilling units [72].

2.5 Benefits and Challenges of Hybrid PEMFC/ESS Systems

In recent years, hybrid PEMFC/ESS systems have gained particular attention as a promising solution for sustainable ship energy systems. Several characteristics make the PEMFC a suitable technology for mobility applications. Firstly, PEMFCs have a quick start-up time, and fast transient response, both in the order of a few seconds [20]. The high degree of maturity already reached in other mobility sectors is a significant advantage for maritime applications, especially in the short-medium term scenario. PEMFCs can achieve high system efficiencies in the range of 40–60%, and specific powers vary from 125 to 750 W/kg, and power density from 50 to 400 W/L at system level [67]. Moreover, PEMFCs offer additional advantages, such as lightweight and more compact sizes when compared to other fuel cell technologies, and they operate at relatively low temperatures, making it easier to contain and reduce thermal losses. In addition, the number of BoP components is low compared to other fuel cell types. Combined with ESS, hydrogen PEMFCs provide a zero-emission solution for ship propulsion plants, that will play a key role in the maritime sector to comply with the stringent upcoming regulatory framework.

On the other side, a series of challenging technical, economic, and safety aspects still hinder the wide spread of marine hybrid PEMFC/ESS systems for a broader range of applications in the sector. First of all, PEMFCs have no fuel flexibility, and hydrogen storage on board is critical for both the intrinsic low energy storage density of the fuel and for its bunkering safety issues. The currently preferred method of hydrogen storage onboard is compressed hydrogen, although liquid hydrogen storage is gaining traction for larger vessels [26]. To complete the same route with hydrogen the required storage will increase by several times compared to conventional fuels, which will reduce cargo space [78]. For this reason, PEMFCs have been mainly considered for small-sized and medium-sized vessels with short or close-to-shore missions [26].

Recently, research interest in alternative hydrogen storage solutions, such as hydrogen carriers and metal hydrides, is growing, and some studies have addressed the possibility of using them for hydrogen-fuelled ships [54, 79]. The research advances on high-density hydrogen storage, and the development of alternative physical and chemical solutions could potentially extend PEMFC use to longer-distance shipping in the future [80]. However, technical, economic, and safety challenging aspects involve not only the onboard storage, but the entire hydrogen supply chain. A global commercially viable green hydrogen supply chain and bunkering infrastructure must be established to allow the widespread use of hydrogen as maritime logistic fuel [78].

Another challenging aspect of hybrid PEMFC/ESS systems is connected to their durability and moderate lifetime. Durability refers to the system performance drop and is associated with the chemical, mechanical, and thermal degradation and aging of its components. The limited durability of PEMFC/ESS systems impacts their reliability, safety, and cost-effectiveness. The degradation of a PEMFC is affected by several internal and external factors: stacks design and assembly, BoP components, operational aspects such as dynamic load demands and controlled operating conditions (e.g., temperature, pressure, humidity), and external factors such as chemical contamination or mechanical stress [81]. PEMFC degradation is a complex and interconnected combination of mechanisms, and a full understanding of the aging process requires a deeper analysis of the application. Currently, most of the knowledge on PEMFC degradation comes from applications in the automotive sector. Maritime applications introduce a novel set of operational and environmental conditions, such as different load demands and operating conditions, influenced by the adopted power control and EMS for the PEMFC/ESS system, new chemical contaminants (e.g., NaCl contamination from marine air), mechanical stresses due to inclinations, ship vibrations, and rough water conditions. The impact of these factors on PEMFC durability remains largely unexplored to date. However, load demand has been identified as one of the most impacting factors on PEMFCs' durability, suggesting that power and EMS could play a crucial role in enhancing the operating lifetime. High load demands, idling, start and stops, and transient loads are recognized as critical operating conditions for PEMFC degradation [82].

Different from PEMFCs, batteries experience degradation both during operations and during resting conditions [83]. Cycle aging occurs when the battery is charged or discharged during operation and depends on battery power/capacity ratio, state of charge (SoC), temperature, number of performed equivalent cycles, and DoD. Calendar aging occurs when no current is flowing in the battery and depends mainly on the battery's SoC and temperature. PEMFC degradation is generally measured in terms of voltage or power loss, while battery degradation results in voltage and power loss, and capacity fade [84]. Similarly to batteries, supercapacitors are also subjected to both cycle and calendar aging, limiting their capacitance and deliverable power [85]. While the power management strategy controls the instantaneous PEMFC/ESS power output to correctly match the power demand of the vessel, the EMS has the additional potential of optimizing the performance and limiting critical operation over time, minimizing degradation, and enhancing the system's lifetime.

2.6 Energy Management Strategies

2.6.1 Intro

As hybrid fuel cell systems are gaining popularity as a zero-emission solution for maritime energy systems, the need to adapt traditional control and EMS to the complex operation of these systems has been raised [86]. Two primary strategies are currently used to manage the complexity of these systems: power management and energy management. While these terms are often used interchangeably, they refer to distinct aspects of system operation. For hybrid fuel cell applications in ships, EMS involves mathematical models to optimize energy allocation between fuel cells, batteries, and other energy sources. Power management refers to the real-time power allocation to meet immediate energy demand. It focuses on distributing power from different sources, such as fuel cells and batteries, at any given moment to meet propulsion and auxiliary loads. Power management ensures that all systems have sufficient power without exceeding operational limits while balancing instantaneous performance and cost. Energy management, on the other hand, involves a more strategic, long-term approach. It is concerned with optimizing the overall energy usage over a period of time, such as an entire voyage. Energy management plans when and how energy should be stored, discharged, or supplied by the fuel cell and auxiliary systems. It aims to minimize total fuel consumption, emissions, and degradation of the power sources, ensuring optimal system performance over time. The key difference lies in the time horizon. Power management operates over short timescales to meet immediate demand, while energy management spans longer periods, focusing on overall efficiency and sustainability. The goal is often to minimize fuel consumption or emissions, subject to operational constraints.

The control hierarchy for vessels using fully electric batteries and fuel cells is structured around three levels: primary, secondary, and tertiary control, each operating at different time granularities and addressing different aspects of the system's operation. These control layers work together to ensure smooth, efficient, and safe operation of the vessel's powertrain. At the heart of the system lies the primary control, which functions at the component level. This layer operates in real-time, reacting within milliseconds to ensure the stability of individual components, such as batteries, fuel cells, inverters, and motors. It is primarily concerned with maintaining the correct voltage, current, and frequency within these systems to ensure that each component performs optimally and safely. For example, it might monitor and adjust the voltage levels within the battery cells or regulate the output of the inverter to keep the electric propulsion running smoothly. The immediacy of primary control makes it the foundation for the vessel's performance, responding in milliseconds (mHz level) to prevent fluctuations that could lead to inefficiencies or component damage.

Building on this is the secondary control, which deals with power management across the entire vessel. This level is responsible for managing the real-time power distribution between the various energy sources—such as batteries and fuel cells—and the electric propulsion system. Power management ensures that the vessel has the right amount of power to meet the propulsion and auxiliary loads at any given time, balancing between energy sources in response to changing operational conditions. Operating on a time scale of seconds to minutes, secondary control is crucial during transitions, such as when the vessel shifts between electric and hybrid modes. By distributing power efficiently, this layer prevents overloads and optimizes the use of available energy, ensuring the vessel runs smoothly even during rapid changes in propulsion demand.

Finally, tertiary control oversees the vessel's EMS. This layer functions over longer timeframes, typically minutes to hours, and optimizes energy use in alignment with the vessel's operational goals. At this level, the focus shifts from moment-to-moment power distribution to a broader, strategic view of energy consumption. The aim here is to make high-level decisions that maximize the overall efficiency of the vessel, considering factors such as fuel economy, battery life, and energy replenishment. For instance, tertiary control might determine when to prioritize using the fuel cells over the batteries to conserve fuel or decide when it's best to charge the batteries when the vessel is docked. The decisions made at this level contribute to long-term sustainability, ensuring that the vessel's energy sources are used in the most efficient manner possible throughout the journey.

Together, these three layers of control create a robust system that balances the immediate demands of component regulation, the dynamic needs of power management, and the long-term energy optimization strategy. The primary, secondary, and tertiary controls, each operating within its own time granularity, ensure that the vessel performs efficiently, safely, and sustainably, regardless of the operational scenario.

A control hierarchy schematic is reported in Fig. 7. At the top, the component-level (primary) control addresses the real-time regulation of individual systems. In the middle, power management (secondary control) deals with power distribution between energy sources and propulsion. At the base of the hierarchy, energy management (tertiary control) focuses on the vessel's long-term operational efficiency. Each level is interconnected, with decisions flowing from fast, momentary adjustments at the component level to slower, strategic decisions about energy use. In the context of EMS for hybrid systems, particularly in fuel cell-battery configurations, there are three main approaches: RB, OB, and LB strategies, as reported in Fig. 8. Each strategy offers unique methods of managing the energy flow within the system, balancing the goals of efficiency, fuel consumption, emissions reduction, and system longevity. In an RB approach, the control of power distribution between components such as fuel cells and batteries is governed by predefined rules that dictate the conditions under which each component operates. These rules often rely on specific thresholds or conditions, such as turning on the battery when power

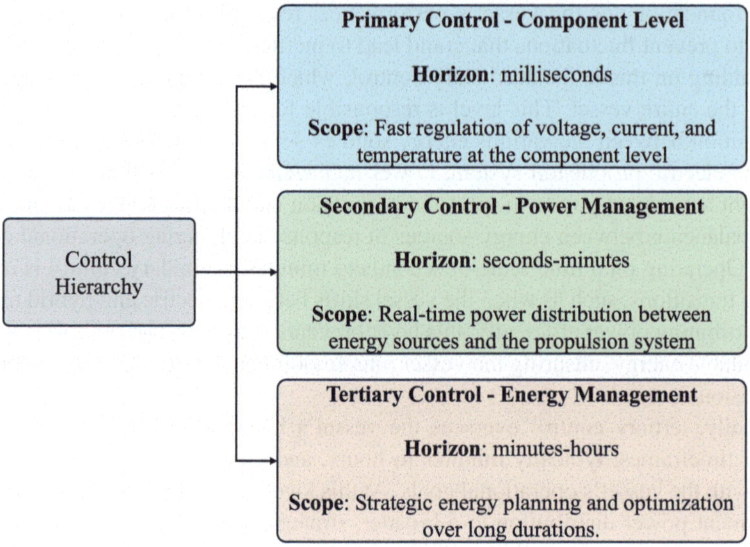

Fig. 7 Control hierarchy for marine energy systems system illustrating the three layers of control: Primary (component-level), secondary (power management), and tertiary (energy management)

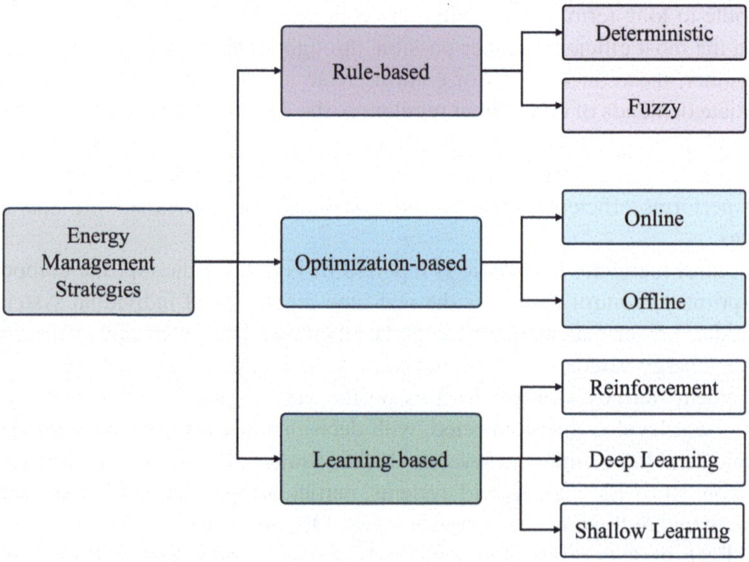

Fig. 8 Overview of EMS categorized into three main approaches: RB (deterministic and fuzzy control), OB (online and offline optimization techniques), and LB (reinforcement, deep learning, and supervised learning methods)

demand exceeds a certain value or activating the fuel cell when the battery's charge drops below a certain level. This kind of approach is relatively straightforward to implement because it follows clear, deterministic logic: if one condition is met, a particular action is triggered. For instance, in a simple system, a rule might state that if power demand goes beyond 100 kW, the battery should kick in to supplement the fuel cell. Or, if the battery's SoC falls below 20%, the fuel cell should become the primary power source. These predefined thresholds create a predictable and reliable system. However, the problem with this deterministic RB approach is its inflexibility. The system operates based on fixed rules, so it cannot adapt dynamically to changing conditions, such as fluctuations in power demand or fuel efficiency. As a result, while easy to understand and implement, it can lead to suboptimal performance because it does not account for real-time changes or optimize energy use effectively over time.

On the other hand, a fuzzy RB approach introduces greater flexibility and adaptability. Unlike deterministic systems, fuzzy logic does not rely on rigid thresholds. Instead, it works with ranges and degrees of truth. For example, rather than having a hard rule that activates the battery only when demand exceeds a specific value, the fuzzy approach might use categories such as low, medium, or high power demand. These categories are not strictly defined by exact numbers but by overlapping ranges. The system then interprets these ranges dynamically, adjusting how much power comes from the fuel cell or battery depending on the situation. In this setup, the battery might gradually start to supplement the fuel cell when power demand is medium, not waiting until it reaches a high, fixed threshold. Fuzzy logic handles this uncertainty and variability much better, allowing smoother transitions between power sources and more efficient management of energy flows. This adaptability is particularly useful in systems where power demand and other factors fluctuate unpredictably. The fuzzy RB approach, therefore, offers better performance by responding in a more nuanced way to the system's real-time needs.

Despite its advantages, the fuzzy RB approach is more complex to design and implement than its deterministic counterpart. The system must evaluate multiple inputs, handle degrees of truth, and continuously adjust decisions based on changing conditions. This requires a more sophisticated control mechanism and a deeper understanding of the system's dynamics. However, the increased flexibility it provides can lead to significant improvements in efficiency, making it a valuable approach for managing energy systems that operate in complex and variable environments.

Although the deterministic RB approach offers simplicity and predictability, it falls short in dynamic settings. With its ability to handle uncertainty and adjust to changing conditions, the fuzzy RB approach can optimize performance much more effectively, though at the cost of increased complexity [87]. The choice between these approaches depends largely on the complexity of the system and the level of adaptability required to meet performance goals.

In contrast to RB strategies, OB energy management techniques approach the control problem with a more sophisticated framework, treating it as a mathematical optimization problem. The goal here is not to rely on predefined rules or thresholds but to dynamically minimize a cost function, including objectives like fuel consumption, emissions, system degradation, or operational costs. By formulating the problem in this way, these techniques can account for various variables and constraints, enabling a more holistic and efficient approach to managing energy flows in complex systems. OB techniques can be implemented in two primary ways: offline and online optimization.

In offline optimization, the control strategy is developed beforehand, using historical data or known conditions to simulate various scenarios and find an optimal solution. This precomputed solution is then implemented during system operation. For instance, before deploying a hybrid power system, the optimization model may be run using expected load profiles, environmental conditions, and system performance data to find the best EMS for a given set of conditions.

Offline optimization has the advantage of being computationally efficient during real-time operation, as the heavy computational work is done in advance. Once the optimal control laws or trajectories are determined, they can be directly applied without requiring intensive calculations. This makes offline optimization particularly suitable for systems where real-time computational power is limited or where conditions are relatively stable and predictable.

However, one major disadvantage is its lack of adaptability. Since offline optimization relies on predefined conditions, it may not perform well if the actual operating conditions deviate significantly from those assumed during the optimization process. In dynamic environments where load demands, fuel costs, or operational constraints can change rapidly, offline optimization may become suboptimal, much like a deterministic RB system. Online optimization, in contrast, involves continuously solving the optimization problem in real-time based on the system's current state and changing conditions. Instead of relying on precomputed solutions, the system dynamically adjusts its control strategy as it operates. This is particularly beneficial in environments where conditions are unpredictable or rapidly changing, as the system can continuously update its strategy to reflect real-time data.

The primary advantage of online optimization is its ability to adapt to real-world fluctuations. For instance, if fuel prices spike, the system can immediately adjust its strategy to prioritize energy sources that minimize costs. Similarly, in a maritime energy system, if the load demand changes unexpectedly or a component experiences degradation, the system can optimize power distribution in real time to maintain performance and efficiency. However, online optimization comes with challenges, most notably computational complexity. Since the optimization problem must be solved continuously, the system requires significant computational resources and must be capable of responding quickly to new inputs. This can be a limiting factor, especially for systems that require fast response times or operate with limited processing power. Additionally, depending on the complexity of the optimization problem and the number of variables involved, solving the problem in real-time can introduce delays, potentially reducing the system's responsiveness. Both offline and online optimization approaches offer clear advantages over traditional RB strategies. While RB methods are straightforward and computationally simple,

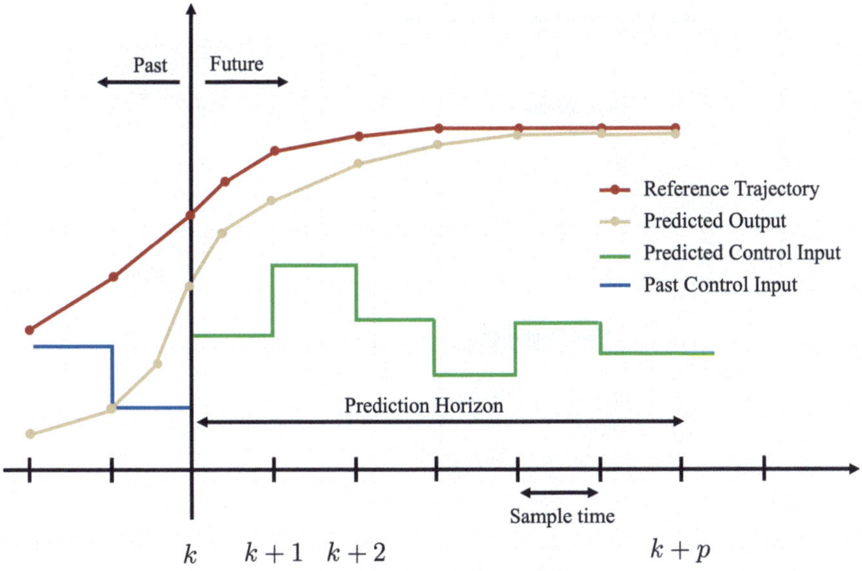

Fig. 9 The reference trajectory (red line) represents the desired system behavior, while the predicted output (tan line) shows the system's future performance based on current control inputs. The predicted control input (green) is optimized over the prediction horizon, balancing power demands and system constraints, with the past control input (blue) indicating previously applied actions. The prediction horizon spans multiple time steps from the current time k to $k + p$ allowing MPC to dynamically adapt to changes in real-time energy demand

they lack the ability to adapt to changing conditions or optimize performance across multiple objectives.

In contrast, OB techniques offer a more flexible and effective way to manage energy flows, especially in complex systems with multiple competing goals, such as reducing emissions, minimizing fuel consumption, and limiting component degradation. One of the most widely adopted methods is model predictive control (MPC), which leverages real-time data and system models to dynamically predict future system behavior and optimize energy and power flow accordingly. At the core of MPC is the concept of optimizing a cost function (as shown in Fig. 9) that typically includes objectives such as minimizing fuel consumption, reducing emissions, and limiting system degradation. In doing so, MPC ensures that operational efficiency is maximized while constraints like power demand, battery SoC, and the longevity of fuel cells and batteries are respected. Figure 9 showcases how MPC uses a reference trajectory (red line) that represents the desired behavior of the system. The predicted output (tan line) shows how the system is expected to behave based on current control inputs. The green line represents the predicted control inputs optimized over a prediction horizon, while the blue line depicts the past control inputs. The core strength of MPC lies in its predictive capability, enabling it to make proactive decisions rather than merely reacting to immediate conditions. For example, if an upcoming surge in power demand is anticipated, the MPC framework can adjust

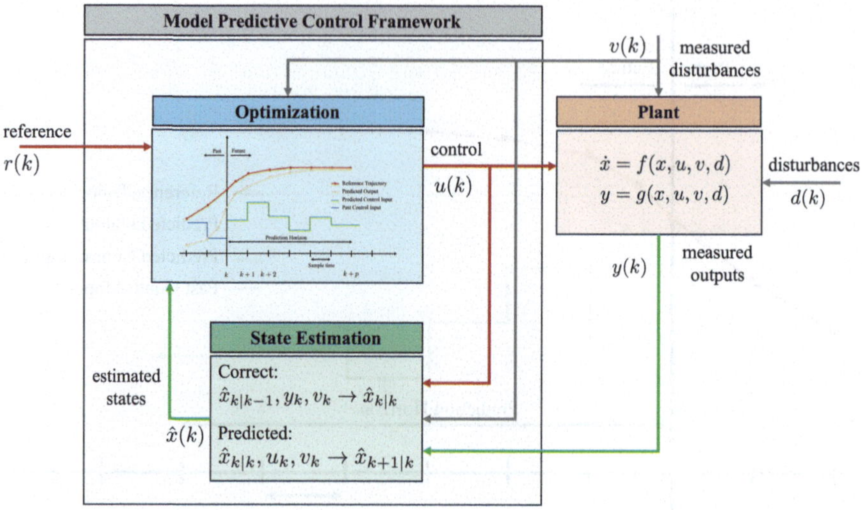

Fig. 10 MPC framework for marine energy systems. The reference input $r(k)$ drives the optimization process, which uses a prediction horizon to balance power distribution and system constraints. The plant model (right) receives the control input $u(k)$, responding to disturbances $d(k)$ and $v(k)$. The state estimation block corrects and predicts future states $\widehat{x}(k)$, ensuring accurate system feedback through real-time measured outputs $y(k)$. The closed-loop system dynamically adjusts control actions to optimize performance and energy efficiency while respecting system constraints and disturbances

power generation or battery charging behavior in advance to meet the load while maintaining system stability.

As illustrated in Fig. 10 the MPC framework consists of three primary components: the optimization block, the plant model, and the state estimation block. The optimization block receives the reference input $r(k)$ and calculates the optimal control actions $u(k)$ based on predicted future states and constraints. The plant model represents the system dynamics, including disturbances $d(k)$ and $v(k)$. At the same time, the state estimation block updates and predicts future states $\widehat{x}(k)$, ensuring that the control actions are continuously refined based on real-time measured outputs $y(k)$.

The core principle of MPC revolves around using a dynamic model of the system—such as those involving fuel cells and batteries—to predict future behavior and make optimal control decisions over a moving time horizon. This allows the controller to not only react to immediate conditions but also to anticipate future changes, leading to more efficient energy management. In practice, MPC solves an optimization problem at each time step. The challenge is that all of these must be achieved while respecting stringent system constraints, such as power demand, battery SoC, and the longevity of system components. To begin, the system model is used to forecast how the system will behave under different control actions. This model considers real-time data inputs, such as current power demand and the SoC of the battery, which ensures the predictions are as accurate as possible.

Based on these predictions, the MPC framework minimizes the cost function by choosing the control inputs that will yield the best performance over the time horizon. For instance, if a surge in power demand is predicted, the MPC can adjust power generation and battery usage in advance to ensure a smooth response. Another key advantage of MPC is its ability to handle constraints naturally. These might include limits on power generation, restrictions on how low the SoC of the battery can fall, or operational boundaries for fuel cells to avoid excessive wear and tear. By considering these constraints directly within the optimization process, MPC can ensure that the system operates within safe limits while still pursuing optimal performance. Once the optimal control strategy is computed, only the first step is implemented in real-time, and the process repeats as new data becomes available. This feedback loop ensures that the system continually adapts to changing conditions, making MPC highly suitable for environments where energy supply and demand can fluctuate unpredictably. MPC's flexibility makes it an ideal choice for multi-objective optimization problems in energy management. It excels at balancing trade-offs, such as minimizing fuel consumption while ensuring the battery remains within its operational limits or optimizing for both short-term efficiency and long-term component health. More advanced optimization techniques, such as Dynamic Programming or Stochastic optimisation optimization [88], can handle more complex systems but often come with increased computational demands. These methods are highly effective for systems requiring real-time adaptability and efficiency.

Finally, LB approaches leverage machine learning algorithms to improve EMSs continuously. Reinforcement Learning, for example, enables the system to learn optimal energy management policies through trial and error by maximizing long-term rewards, making it well-suited for dynamic environments like hybrid power systems. Reinforcement LB models can adaptively control the distribution of energy between components such as batteries and fuel cells, improving operational efficiency over time [89].

Both shallow and deep learning models can be used to predict future power demands or battery states, enabling the system to make more informed decisions [90]. LB methods are particularly powerful in complex, dynamic environments where traditional strategies may fall short, as they adapt based on past performance and new data, offering a robust solution for managing uncertainty [90, 91, 92].

2.6.2 Rule-Based Approaches

A basic strategy is an RB approach in which predefined control rules determine the power distribution between the fuel cell and other energy sources. For example, when the power demand exceeds a certain threshold, the battery might be used to supplement the fuel cell output. While this approach is straightforward to implement, it may not always be optimal.

For example, when focusing on fuel cells, RB strategies can be mathematically formulated using a series of conditional statements

$$P_{\rm fc}(t) = \begin{cases} P_{\rm max,fc} & \text{if } P_{\rm demand}(t) > P_{\rm fc,thresh} \\ P_{\rm demand}(t) & \text{otherwise} \end{cases} \tag{11}$$

where $P_{fc}(t)$ is the power supplied by the fuel cell at a time t, and $P_{demand}(t)$ is the total power demand. The system switches between using full fuel cell power and supplementing with auxiliary sources based on predefined thresholds.

Table 3 provides a structured strategy for energy management in a hybrid fuel cell-battery system for ships, considering both the SoC of the battery and the power demand. The table separates the power demand scenarios into three categories. The fuel cell's primary role is highlighted, particularly for providing base power, while the battery compensates for variable or excess demand. In high-demand situations, the system leverages the combined output of both the fuel cell and the battery to meet total power requirements. The table also reports the system's behavior under three conditions of demand power. When the power demand is equal to zero$P_{demand}(t) = 0$, we are in idle or zero-load conditions. When the battery's SoC is low $SoC(t) < SoC_{min}$, the fuel cell charges the battery until the SoC reaches a safe level. In cases where the battery is sufficiently charged, both the fuel cell and battery are turned off to avoid unnecessary energy consumption. When the battery was previously discharged, but the SoC is still within acceptable limits, the system shuts off to preserve energy. A power demand where $0 < P_{demand}(t) \leq P_{opt}$ represents a moderate load where the system can rely primarily on the fuel cell and battery working together. When the SoC is low, the fuel cell works to both meet demand and charge

Table 3 RB EMS for hybrid fuel cell and battery systems

	$P_{demand}(t) = 0$	$0 < P_{demand}(t) \leq P_{opt}$	$P_{demand}(t) > P_{opt}$
$SoC(t) < SoC_{min}$	Fuel-cell: ON Battery: Charging (until SoC reaches SoC_{min}) $P_{fc} = -P_{bat} = P_{opt}$	Fuel-cell: ON Battery: Charging (until SoC reaches SoC_{min}) $P_{fc} = P_{opt}$ $P_{bat} = P_{dem} - P_{fc}$	Fuel-cell: ON Battery: OFF $P_{fc} = P_{dem}$
Previous battery mode: Charging $SoC_{min} \leq SoC(t) < SoC_{max}$	Fuel-cell: ON Battery: Charging (until SoC reaches SoC_{min}) $P_{fc} = P_{opt}$	Fuel-cell: ON Battery: Charging (until SoC reaches SoC_{min}) $P_{fc} = P_{opt}$ $P_{bat} = P_{dem} - P_{fc}$	Fuel-cell: ON Battery: Discharging $P_{fc} = P_{opt}$ $P_{bat} = P_{dem} - P_{fc}$
Previous battery mode: Discharging $SoC_{min} \leq SoC(t) < SoC_{max}$	Fuel-cell: OFF Battery: OFF	Fuel-cell: OFF Battery: Discharging $P_{bat} = P_{dem}$	Fuel-cell: ON Battery: Discharging$P_{fc} = P_{opt}$ $P_{bat} = P_{dem} - P_{fc}$
$SoC(t) \geq SoC_{max}$	Fuel-cell: OFF Battery: OFF	Fuel-cell: OFF Battery: Discharging $P_{bat} = P_{dem}$	Fuel-cell: ON Battery: Discharging $P_{fc} = P_{opt}$ $P_{bat} = P_{dem} - P_{fc}$

the battery. When the battery is in its optimal SoC range, the fuel cell continues to charge the battery if needed or meets the demand and discharges the battery if the SoC is stable. If the battery was discharging in the previous step and the SoC remains within bounds, the battery continues to discharge to meet demand. In high power demand conditions, $P_{demand}(t) > P_{opt}$, the fuel cell operates at maximum efficiency, providing P_{opt} while the battery discharges to meet the excess demand. If the SoC is low, the battery is not used, and the fuel cell alone powers the system. For batteries within their operational range, the system effectively balances the power supply by using both the fuel cell and battery, keeping the system running efficiently. Finally, when the battery is full $SoC(t) \geq SoC_{max}$, the battery discharges to meet excess power demand, ensuring that the fuel cell is used optimally without overcharging the battery.

The battery's SoC is critical for determining whether the fuel cell should operate alone or in conjunction with the battery. When the battery's SoC is below a minimum threshold, the system prioritizes recharging the battery to avoid depleting it further. When the SoC is high ($SoC(t) \geq SoC_{max}$), the battery is used for discharging rather than charging, ensuring efficient power use.

This exemplificatory strategy ensures that the fuel cell is optimally used by charging the battery when the power demand is low or moderate, and the battery is used primarily to meet excess demand and discharge more when power requirements exceed the fuel cell's optimal output.

2.6.3 Optimization-Based Energy Management

A more sophisticated approach is OB energy management, where the objective is to minimize a comprehensive cost function that accounts for fuel consumption, emissions, system degradation (including both fuel cells and batteries), and operational efficiency, all while respecting system constraints such as power demand, SoC, and component longevity. The most commonly used optimization technique is MPC, which uses real-time data and system models to forecast future energy demand and dynamically optimize power flow over a moving time horizon. MPC has been widely applied due to its adaptability to real-time conditions and ability to handle system constraints such as power limits and SoC for batteries.

However, there are several other optimization strategies that can be applied depending on the objectives and complexity of the system. Dynamic Programming is another well-known technique that solves problems by breaking them into simpler subproblems. Dynamic Programming can find the globally optimal solution over a defined time horizon, making it suitable for systems with multi-stage decision processes, such as energy management in hybrid systems. However, Dynamic Programming suffers from the curse of dimensionality, making it computationally expensive for large-scale systems [93]. To overcome the computational complexity of Dynamic Programming, Approximate Dynamic Programming uses approximation methods to estimate the cost-to-go function, making it feasible for large-scale energy management problems. Approximate Dynamic Programming is an attractive option for complex systems with high-dimensional state spaces where traditional Dynamic Programming is not computationally feasible [94]. Stochastic optimization techniques explicitly account for uncertainties in the system, such as varying weather conditions or uncertain power demand. These techniques, including stochastic Dynamic Programming and stochastic MPC, optimize energy management by modeling probabilistic scenarios and determining the best strategy under these uncertainties. Stochastic optimization is particularly useful when dealing with unpredictable renewable energy sources [95]. Heuristic and metaheuristic algorithms, such as Genetic Algorithms and Particle Swarm optimization, provide a non-deterministic approach to optimization. These methods can quickly find near-optimal solutions for complex, non-linear problems. Genetic Algorithms evolve potential solutions over generations by mimicking natural selection [96, 97], while Particle Swarm optimization simulates the social behavior of animals to find optimal solutions [98]. These methods are computationally efficient and can be applied to large-scale problems, though they do not guarantee finding the global optimum. Mixed-Integer Linear Programming is an optimization technique used when the problem involves both continuous and discrete variables, such as the on/off states of fuel cells or batteries. Mixed-Integer Linear Programming solves problems where relationships between variables are linear, making it ideal for hybrid systems with discrete switching controls. However, Mixed-Integer Linear Programming is limited to linear problems and may not handle non-linear system behaviors efficiently [99]. In cases where the relationships between system variables are non-linear, Nonlinear Programming can be applied. Nonlinear Programming optimizes the cost function and system constraints for systems where efficiency, degradation, and power output are non-linear, such as in hybrid fuel cell systems. Though more computationally intensive, Nonlinear Programming offers better modeling of real-world systems where non-linearities are prevalent [100]. Bi-level optimization is a hierarchical approach where one optimization problem is nested within another. This is suitable for EMS, where high-level decisions, such as overall power distribution, depend on lower-level decisions, such as device-level operational efficiency. Bi-level optimization is often used in large-scale systems like smart grids [101]. In hybrid energy systems, there are typically competing objectives, such as minimizing fuel consumption, emissions, and system degradation. Multi-Objective optimization allows for the simultaneous optimization of these competing objectives, resulting in a set of Pareto-optimal solutions that offer trade-offs between the different objectives.

Multi-Objective optimization is particularly useful for balancing the conflicting goals of efficiency and environmental sustainability [102, 103]. Receding Horizon Control is similar to MPC but is more flexible in handling real-time changes in system constraints and objectives. Receding Horizon Control continually adjusts the optimization horizon, effectively managing systems where demand and operational constraints frequently change. Receding Horizon Control is particularly suitable for hybrid systems that must adapt to fluctuating power demands [104]. In addition to these standalone methods, hybrid approaches that combine different optimization techniques are often used to leverage the advantages of multiple methods. For example, combining MPC with metaheuristic algorithms like Genetic Algorithm or Particle Swarm optimization can improve performance by speeding up the search for optimal solutions while retaining real-time adaptability [103].

In a hybrid fuel cell system, the energy management problem can be modeled as a multi-objective optimization problem. The goal is to minimize several competing objectives, such as fuel consumption, emissions, and system degradation, and maximize operational efficiency. We seek to minimize the following set of objectives over time $t \in [0, T]$

$$\min_{P_{\text{fc}}(t),\ P_{\text{batt}}(t)} \begin{cases} f_{\text{fuel}}(P_{\text{fc}}(t)), & \text{Fuel consumption} \\ f_{\text{degr,fc}}(P_{\text{fc}}(t)), & \text{Degradation of the fuel cell} \\ f_{\text{degr,batt}}(P_{\text{batt}}(t)), & \text{Degradation of the battery} \\ -f_{\text{eff}}(P_{\text{fc}}(t), P_{\text{batt}}(t)), & \text{System efficiency} \end{cases} \tag{12}$$

where $P_{\text{fc}}(t)$ and $P_{\text{batt}}(t)$ are the power supplied by the fuel cell and battery, respectively, at the time t, $f_{\text{fuel}}(P_{\text{fc}}(t))$ is the fuel consumption cost of the fuel cell, $f_{\text{degr, fc}}(P_{\text{fc}}(t))$ is the degradation cost of the fuel cell as a function of its power output over time, $f_{\text{degr, batt}}(P_{\text{batt}}(t))$ is the degradation cost of the battery, typically modeled based on charge/discharge cycles or usage patterns, $f_{\text{eff}}(P_{\text{fc}}(t), P_{\text{batt}}(t))$ is a term representing the system's operational efficiency. The optimization is subject to constraints

$$\begin{array}{ll} \mathrm{P_{fc}(t) + P_{batt}(t) = P_{load}(t)} & \text{Power Balance} \\ \mathrm{P_{fc}^{min} \leq P_{fc}(t) \leq P_{fc}^{max}} & \text{Fuel Cell Power Limits} \\ \mathrm{P_{batt}^{min} \leq P_{batt}(t) \leq P_{batt}^{max}} & \text{Battery Power Limits} \\ \mathrm{SoC^{min} \leq SoC(t) \leq SoC^{max}} & \text{Battery SoC Limits} \\ \mathrm{D_{fc}^{min} \leq D_{fc}(t) \leq D_{fc}^{max}} & \text{Fuel Cell Degradation Limits} \\ \mathrm{D_{batt}^{min} \leq D_{batt}(t) \leq D_{batt}^{max}} & \text{Battery Degradation Limits} \\ \mathrm{SoH_{fc}(t) \geq SoH_{fc}^{min}} & \text{Fuel Cell SoH Limits} \\ \mathrm{SoH_{batt}(t) \geq SoH_{batt}^{min}} & \text{Battery SoH Limits} \end{array} \tag{13}$$

Where $P_{\text{load}}(t)$ represents the load power requested by the system at a time t, which should be provided by the fuel cell and the battery. $P_{\text{fc}}(t)$ is constrained by minimum $\left(P_{\text{fc}}^{\text{min}}\right)$ and maximum $(P_{\text{fc}}^{\text{max}})$ limits, which could be based on the operating characteristics of the fuel cell to prevent overloading or operating below an inefficient threshold. $P_{\text{batt}}(t)$ is limited by the minimum $(P_{\text{batt}}^{\text{min}})$ and maximum $(P_{\text{batt}}^{\text{max}})$ values to ensure the battery operates within safe limits, considering charge/discharge rates. SoC(t) represents the battery's SoC at a time t, indicating how much charge is left in the battery as a percentage of its total capacity. It is constrained between minimum (SoC^{min}) and maximum (SoC^{max}) values to prevent overcharging or deep discharging, which could harm the battery's lifespan. $D_{\text{fc}}(t)$ represents the degradation level of the fuel cell at a time t, capturing the performance loss due to aging, operational stress, or wear. The degradation is kept within limits, $D_{\text{fc}}^{\text{min}}$ and $D_{\text{fc}}^{\text{max}}$, to ensure the fuel cell operates reliably and predictably without excessive wear. $D_{\text{batt}}(t)$ refers to the degradation of the battery at a time t, which reflects the reduction in battery performance or capacity due to cycles of charging and discharging. The degradation is constrained between $D_{\text{batt}}^{\text{min}}$ and $D_{\text{batt}}^{\text{max}}$, ensuring the battery remains operational within a desired degradation range. $\text{SoH}_{\text{fc}}(t)$ is the state of health of the fuel cell and is a measure of the remaining useful life or efficiency of the fuel cell, where 100% (or 1) represents optimal health and 0% represents complete failure. This value is constrained to ensure that the fuel cell does not drop below a minimum acceptable health level ($\text{SoH}_{\text{fc}}^{\text{min}}$) to maintain functionality. $\text{SoH}_{\text{batt}}(t)$ indicates the health of the battery, reflecting the remaining capacity relative to its original state. The constraint $\text{SoH}_{\text{batt}}(t) \geq \text{SoH}_{\text{batt}}^{\text{min}}$ ensures that the battery's state of health remains above a critical level to prevent failure or underperformance.

In order to solve this multi-objective problem, we can combine the objectives into a single scalar function using the weighted sum approach

$$\min_{P_{\text{fc}}(t),P_{\text{batt}}(t)} \int_0^T \Big(\lambda_1 f_{\text{fuel}}(P_{\text{fc}}(t)) + \lambda_2 f_{\text{emissions}}(P_{\text{fc}}(t)) + \lambda_3 f_{\text{degr, fc}}(P_{\text{fc}}(t)) \\ + \lambda_4 f_{\text{degr, batt}}(P_{\text{batt}}(t)) - \lambda_5 f_{\text{eff}}(P_{\text{fc}}(t), P_{\text{batt}}(t))\Big) dt \tag{14}$$

where λ_1, λ_2, λ_3, λ_4, λ_5 are weights representing the relative importance of each objective in the system. This leads to a set of trade-off solutions known as the Pareto front, where improvements in one objective (e.g., minimizing fuel consumption) can only be achieved by worsening another (e.g., increasing degradation).

2.7 Learning-Based Energy Management

LB approaches have emerged as powerful tools to enhance the EMS by enabling accurate prediction of various system states and environmental conditions. These predictions are subsequently used in OB EMS, which plays a key role in minimizing fuel consumption, maximizing efficiency, and reducing emissions. This section discusses how LB methods contribute to energy management in hybrid fuel cell applications and the differences between shallow and deep learning approaches in this domain.

In hybrid fuel cell applications, the role of energy management is to balance the power demands between different energy sources, such as batteries, fuel cells, and auxiliary systems, while ensuring efficient energy use and extending the lifetime of key components. LB methods facilitate this task by accurately predicting power demand, fuel cell behavior, battery SoC, and environmental variables like weather and sea conditions. These predictions feed into the OB EMS, which continuously adjusts energy distribution to meet operational requirements while adhering to performance and sustainability goals.

LB prediction models have proven particularly effective in hybrid fuel cell systems due to their ability to process large amounts of data and recognize complex patterns over time. The models take historical data and system measurements as input and learn to predict future states or energy demand, which are subsequently optimized using mathematical techniques such as linear programming, dynamic programming, or MPC. The success of these predictions directly impacts overall energy efficiency; choosing a learning approach is critical for achieving optimal energy management.

In the context of LB energy management for ships, two main categories of machine learning approaches are widely used for making predictions: shallow and deep learning [105, 106, 107, 108]. Each has its strengths and appropriate application scenarios, depending on the nature of the data and the complexity of the prediction task.

Shallow learning methods [108] include kernel methods (e.g., support vector machines [108]), ensemble methods (e.g., random forests [109], gradient boosting machines [110]), and Gaussian processes [111]. These models are typically characterized by their relatively simple architectures and a limited number of layers. Their primary advantage lies in their ability to handle smaller datasets and structured tabular data effectively, which makes them particularly suitable for certain applications in energy management. Kernel-based methods [108], such as Support Vector Machines, map data into higher-dimensional spaces where linear separations can be applied. They are effective when dealing with non-linear relationships in the data and can produce high-quality predictions with small datasets. Ensemble Methods [109, 110] combine multiple weak learners (such as decision trees) to form a strong predictor. Random Forests and Gradient Boosting Machines are examples of ensemble methods widely used in energy prediction tasks, as they excel at capturing variance and avoiding overfitting, especially in tabular data. Gaussian Processes [111] are a non-parametric approach that provides probabilistic predictions with uncertainty quantification. This characteristic is useful in energy management applications, where uncertainty plays a key role in decision-making under variable environmental conditions. Shallow learning methods are generally well-suited for tabular data, where each input variable is explicitly defined, and relationships between variables can be modeled with relatively simple algorithms. For example, shallow models can efficiently handle historical fuel consumption, power load, or sea state data stored in tabular form due to their ability to deal with structured input.

Deep learning [105, 106, 107], on the other hand, is better suited for more complex and unstructured data types, such as time series data, where sequential dependencies must be captured. Deep learning methods include Long Short-Term Memory networks [112], Temporal Convolutional Networks [113], and Transformers [114], all of which are designed to model long-range dependencies in data and capture dynamic relationships that evolve over time. Long Short-Term Memory Networks [112] are a type of recurrent neural network designed to handle time series data with long-term dependencies. They maintain memory over extended periods, allowing them to predict energy demand or fuel cell behavior based on previous time steps in the data. This makes Long Short-Term Memory Networks particularly useful for managing power flows in hybrid systems, where temporal patterns, such as fuel cell degradation over time, play a crucial role. Temporal Convolutional Networks [113] use convolutional layers to model temporal data, offering a more efficient way to capture dependencies than traditional Recurrent Neural Networks like Long Short-Term Memory Networks. They allow for parallelization during training and are often better at handling longer sequences, making them well-suited for complex, time-dependent energy management tasks in ships, such as forecasting power demands or weather conditions over extended periods. Transformers [114], originally developed for natural language processing, transformers have proven to be highly effective in handling sequential data by using self-attention mechanisms that capture relationships between different time steps without relying on traditional recurrence. Transformers are particularly powerful when dealing with long sequences, where attention to both local and global time steps is critical, such as in predicting fuel consumption over long voyages.

The choice between shallow and deep learning methods depends on the type of data being used for prediction in the EMS. Shallow learning methods typically outperform deep learning approaches when dealing with structured tabular data, as they can efficiently capture relationships between features with relatively low computational cost. For example, predicting short-term energy demands based on current operational conditions (e.g., fuel cell and battery temperature, fuel flow rates, battery SoC) can be effectively handled by ensemble methods or Gaussian processes. However, when the data is more complex and unstructured, such as time series data that includes temporal patterns (e.g., power demand forecasts over time or environmental conditions during voyages), deep learning methods like Long Short-Term Memory networks, Temporal Convolutional Networks, or transformers become more advantageous. These models can learn from past time steps and make predictions that account for long-range dependencies, which is crucial in dynamic systems like ships.

LB energy management for hybrid fuel cell applications in ships benefits from the strengths of both shallow and deep learning methods, with the choice of approach depending on the type and complexity of the data. Shallow learning models are optimal for structured data and straightforward relationships, while deep learning models excel in capturing temporal dependencies and handling unstructured, time-series data in maritime energy management applications.

Both shallow and deep learning approaches have been successfully applied in the context of hybrid fuel cell systems. Tang et al. [115] demonstrated the potential of deep reinforcement learning in optimizing EMS for hybrid fuel cell systems, significantly improving fuel efficiency and reducing emissions in varying operational conditions. On the other hand, shallow models, while simpler, can be used for more basic tasks like short-term power demand forecasting or as components of larger, more complex systems.

Recent advancements in LB EMS also highlight the importance of deep learning in handling real-time optimization and prediction tasks. Hu et al. [116] applied a deep learning framework to predict energy demand and optimize performance in hybrid fuel cell systems, leading to improved system efficiency and lower emissions. Such applications underscore the versatility and power of deep learning in dealing with the complexities of modern energy systems.

Shallow and deep learning offer distinct advantages for EMSs, depending on the complexity of the system and the desired outcomes. Shallow learning models are computationally efficient and suitable for simpler systems or when immediate, straightforward predictions are needed. Deep learning, however, excels in capturing complex interactions and adapting to changing conditions, making it a superior choice for large-scale hybrid fuel cell systems where optimizing energy efficiency, sustainability, and operational reliability is critical. Both shallow and deep learning approaches are essential tools in the evolving landscape of LB energy management, enabling EMS strategies that continuously improve and adapt to meet the challenges of dynamic energy environments.

Reinforcement Learning [117] is a powerful machine learning technique particularly well-suited for EMS due to its ability to learn optimal control strategies through interaction with dynamic environments. In Reinforcement Learning, an agent (in this case, the EMS) makes decisions sequentially, taking actions that influence the system's state, intending to maximize long-term rewards. This approach is highly effective for managing hybrid energy systems, where the optimal strategy may change based on varying operational conditions such as fluctuating energy demands, fuel availability, or battery SoC.

The Reinforcement Learning framework can be mathematically modeled as a Markov Decision Process, where the system is described by a tuple $(\mathcal{S}, \mathcal{A}, \mathcal{P}, \mathcal{R}, \gamma)$, consisting of:

- $\mathcal{S}$: the set of possible *states* the system can be in. For an EMS, states might include the current load demand, SoC of the battery, fuel consumption rate, or energy production capacity.
- $\mathcal{A}$: the set of *actions* the agent can take. Actions in an EMS could include decisions like adjusting the fuel cell output, charging or discharging the battery, or selecting which energy source to use at a given moment.
- $\mathcal{P}$: the *state transition probability* function $\mathcal{P}(s_{t+1}|s_t, a_t)$, which defines the probability of transitioning from one state s_t to the next state s_{t+1} after taking action a_t.
- $\mathcal{R}$: the *reward function* $r(s_t, a_t)$, which gives a numerical reward for the agent's action a_t in state s_t. In EMS, this reward can represent objectives such as minimizing fuel consumption, reducing emissions, or optimizing battery usage.
- γ: the *discount factor* $\gamma \in [0, 1]$, which determines the importance of future rewards compared to immediate rewards.

The objective of the RL agent is to learn an optimal policy π^* that maps states to actions, i.e., $\pi^* : \mathcal{S} \rightarrow \mathcal{A}$, in a way that maximizes the expected cumulative reward over time. This can be formalized as

$$\max_{\pi} \ \mathbb{E}_{\pi}\left[\sum_{t=0}^{T} \gamma^t r(s_t, a_t)\right] \tag{15}$$

where $\pi(a|s)$ is the policy that defines the probability of taking action a in state s, and the expectation is taken over the stochastic process governing the state transitions and actions.

In hybrid EMS, the reward function can be designed to capture various system performance metrics, such as minimizing fuel consumption, reducing wear on system components, maintaining an optimal SoC for the battery, or minimizing GHG emissions. This multi-objective nature of the reward function allows the RL framework to balance competing objectives dynamically, enabling the system to operate optimally under varying and uncertain conditions.

A central concept in RL is the exploration-exploitation dilemma. The agent must balance exploration, where it tries new actions to discover potentially better strategies, and exploitation, where it applies the best-known strategy based on past experience to maximize immediate reward.

In the context of an EMS, exploration could involve trying different energy distribution strategies (e.g., prioritizing battery over fuel cell or vice versa) to discover more efficient management policies. However, too much exploration may result in suboptimal decisions in the short term, such as unnecessary battery degradation or excessive fuel consumption. On the other hand, excessive exploitation might cause the agent to miss out on discovering better strategies for energy management in the long run.

To manage this trade-off, RL algorithms typically employ techniques such as ϵ-greedy policies, where the agent selects a random action with probability ϵ (exploration) and the best-known action with probability $1 - \epsilon$ (exploitation).

More advanced strategies, such as Upper Confidence Bound [118] or Boltzmann exploration [119], adjust the exploration rate dynamically based on the uncertainty of the agent's knowledge about certain actions.

RL algorithms rely on estimating value functions to guide decision-making:

The *state − value function* $V^{\pi}(s)$ represents the expected cumulative reward starting from the state s and the following policy π

$$V^{\pi}(s) = \mathbb{E}_{\pi}\left[\sum_{t=0}^{T} \gamma^t r(s_t, a_t) | s_0 = s\right]. \tag{16}$$

The *action − value function* or Q-learning $Q^{\pi}(s, a)$ gives the expected cumulative reward after taking action a in state s, and then the following policy π

$$Q^{\pi}(s,a)=\mathbb{E}_{\pi}\left[\sum_{t=0}^{T}\gamma^{t}r(s_t,a_t)|s_0=s,a_0=a\right]. \tag{17}$$

The agent's goal is to learn the optimal Q-function $Q^{*}(s,a)$ that maximizes the expected cumulative reward. Once the optimal Q-function is learned, the optimal policy π^{*} is simply to choose the action that maximizes the Q-value for any given state

$$\pi^{*}(s)=\arg\max{}_{a}Q^{*}(s,a), \tag{18}$$

Q-learning is one of the most widely used RL algorithms for learning the optimal Q-function through interaction with the environment [120]. It updates the Q-function using the Bellman equation [121].

$$Q(s_t,a_t)\leftarrow Q(s_t,a_t)+\alpha[r(s_t,a_t)+\gamma\max{}_{a'}Q(s_{t+1},a')-Q(s_t,a_t)], \tag{19}$$

where α is the learning rate that controls how much new information is incorporated into the Q-value estimate.

Traditional Q-learning methods may struggle with complex systems like hybrid energy management, where the state and action spaces are large or continuous. This is where Deep Reinforcement Learning comes in, which combines RL with deep neural networks to approximate the value function, policy, or both. Deep Reinforcement Learning is particularly effective in energy management, as it can handle large-scale and highly non-linear systems [122].

In Deep Q-Networks, a neural network is used to approximate the Q-function

$$Q(s_t,a_t;\theta)\approx Q^{*}(s_t,a_t), \tag{20}$$

where θ represents the network parameters (weights and biases). The Deep Q-Networks algorithm uses experience replay to store past transitions (s_t, a_t, r_t, s_{t+1}) in a replay buffer and sample mini-batches to update the Q-network, reducing the correlation between consecutive updates and improving the stability of training.

Deep Reinforcement Learning techniques, such as Deep Q-Networks [123] and Policy Gradient methods [124], have been successfully applied to hybrid fuel cell EMS to optimize long-term energy efficiency while balancing multiple objectives, such as fuel consumption, emissions, and system longevity. For instance, in a recent study, Tang et al. [115] used Deep Reinforcement Learning to manage the energy flow in a hybrid fuel cell-battery system, achieving significant improvements in both energy efficiency and emission reduction compared to traditional methods. One of the key advantages of RL, especially in energy management systems, is the ability to learn without a detailed model of the environment. Model-free RL methods, such as Q-learning and Policy Gradient, learn the optimal policy directly from interactions with the environment, bypassing the need for explicit modeling of the system dynamics. This is particularly useful in hybrid fuel cell systems where the dynamics can be highly complex and difficult to model accurately.

3 State-of-the-Art on Energy Management Strategies

This section provides an overview of various EMS (RB, OB, LB) in marine energy systems, highlighting their advantages and disadvantages. EMS play a crucial role in distributing power among the different power-generating components of hybrid marine power plants. This task becomes increasingly complex when the EMS must satisfy multiple, often conflicting objectives, including i) ensuring the vessel can complete its mission, ii) minimizing energy consumption, and iii) reducing component degradation [6].

While this chapter reviews these three EMS strategies, it is important to acknowledge that several comprehensive reviews on energy management systems have been published in the literature. These reviews provide valuable insights into the evolution of EMS methodologies and their specific applications. Readers are encouraged to consult these existing reviews to gain a holistic understanding of EMS strategies in maritime applications.

The growing interest in EMS approaches within the literature reflects a dynamic and evolving landscape for review papers. This development aligns with the themes explored in various book chapters [125, 126, 127]. Initial discussions in these reviews focus on the motivations driving the energy transition and zero-emission goals [23, 125, 128]. They briefly address the economic and environmental imperatives behind the necessity for innovative EMS designs in the maritime sector.

It is also essential to highlight the applicable fuel cell technologies and their configurations in maritime applications. The authors of [23] provide a clear narrative regarding the working principles of various fuel cell types, including PEMFC, SOFCs, and MCFCs. The type of fuel utilized is often influenced by the topology and design of the EMS, with a significant focus on their environmental impacts. PEMFCs are the predominant contributors in this sector; however, HT-PEMFCs, SOFCs, and MCFCs are also gaining attention for their promising capabilities [20, 126].

Additionally, the literature covers other critical system components, such as batteries, DC-DC converters, inverters, and motors [23, 129]. Control approaches are discussed in detail [125, 126, 130, 131] and categorized under RB, OB, and LB methods. RB approaches stand out for their simplicity and ease of implementation, often relying on straightforward engineering operation schemes for decision-making and power allocation [125]. In contrast, OB control strategies are extensively examined, typically divided into online and offline methods. Offline methods are further categorized into classical optimization approaches and heuristic methods. Notable strategies include Dynamic Programming, Linear Programming, Nonlinear Programming, Pontryagin's Minimum Principle, Mixed Integer Linear Programming, and Mixed Integer Nonlinear Programming [125, 126]. Heuristic algorithms, often referred to as intelligent algorithms, include Particle Swarm optimization, Genetic Algorithms, Grey Wolf optimizer, and Ant Colony optimization.

While offline optimization methods can be powerful, they require prior knowledge of load profiles. In contrast, online or real-time optimization strategies focus on smaller segments of the optimization problem, allowing for more efficient computations. Common approaches include the Equivalent Consumption Minimization Strategy (ECMS) and MPC. The selection of the optimization objective function and the application of desired costs in each scenario are crucial components of this approach.

LB strategies, increasingly present in the literature, utilize artificial intelligence methods to explore system dynamics, providing advantages at various stages of control. For instance, regression models can play a significant role in parametric identification within power plant models. The combination of MPC with artificial intelligence forecasting models presents a viable option for enhancing EMS effectiveness [125].

3.1 Rule-Based Energy Management Strategies

RB control systems have long been favored in EMS due to their simplicity, robustness, and real-time performance capabilities [16]. These systems rely on predefined sets of rules to manage the distribution of power between components, allowing for immediate responses to changes in system conditions. However, despite their strengths, RB strategies also present several limitations [132]. Their responses are often suboptimal when compared to more sophisticated OB strategies, as the rules may not align with the broader vehicle design or control objectives [133]. Additionally, these systems tend to have weaker dynamic responses to rapid changes in operating conditions [16], and they require significant expert knowledge during the design phase to ensure the system functions optimally [134]. RB control systems can be categorized into two primary strategies: fuzzy RB and deterministic RB approaches.

Deterministic RB strategies follow predefined rules based on tables or flow charts to determine which power source will be used and to what extent. These deterministic strategies can be further subdivided into several methods. The thermostat control approach is characterized by a simple on/off control method that turns system components on or off based on the current state of the power plant. For example, the system may turn on a diesel generator when the battery's SoC falls below a certain threshold. Xie et al. [135] present a two-layer thermostat-based energy management system where the outer layer controls the activation of the diesel engine based on the SoC and load. However, while thermostat control is common in the automotive industry, it is less suitable for maritime applications, where starting and stopping large engines, such as diesel generators, requires significant energy and increases emissions due to start-up inefficiencies.

The power follower method approach ensures that the power distribution between components follows the highest efficiency levels for each given moment, as identified in pre-established efficiency maps. Trovao et al. [136] applied the power follower method to a hybrid electric vehicle EMS. While this method ensures optimal efficiency at any particular time, it does not guarantee the highest efficiency for the entire operational cycle, making it less effective over extended periods.

Bassam et al. [137] developed an RB PI-control-based EMS for a hybrid fuel cell/battery passenger vessel, which uses predefined rules to determine the power distribution between the fuel cell and the battery system based on various operational states such as battery SoC, required power, and other operational limits.

Finally, the gliding average control approach, also known as the moving average method, smooths the load demand on power sources by calculating the load's moving average over time. Tritschler et al. [138] applied this method to a fuel cell-based system, finding that it successfully reduced load fluctuations. However, while this method can help extend component lifespan by reducing strain, it can also result in delayed system responses, affecting real-time performance. This trade-off between dynamic response and system longevity is a key consideration for applying gliding average control in hybrid energy systems.

Fuzzy RB strategies utilize fuzzy logic to imitate human reasoning and handle uncertainty in decision-making [134]. Fuzzy logic translates non-binary inputs into outputs that a binary system can process, creating a black-box model that mimics human logic. One of this method's main advantages is its adaptability; the fuzzy rules can be optimized for different drive cycles without requiring deep expertise in the design phase. Studies such as [139] demonstrate fuzzy rule strategies' robustness and anti-disturbance capabilities, showing improved energy efficiency compared to conventional RB approaches. Fuzzy rules' adaptability also allows them to handle a wide range of operating conditions, making them particularly suitable for hybrid energy systems with fluctuating loads.

As research progresses, the limitations of traditional RB strategies are becoming more evident, particularly in the context of complex energy systems [140], particularly in dynamic environments, necessitating continued innovation. Recent trends indicate a shift toward machine LB optimization of RB strategies, reducing dependence on human expertise and allowing for real-time adaptation to changing system conditions [141].

The increasing reliance on machine learning techniques to optimize RB strategies holds promise for the future of energy management in hybrid systems. Several studies that demonstrate improved efficiency and adaptability of RB systems through automated rule optimization [142, 143] highlight this growing integration of machine learning. This approach offers a blend of simplicity, robustness, and adaptability that could redefine the role of RB strategies in complex energy systems.

Machine learning algorithms, such as RL and genetic algorithms, are increasingly being used to optimize rule sets based on large datasets. These approaches enhance the efficiency of EMS by improving adaptability, reducing energy consumption, and increasing system reliability while maintaining the simplicity and responsiveness of RB strategies. Research indicates that RL's ability to optimize EMS in dynamic and

Table 4 Summary of RB EMS in literature

Ref.	Approach	Objective
[16]	RB	Simplicity, real-time performance
[132]	RB	Limitations in dynamic response, suboptimal control
[133]	RB	Suboptimal when compared to OB strategies
[134]	Fuzzy RB	Adaptability, robustness, anti-disturbance capabilities
[135]	Thermostat	SoC-based control, startup inefficiencies
[136]	Power follower	Component efficiency optimization
[137]	RB PI-control	Fuel Consumption optimization
[138]	Gliding average	Load smoothing, reduced strain at delayed response
[139]	Fuzzy RB	Energy efficiency, adaptability for fluctuating loads
[140]	RB	Highlighted limitations in complex systems
[141]	ML-enhanced RB	Real-time adaptation, reduced expert dependence
[142]	RB+GA	Optimization of rule sets for enhanced performance
[143]	RB+GA	Optimization of rule sets for enhanced performance
[144]	RB+RL&GA	Non-convex problems, improved efficiency, and reliability
[145]	RB+RL&GA	Non-convex problems, improved efficiency, and reliability

unpredictable environments, such as grids with renewable energy sources, is highly promising. Meanwhile, genetic algorithms are leveraged for their capacity to handle non-convex optimization problems within EMS frameworks and refine rule sets for better performance in real-world applications [144, 145].

The summary of the referenced works is reported in Table 4.

3.2 *Optimization-Based Energy Management Strategies*

OB energy management differs inherently from RB energy management. While the decision of RB control systems is, to an extent, predefined by an applied set of rules that aim to describe the system's behavior, OB energy management bases its control decision on the minimization of an objective function. The objective function serves as a mathematical representation of the system's behavior with respect to one or more performance criteria that are deemed significant for the control application. The solution to the resulting optimization problem is contingent upon the selected algorithm. Importantly, the choice of algorithm is not arbitrary; its performance cannot be universally guaranteed to be optimal for all problems, as stated by the no-free-lunch theorem [146]. For energy management, OB strategies can be split up into two different categories: offline methods and online methods, as reported in Fig. 8.

Offline methods solve the problem globally by considering the full operational profile, which allows for an optimization of the complete operation beforehand. However, this includes extensive knowledge of the system and the planned operation, such as route, vessel speed, weather conditions, and power profile. Considering all that information, the optimal control sequence for the operation can be determined using a minimization of the objective function. A variety of different global optimization algorithms can be used, whereby Dynamic Programming [93] is worthy of mentioning in specific. In reality, this method is hard to apply, as much of the required information cannot be accurately determined pre-operation. This limits the applicability of such methods to short time horizons or commonly known load profiles, such as a passenger ferry with a repeating route and power profile. An example of an offline EMS is shown by Xie et al. [147, 148], with a focus on improving fuel efficiency. For this, the EMS is split into two layers, whereby the first is an offline optimization to decide on the power generation plan for the next day, while the second layer allocates the power generation strategy between the fuel cell modules.

Han et al. [149] present a hybrid fuel cell and battery system for a low-power boat, where the EMS aims to optimize system efficiency by managing power flow between the fuel cell and battery. The EMS determines the operating point of each component based on the system's operational states, which maximizes efficiency under various load conditions. The proposed EMS utilizes predefined operating states that consider the battery's SoC, load demands, and fuel cell efficiency. The system dynamically adjusts the power output from the fuel cell and battery based on these states. For example, when the battery's SoC is high, the EMS prioritizes discharging the battery, while at lower SoC levels, it prioritizes recharging the battery using the fuel cell.

Online methods include all approaches to optimize the cost function in real-time. To achieve this, the continuous operation is broken down into singular discrete time steps. With an appropriate sampling rate between the time steps, the system behavior between the time steps can be safely ignored [150]. The operation can be optimized, minimizing the cost function repeatedly for each time step of the operation, and implemented in the system in real-time. The major advantage of those methods compared to the offline methods is the reduced requirements of system and operational knowledge, as well as a reduction of computational time due to the reduction of the problem size to singular time steps rather than a complete operation. On the other hand, online methods are limited by the hardware available on board the vessel, as well as the chosen sampling rate, to ensure the accuracy of the real-time optimization. The latter is of interest as it is a hard limit requirement for the solution time of the chosen algorithm, which generally increases with the system's complexity. However, as energy management is situated on the tertial control layer (see Fig. 7), the time step size is large enough for OB control methods.

The spectrum of online methods can be divided into instantaneous and predictive optimization approaches. Instantaneous optimization approaches address only one of the discrete time steps of the problem and solve the objective function for this respective time step without any knowledge about the future operation. The control problem is set up with an objective function for each time step and solved using an adequate optimization algorithm. The setup of the control problem itself can vary and incorporate various control approaches, such as reference tracking or the introduction of additional Key Performance Indicators to help determine the optimal control sequence. A notable concept for hybrid fuel cell/battery systems is ECMS.

ECMS introduces an equivalent factor to quantify the usage of battery power in comparison to fuel usage from the fuel cell. This approach allows the connection of two inherently different energy sources to the same factor, for example, the financial cost of a kWh, and therefore single out the global minimum for the power split. Further, it is also possible to solve the optimization problem using a variety of algorithms. One example is Particle Swarm optimization. This is demonstrated by Peng et al. [151], who use Particle Swarm optimization to improve the power quality, as well as the battery lifetime.

Predictive optimization approaches combine multiple control steps to solve the control problem. This is also commonly referred to as MPC (see Sect. 2.6.3) or receding horizon control. In this setup of the optimization problem, the optimal control sequence is based on the expected behavior of the system, taking into account a prediction for the future. This allows for the optimization to consider upcoming system changes beforehand, such as upcoming power fluctuations of significant scale. Commonly, only the first step of the optimal control sequence found as a solution to the optimization problem is implemented in the system. Afterward, the control horizon is moved one step forward and optimized again. While this method promises an accurate, optimal control solution because it can adjust to upcoming changes early, it also has some requirements. The first one is a model of the controlled system, which allows for an accurate prediction of the system behavior under the upcoming system conditions. The second one is an accurate forecast of the external conditions, such as the load profile and weather conditions. Both the model of the system and the external prediction model need to fulfill requirements on the accuracy of their forecast, as otherwise, the added uncertainty invalidates the control performance. In addition, predictive OB approaches also increase the size of the control problem as they add multiple steps together, which leads to an increase in computational requirements and solution time. A first example of predictive control of hybrid fuel cell/battery systems is shown by Banaei et al. [152]. The authors propose a stochastic MPC for the control of a fuel cell/battery/cold-ironing vessel for the optimization of operational cost, as well as considering aging factors for the fuel cell to account for component degradation. The summary of the referenced works is reported in Table 5.

Table 5 Summary of OB EMS

Ref.	Approach	Objective
[93]	Dynamic programming—Offline	Global optimization
[147]	Two layer—Offline	Fuel efficiency
[148]	Two layer—Offline	Fuel efficiency
[149]	PSO—Offline	Power quality, battery lifetime
[150]	Two layer—Offline	Fuel efficiency, power quality, system efficiency
[151]	MPC—Online	Real-time optimization
[152]	Stochastic MPC—Online	Cost, aging

3.3 Learning-Based Energy Management Strategies

LB EMS are becoming increasingly essential for optimizing energy distribution and minimizing operational costs in hybrid fuel cell and battery systems within maritime applications. These strategies leverage operational data from the ship's energy systems to inform control decisions, dynamically adapting to real-time conditions. Compared to traditional RB methods, LB approaches often outperform by learning from data and utilizing predictive capabilities that are not restricted by predefined rules [153]. Furthermore, they avoid the extensive computational demands and modeling complexities associated with offline OB strategies.

In the context of hybrid fuel cell systems for ships, operational data such as power demand, load profiles, fuel consumption, and battery states can be harnessed to train machine learning models that predict energy requirements over short time horizons. This predictive capability enables LB strategies to effectively manage the dynamic and variable nature of ship energy management. However, challenges persist regarding data availability, as hybrid fuel cell/ESS systems are not yet widely adopted in the maritime industry, which limits access to large, high-quality datasets.

LB EMS is becoming increasingly critical for optimizing energy distribution and reducing operational costs in hybrid fuel cell and battery systems used in maritime applications. These strategies leverage real-time operational data from the ship's energy systems to inform dynamic control decisions, adapting continuously to changing conditions. Compared to traditional RB methods, LB approaches often demonstrate superior performance by learning from historical data and utilizing predictive capabilities, which are not limited by predefined rules [153]. Additionally, they avoid the significant computational demands and modeling complexities inherent in offline OB strategies. In the context of hybrid fuel cell systems for ships, operational data—including power demand, load profiles, fuel consumption, and battery states—can be used to train machine learning models that predict energy requirements over short timeframes. This predictive capability enables LB strategies to manage the dynamic and variable nature of shipboard energy systems more effectively. However, challenges remain, particularly in terms of data availability, as hybrid fuel cell/ESS systems are not yet widely deployed in the maritime industry, limiting access to large, high-quality datasets.

Among the promising approaches within LB energy management is reinforcement RL, which has demonstrated the ability to optimize energy consumption, reduce fuel usage, and enhance overall system efficiency. RL frameworks, particularly deep RL, have shown success in real-time decision-making for hybrid fuel cell systems. For example, Jung and Chang [154] applied Deep RL to manage the energy flow in a hybrid fuel cell/battery-powered ship, achieving significant reductions in fuel consumption and operational expenditures.

In another noteworthy application, Jung et al. [154] propose a Deep RL-based EMS tailored for hybrid electric ship propulsion systems powered by liquid hydrogen. This system integrates PEMFCs, a fuel gas supply system, and lithium-ion battery systems. The Deep RL EMS was evaluated against benchmark algorithms such as dynamic programming and sequential quadratic programming, demonstrating superior real-time optimization capabilities across various operational scenarios.

Moreover, Wu et al. [155] present a Double Q-learning-based RL EMS for plug-in PEMFC and battery-powered ships. Their method optimizes cost-effectiveness in operating these hybrid ships without requiring prior knowledge of future power demands. The study focuses on a coastal ferry, utilizing continuous monitoring data from real-world voyages to train and validate the RL agent. The energy management system is designed to efficiently allocate power among the fuel cell, battery, and shore power, taking into account factors like fuel cell degradation and battery charging/discharging dynamics.

Gan et al. [156] provide a comprehensive examination of machine learning algorithms to approximate MPC for real-time energy management in hybrid energy ships. Their research introduces an approximate MPC strategy aimed at minimizing operating costs and deviations from the reference SoC of a ship's battery. Various machine learning algorithms, including Random Forest, Gradient Boosting Decision Trees, and Support Vector Machines, were evaluated for their performance in real-time energy management tasks on a ferry equipped with photovoltaic systems. The results indicate that the Random Forest algorithm strikes the best balance between training accuracy and computational efficiency, making it well-suited for real-time applications in hybrid energy ship systems.

Neural networks are also a powerful tool in LB EMS due to their capability to model complex, nonlinear relationships within data, making them particularly suitable for the intricate dynamics of hybrid fuel cell/battery systems. Geng et al. [157] demonstrate an accurate method for estimating the SoC of batteries for all-electric ships, employing a Long Short Term Memory recurrent neural network. This approach improves estimation accuracy by considering nonlinear dependencies in input variables like voltage, current, and battery temperature, achieving a maximum absolute SoC estimation error of 2.0%.

Walker et al. [92] explore the application of shallow and deep machine learning models for power demand forecasting in a hybrid marine energy system. The study focuses on predicting the total power demand of a vessel that operates using electric motors, batteries, and diesel engines, which is increasingly relevant in optimizing energy efficiency and reducing emissions in maritime transport. By comparing shallow learning models (such as kernel methods and ensemble techniques) with a deep learning model (Temporal Convolutional Network), the paper highlights both the potential and challenges of applying machine learning techniques in energy and power management systems.

In addition, Zhou et al. [158] introduce a Soft Actor-Critic RL framework aimed at optimal energy management in electric vehicles equipped with hybrid ESS, including batteries and supercapacitors. The proposed energy management system overcomes the limitations faced by traditional RB and OB control strategies, improving energy efficiency, battery longevity, and overall performance by dynamically allocating power between the battery and supercapacitor during peak power demands. This approach has been validated through extensive experiments, demonstrating superior performance compared to other control strategies, including deep deterministic policy gradient and RB methods.

Kernel methods, such as Support Vector Machines and Kernel Ridge Regression, are increasingly applied in maritime energy management tasks due to their effectiveness in modeling nonlinear relationships, even with small datasets. This characteristic is particularly beneficial in the maritime domain, where operational data for hybrid fuel cell systems may often be sparse. Kernel methods can predict energy usage and optimize system control by transforming the input space into a higher-dimensional feature space using the kernel trick, thus enabling the algorithm to manage complex relationships among operational parameters without requiring extensive data or computational resources [91].

For example, Valchev et al. [91] employed kernel-based learning methods to optimize energy flow in hybrid propulsion systems, finding that Kernel Ridge Regression effectively predicted power demand for controlling and allocating energy between fuel cells and batteries. This demonstrates the potential of kernel methods in scenarios with limited data, which are common during the early stages of hybrid fuel cell technology deployment in maritime applications.

Ensemble methods, such as Random Forests and Gradient Boosting Machines, are also powerful in maritime energy management because they combine multiple models to enhance predictive accuracy and robustness. These ensemble methods have been utilized to predict energy consumption, optimize fuel usage, and manage power distribution in hybrid fuel cell systems. For instance, Wen et al. [159] implemented a hybrid ensemble model designed for interval prediction of solar power output in shipboard power systems. Their model integrates various machine learning techniques—including neural networks, Radial Basis Function Neural Networks, Extreme Learning Machines, and Elman Neural Networks—with Particle Swarm optimization to address the stochastic nature of solar power generation on moving ships. This hybrid approach was validated using real-world data from a large oil tanker operating along a route from China to Yemen, demonstrating improved prediction accuracy compared to conventional methods.

LB EMS present significant potential to enhance energy efficiency and reduce operational costs in hybrid fuel cell/battery systems for maritime applications. Approaches such as reinforcement learning, neural networks, and ensemble methods are particularly well-suited for managing maritime energy systems' dynamic and complex nature. Nevertheless, the success of these methods hinges on the availability of quality operational data, careful model tuning, and effective management of computational constraints aboard ships. As the adoption of hybrid fuel cell systems in maritime applications continues to grow, the integration of advanced LB methods will likely play a pivotal role in optimizing EMS. Despite their advantages, machine LB methods face challenges related to the quantity and quality of training data required and the substantial computational power needed to train these models effectively. The summary of the referenced works is reported in Table 6.

Table 6 Summary of research on LB EMS

Ref.	Approach	Application
[154]	DRL	Hybrid fuel cell/battery-powered ships
[154]	DRL	Hybrid electric propulsion systems powered by liquid hydrogen
[155]	Double Q-learning	EMS for PEMFC and battery-powered ships
[156]	Approximate MPC	Real-time energy management in hybrid energy ships
[157]	LSTM	SoC estimation for all-electric ships
[92]	TCN	Power demand forecasting in hybrid marine energy systems
[158]	SAC-RL	EMS in electric vehicles with hybrid ESS
[91]	KRR	Energy flow optimization in hybrid propulsion systems
[159]	Hybrid	Solar power output prediction in shipboard power systems

3.4 Comparative Analysis of Energy Management Strategies

In their comparative study, the authors of [153] between Dynamic Programming and ECMS indicate that both the Partially Observable Markov Decision Process and Dynamic Programming deliver equivalent optimal solutions. This study underscores the critical relationship between the performance of ECMS and the Partially Observable Markov Decision Process, highlighting their dependence on the precise definitions of co-states used in their calculations.

A key distinction exists between EMS in the automotive and maritime sectors, particularly regarding the utilization of regenerative braking and the steady load demands of marine power plants. As observed by [160], the energy savings realized through ECMS in shipping are less significant than those documented in the automotive industry, largely due to the limitations of ship propellers in capturing energy via regenerative braking. The literature indicates that RB systems often incur higher costs compared to OB strategies [160, 161, 162]. Historically reliant on the expertise of engineers, there is a discernible shift toward offline-optimized RB strategies in recent research [161]. Notably, the Partially Observable Markov Decision Process strategy consistently yields optimal solutions while maintaining lower computational complexity compared to other methods [163]. Furthermore, the cost differential between RB and OB strategies tends to vary with the load profile; more variable loads typically benefit more from non-RB strategies than constant loads. While OB strategies often provide superior fuel efficiency relative to RB methods, they frequently present complexities that hinder real-time application or are overly reliant on precise tuning of co-states in online methods. Literature commonly notes that strategies such as ECMS and Partially Observable Markov Decision Process are optimized for specific drive cycles but are rarely tested under unpredictable, random cycles, which is a crucial factor for both shipping and automotive applications. This observation raises questions regarding their effectiveness in random operational scenarios and whether certain shipping operations—such as ferry routes along the Norwegian coast, as studied by [164] might experience more consistent load cycles. Additionally, while numerous studies acknowledge that hybrid propulsion systems could contribute to addressing global GHG challenges, most tend to concentrate primarily on optimizing fuel consumption. Surprisingly, only a limited number of studies explicitly integrate GHG considerations into their findings, indicating a gap in the comprehensive evaluation of EMS effectiveness in relation to environmental impacts.

4 Challenges and Future Directions

The integration of state-of-the-art EMS for hybrid PEMFC/ESS applications in ships faces a variety of technical, economic, and regulatory challenges. Technical challenges include the complexity of optimizing power flow in real-time, especially with fluctuating energy demands and the integration of renewable energy sources like solar. Critical obstacles remain to achieve seamless transitions between energy sources and maintain system stability while respecting operational constraints such as fuel cell efficiency, battery SoC, and thermal management.

The performance degradation of PEMFC over time, as well as the ESS aging, adds to the complexity of designing predictive models that accurately capture the system dynamics. Currently, most of the existing health-conscious EMS set upper or lower boundaries to the fuel cell and ESS operating ranges, or battery' SoC. Developing models able to quantify the degradation or the lifetime of these energy sources and integrating them into EMS remains an open challenge.

On the economic front, the high capital cost of fuel cell technologies and advanced EMS is a significant barrier to widespread adoption. Regulatory challenges also hinder adoption, as international maritime regulations are still evolving to include standards for the safe and efficient operation of hydrogen-based systems. Compliance with environmental regulations, such as the IMO's goals for carbon reduction, requires further harmonization of policies and incentives to facilitate green shipping technologies.

Looking ahead, future research directions should focus on improving the longevity and efficiency of fuel cells through advances in material science. Developing practical models to estimate fuel cells and batteries degradation, and their interaction and combined effect on the hybrid systems' lifetime is crucial for the integration of health-conscious approaches into EMS. Enhanced optimization algorithms, such as machine learning-driven predictive control, could offer more robust energy management solutions that can adapt to the variability in marine operational profiles. Furthermore, research into integrated systems that combine hydrogen fuel cells with renewable energy sources will be key to reducing dependency on fossil fuels. Addressing economic challenges will require continued efforts to scale up hydrogen production and infrastructure and improve energy storage density through alternative storage solutions, as well as innovations in reducing the cost of fuel cells. On the regulatory side, establishing clear, global guidelines for hydrogen safety, handling, and emissions reporting will be critical to accelerating the adoption of hybrid fuel cell systems in the maritime sector.

5 Conclusion

The adoption of state-of-the-art EMS for hybrid PEMFC/ESS applications in ships represents a critical step toward reducing emissions and enhancing energy efficiency in the maritime sector. Among these strategies, RB, OB, and LB approaches each offer unique benefits and face specific limitations. RB methods, while straightforward and reliable, lack the flexibility needed for dynamic marine environments where operational conditions fluctuate. In contrast, OB techniques, particularly MPC, provide advanced real-time adaptability, making them well-suited for optimizing energy distribution in complex scenarios. While LB strategies that leverage machine learning hold the promise of continuous system improvement through adaptation to new data, they require further development to be widely adopted in practical marine applications. Despite the clear advantages of these approaches, several challenges impede widespread adoption. Technical challenges persist, including modeling the degradation of fuel cells and batteries over time, the complexity of real-time optimization, and the need for more accurate predictive models capable of handling diverse operational conditions. Additionally, the high capital costs associated with hybrid PEMFC/ESS systems and the evolving nature of regulatory frameworks in the maritime industry present substantial economic and regulatory hurdles. These challenges must be addressed to utilize and exploit the full potential of these technologies.

Looking to the future, research must prioritize efforts to improve the durability and performance of fuel cells and batteries and focus on their degradation modeling. The development of more computationally efficient optimization algorithms will also be essential, particularly as MPC, though powerful, remains computationally intensive for practical real-time applications in the marine sector. Exploring more scalable algorithms, especially those driven by machine learning, could offer more effective solutions for managing multiple competing objectives, such as balancing power demand and preserving system longevity.

Another crucial area for future development is the integration of renewable energy sources, such as wind-assisted propulsion and solar power, into hybrid fuel cell systems. This will reduce reliance on hydrogen and improve the sustainability of marine energy systems. Hybrid configurations that blend renewable energy with hydrogen-based systems will be vital for optimizing fuel efficiency, thus aligning with the broader goals of maritime decarbonization. However, the integration of other renewable sources into marine hybrid systems will increase the complexity of their operation, control, and EMS.

Addressing the economic challenges will require targeted financial strategies to lower the barriers to adoption. Governments and industry stakeholders should collaborate to create economic incentives that support the transition to hybrid systems. Subsidies for hydrogen production, grants for infrastructure development, and tax incentives for retrofitting ships with hybrid systems can help mitigate the high upfront costs. Additionally, scaling up production and standardizing fuel cell components will be key to driving down costs and making hybrid systems more financially viable. With the potential to optimize operational efficiency, minimize fuel consumption, and increase components' operational lifetime, EMS can significantly contribute to enhancing the cost-effectiveness of marine hybrid fuel cell systems.

From a regulatory perspective, the maritime industry will need clear, consistent guidelines to facilitate the adoption of hybrid fuel cell technologies. Regulatory bodies should continue to lead the effort in establishing global standards for hydrogen-based systems, addressing critical issues such as safety protocols, emissions reporting, and operational guidelines. A harmonized regulatory framework will reduce uncertainty for shipping companies and incentivize investment in clean technologies.

Finally, to support the successful deployment of these advanced systems, specialized training programs are essential. These programs should equip ship operators and engineers with the necessary skills to safely operate and maintain hybrid fuel cell systems, as well as provide training on the use of advanced energy management tools. As these technologies become more complex, building a skilled workforce will be critical to ensuring their effective use and long-term success.

References

1. I. M. O. (2020). Fourth IMO GHG Study 2020 Full Report: Tech. Rep. [Online]. Available: https://www.imo.org/en/OurWork/Environment/Pages/Fourth-IMO-Greenhouse-Gas-Study-2020.aspx
2. H. Council. (2021). Hydrogen for net zero-a critical cost-competitive energy vector.
3. Whitehead, J., Newman, P., Whitehead, J., & Lim, K. L. (2023). Striking the right balance: Understanding the strategic applications of hydrogen in transitioning to a net zero emissions economy. *Sustainable Earth Reviews, 6*(1), 1.
4. Yrjänäinen, A., Johnsen, T., Dæhlen, J. S., Kramer, H., & Monden, R. (2018). Requirements and operational profiles. In *A holistic approach to ship design: Volume 1: Optimisation of ship design and operation for life cycle* (p. 75).
5. Sorensen, A. J., et al. (2017). Toward safer, smarter, and greener ships: Using hybrid marine power plants. *IEEE Electrification Magazine, 5*(3), 68–73.
6. Geertsma, R. D., Negenborn, R. R., Visser, K., & Hopman, J. J. (2017). Design and control of hybrid power and propulsion systems for smart ships: A review of developments. *Applied Energy, 194*, 30–54.
7. Sürer, M. G., & Arat, H. T. (2022). Advancements and current technologies on hydrogen fuel cell applications for marine vehicles. *International Journal of Hydrogen Energy, 47*(45), 19865–19875.
8. Gianni, M., Pietra, A., Coraddu, A., & Taccani, R. (2022). Impact of SOFC power generation plant on carbon intensity index (CII) calculation for cruise ships. *Journal of Marine Science and Engineering, 10*(10), 1478.
9. Baldi, F., Coraddu, A., & Mondejar, M. E. (2022). *Sustainable energy systems on ships: Novel technologies for low carbon shipping*. Elsevier.
10. Baldi, F., & Coraddu, A. (2022). Towards halving shipping GHG emissions by 2050: The IMO introduces the CII and the EEXI. In *Sustainable energy systems on ships* (pp. 513–517). Elsevier.

11. Vasilikis, N., Geertsma, R., & Coraddu, A. (2023). A digital twin approach for maritime carbon intensity evaluation accounting for operational and environmental uncertainty. *Ocean Engineering, 288*, 115927.
12. Soares, C. G., & Santos, T. A. R. (2022). *Trends in maritime technology and engineering*. CRC Press Taylor & Francis Group.
13. Kopka, T., Mylonopoulos, F., Coraddu, A., & Polinder, H. (2023). Decentralized power sharing with frequency decoupling for a Fuel Cell-Battery DC Shipboard Power System. *Modelling and Optimisation of Ship Energy Systems 2023*. https://doi.org/10.59490/moses.2023.670
14. Kopka, T., Löffler, C., Coraddu, A., & Polinder, H. (2023). Hierarchical control strategy for fuel cell-battery shipboard power system utilizing a modular control architecture. *IEEE*, 1–8.
15. Nguyen, H. P., et al. (2021). The electric propulsion system as a green solution for management strategy of CO2 emission in ocean shipping: A comprehensive review. *International Transactions on Electrical Energy Systems, 31*(11).
16. Inal, O. B., Charpentier, J. F., & Deniz, C. (2022). Hybrid power and propulsion systems for ships: Current status and future challenges. *Renewable and Sustainable Energy Reviews, 156*, 111965.
17. Yang, Z., Huang, H., & Lin, F. (2022). Sustainable electric vehicle batteries for a sustainable world: Perspectives on battery cathodes, environment, supply chain, manufacturing, life cycle, and policy. *Advanced Energy Materials, 12*(26), 2200383.
18. Editorial. (2021). Lithium ion batteries need to be greener and more ethical. *Nature, 595*, 7.
19. Peters, J. F., Baumann, M., Zimmermann, B., Braun, J., & Weil, M. (2017). The environmental impact of Li-Ion batteries and the role of key parameters – A review. *Renewable and Sustainable Energy Reviews, 67*, 491–506.
20. van Biert, L., Godjevac, M., Visser, K., & Aravind, P. V. (2016). A review of fuel cell systems for maritime applications. *Journal of Power Sources, 327*, 345–364.
21. Kandidayeni, M., Trovão, J. P., Soleymani, M., & Boulon, L. (2022). Towards health-aware energy management strategies in fuel cell hybrid electric vehicles: A review. *International Journal of Hydrogen Energy, 47*(17), 10021–10043.
22. Mallouppas, G., & Yfantis, E. A. (2021). Decarbonization in shipping industry: A review of research, technology development, and innovation proposals. *Journal of Marine Science and Engineering, 9*(4), 415.
23. Xing, H., Stuart, C., Spence, S., & Chen, H. (2021). Fuel cell power systems for maritime applications: Progress and perspectives. *Sustainability, 13*(3), 1213.
24. Tronstad, T., Åstrand, H. H., Haugom, G.-P., & Langfeldt, L. (2017). *Study on the use of fuel cells in shipping*. European Maritime Safety Agency.
25. Elkafas, A. G., Rivarolo, M., Gadducci, E., Magistri, L., & Massardo, A. F. (2022). Fuel cell systems for maritime: A review of research development, commercial products, applications, and perspectives. *PRO, 11*(1), 97.
26. Dall'Armi, C., Pivetta, D., & Taccani, R. (2023). Hybrid PEM fuel cell power plants fuelled by hydrogen for improving sustainability in shipping: State of the art and review on active projects. *Energies (Basel), 16*(4), 2022.
27. Chen, W., et al. (2021). DC-distributed power system modeling and hardware-in-the-loop (HIL) evaluation of fuel cell-powered marine vessel. *IEEE Journal of Emerging and Selected Topics in Industrial Electronics, 3*(3), 797–808.
28. Maurya, R., Prakash, S., & Singh, A. K. (2023). Challenges, configuration, control, and scope of DC microgrid systems: A review. *Journal of Electrical Engineering & Technology, 18*(3), 1655–1674.
29. Chen, D., & Xu, L. (2017). AC and DC microgrid with distributed energy resources. In *Technologies and applications for smart charging of electric and plug-in hybrid vehicles* (pp. 39–64). Springer Nature.
30. Pires, V. F., Pires, A., & Cordeiro, A. (2023). DC microgrids: Benefits, architectures, perspectives and challenges. *Energies (Basel), 16*(3), 1217.

31. Aghmadi, A., & Mohammed, O. A. (2024). Energy storage systems: Technologies and high-power applications. *Batteries, 10*(4), 141.
32. Abo-Khalil, A. G., & Alobaid, M. (2023). A guide to the integration and utilization of energy storage systems with a focus on demand resource management and power quality enhancement. *Sustainability, 15*(20), 14680.
33. Oliveira, D. Q., et al. (2021). A critical review of energy storage technologies for microgrids. *Energy Systems*, 1–30.
34. Alkaner, S., & Zhou, P. (2006). A comparative study on life cycle analysis of molten carbon fuel cells and diesel engines for marine application. *Journal of Power Sources, 158*(1), 188–199.
35. Trondstad, T., & Endresen, Ø. (2005). FellowSHIP-Fuel cells for low emission ships. In *Risoe International Energy Conference* (pp. 141–149). Risoe National Laboratory, Systems Analysis Dept. https://doi.org/10.11581/20623166
36. Bensaid, S., Specchia, S., Federici, F., Saracco, G., & Specchia, V. (2009). MCFC-based marine APU: Comparison between conventional ATR and cracking coupled with SR integrated inside the stack pressurized vessel. *International Journal of Hydrogen Energy, 34*(4), 2026–2042.
37. Nehter, P., Wildrath, B., Bauschulte, A., & Leites, K. (2017). Diesel based SOFC demonstrator for maritime applications. *ECS Transactions, 78*(1), 171.
38. McConnell, V. P. (2010). Now, voyager? The increasing marine use of fuel cells. *Fuel Cells Bulletin, 2010*(5), 12–17.
39. Chakraborty, S., Dzielendziak, A. S., Koroglu, T., & Yang, K. (2013). Evaluation of smart eco-friendly public transport options in coastal cities: Towards a green future for the city of Southampton (Vol. 2). In: R. A. Shenoi, P. A. Wilson, S. S. Bennett (eds.) University of Southampton. https://eprints.soton.ac.uk/359323/
40. German e4ships project reports on fuel cell maritime demos. *Fuel Cells Bulletin, 2016*(10), 5 (2016). https://doi.org/10.1016/S1464-2859(16)30271-1.
41. Huerta, G. V., Jordán, J. Á., Dragon, M., Leites, K., & Kabelac, S. (2018). Exergy analysis of the diesel pre-reforming solid oxide fuel cell system with anode off-gas recycling in the SchIBZ project. Part I: Modeling and validation. *International Journal of Hydrogen Energy, 43*(34), 16684–16693.
42. Borgogna, G., et al. (2019). Design and development of a laboratory for the study of PEMFC system for marine applications. *EDP Sciences*, 02020.
43. Rupo, D., Perano, M., Centorrino, G., & Sanchez, A. V. (2018). A framework based on sustainability, open innovation, and value cocreation paradigms—A case in an Italian maritime cluster. *Sustainability, 10*(3), 729.
44. A. Energy. (2024). RIVERCELL: SERENE to Maritime. [Online]. Available: https://serene.advent.energy/cases/rivercell-serene-to-maritime/
45. E. R. M. (2021). The Fuel Cell Industry Review 2021. [Online]. Available: https://www.erm.com/globalassets/documents/fuel-cell-industry-review/thefuelcellindustryreview2021.pdf
46. S. Maritime. (2024). SeaChange Project. [Online]. Available: https://www.switchmaritime.com/projects
47. M. Project. (2024). *Marine application of a new fuel cell powertrain validated in demanding Arctic conditions*. MARANDA.
48. Vafamand, N., Boudjadar, J., & Khooban, M. H. (2020). Model predictive energy management in hybrid ferry grids. *Energy Reports, 6*, 550–557.
49. Igder, M. A., Rafiei, M., Boudjadar, J., & Khooban, M.-H. (2020). Reliability and safety improvement of emission-free ships: Systemic reliability-centered maintenance. *IEEE Transactions on Transportation Electrification, 7*(1), 256–266.
50. Rafiei, M., Boudjadar, J., & Khooban, M.-H. (2020). Energy management of a zero-emission ferry boat with a fuel-cell-based hybrid energy system: Feasibility assessment. *IEEE Transactions on Industrial Electronics, 68*(2), 1739–1748.

51. Khooban, M. H., Gheisarnejad, M., Farsizadeh, H., Masoudian, A., & Boudjadar, J. (2019). A new intelligent hybrid control approach for DC–DC converters in zero-emission ferry ships. *IEEE Transactions on Power Electronics, 35*(6), 5832–5841.
52. Banaei, M., Boudjadar, J., & Khooban, M.-H. (2021). *Optimal energy resources scheduling of hybrid diesel/battery ships in shallow waters* (pp. 369–378). Springer.
53. Trillos, J. C. G., Wilken, D., Brand, U., & Vogt, T. (2021). Life cycle assessment of a hydrogen and fuel cell ropax ferry prototype. *Progress in Life Cycle Assessment, 2019*, 5–23.
54. Cavo, M., Gadducci, E., Rattazzi, D., Rivarolo, M., & Magistri, L. (2021). Dynamic analysis of PEM fuel cells and metal hydrides on a zero-emission ship: A model-based approach. *International Journal of Hydrogen Energy, 46*(64), 32630–32644.
55. Cavo, M., et al. (2022). Thermal integration of PEM Fuel Cells and metal hydrides storage system for Zero Emission Ultimate Ship (ZEUS). *EDP Sciences*, 04004.
56. Di Ilio, G., Di Giorgio, P., Tribioli, L., Cigolotti, V., Bella, G., & Jannelli, E. (2021). *Assessment of a hydrogen-fueled heavy-duty yard truck for roll-on and roll-off port operations*. SAE Technical Paper.
57. e4ships. (2024). PA-X-ELL 2: Entwicklung und Erprobung eines emissionsfreien und nachhaltigen Energieversorgungssystems für Seeschiffe. [Online]. Available: https://www.e4ships.de/deutsch/seeschifffahrt/pa-x-ell-2/
58. Danfoss. (2024). Hydrogenia Vessel Case Story. [Online]. Available: https://www.danfoss.com/en/service-and-support/case-stories/dps/hydrogenia-vessel-case-story/
59. Meca, V. L., Villalba-Herreros, A., d'Amore-Domenech, R., & Leo, T. J. (2022). Zero emissions wellboat powered by hydrogen fuel cells hybridised with batteries. *Proceedings of the Institution of Mechanical Engineers, Part M: Journal of Engineering for the Maritime Environment, 236*(2), 525–536.
60. Trivyza, N. L., Cheliotis, M., Boulougouris, E., & Theotokatos, G. (2021). Safety and reliability analysis of an ammonia-powered fuel-cell system. *Safety, 7*(4), 80.
61. CORDIS. (2024). *Nautical integrated hybrid energy system for long-haul cruise ships*. CORDIS.
62. H. Project. (2024). HyShip Project. [Online]. Available: https://hyship.eu/
63. E. M. E. C. (EMEC). (2024). HIMET: Hydrogen in an Integrated Maritime Energy Transition. [Online]. Available: https://www.emec.org.uk/projects/hydrogen-projects/himet/
64. CORDIS. (2024). *High efficiency Low emission nautical solid oxide fuel cell*. CORDIS.
65. Wang, Y., Diaz, D. F. R., Chen, K. S., Wang, Z., & Adroher, X. C. (2020). Materials, technological status, and fundamentals of PEM fuel cells–a review. *Materials Today, 32*, 178–203.
66. Kandlikar, S. G., Garofalo, M. L., & Lu, Z. (2011). Water management in a pemfc: Water transport mechanism and material degradation in gas diffusion layers. *Fuel Cells, 11*(6), 814–823.
67. van Biert, L., & Visser, K. (2022). Fuel cells systems for sustainable ships. In *Sustainable energy systems on ships* (pp. 81–121). Elsevier.
68. Rosli, R. E., et al. (2017). A review of high-temperature proton exchange membrane fuel cell (HT-PEMFC) system. *International Journal of Hydrogen Energy, 42*(14), 9293–9314.
69. Wang, H., & Turner, J. A. (2010). Reviewing metallic PEMFC bipolar plates. *Fuel Cells, 10*(4), 510–519.
70. Pietra, A., Gianni, M., Zuliani, N., Malabotti, S., & Taccani, R. (2022). Experimental characterization of a PEM fuel cell for marine power generation. *EDP Sciences*, 05002.
71. Larminie, J., Dicks, A., & McDonald, M. S. (2003). *Fuel cell systems explained* (Vol. 2). J. Wiley.
72. E. M. S. Agency. (2020). Study on Electrical Energy Storage for Ships. [Online]. Available: https://www.emsa.europa.eu/publications/item/3895-study-on-electrical-energy-storage-for-ships.html
73. Reusser, C. A., & Pérez Osses, J. R. (2021). Challenges for zero-emissions ship. *Journal of Marine Science and Engineering, 9*(10), 1042.

74. Zhou, Z., Benbouzid, M., Charpentier, J. F., Scuiller, F., & Tang, T. (2013). A review of energy storage technologies for marine current energy systems. *Renewable and Sustainable Energy Reviews, 18*, 390–400.
75. Ongaro, F., Saggini, S., & Mattavelli, P. (2012). Li-ion battery-supercapacitor hybrid storage system for a long lifetime, photovoltaic-based wireless sensor network. *IEEE Transactions on Power Electronics, 27*(9), 3944–3952.
76. Chen, W., Ådnanses, A. K., Hansen, J. F., Lindtjørn, J. O., & Tang, T. (2010). Supercapacitors based hybrid converter in marine electric propulsion system. *IEEE*, 1–6.
77. Kopka, R., & Tarczyński, W. (2017). A fractional model of supercapacitors for use in energy storage systems of next-generation shipboard electrical networks. *Journal of Marine Engineering & Technology, 16*(4), 200–208.
78. Stark, C., Xu, Y., Zhang, M., Yuan, Z., Tao, L., & Shi, W. (2022). Study on applicability of energy-saving devices to hydrogen fuel cell-powered ships. *Journal of Marine Science and Engineering, 10*(3), 388.
79. Lázár, M., et al. (2023). A newly proposed method for hydrogen storage in a metal hydride storage tank intended for maritime and inland shipping. *Journal of Marine Science and Engineering, 11*(9), 1643.
80. Taccani, R., Malabotti, S., Dall'Armi, C., & Micheli, D. (2020). High energy density storage of gaseous marine fuels: An innovative concept and its application to a hydrogen powered ferry. *International Shipbuilding Progress, 67*, 29–52.
81. Wu, J., et al. (2008). A review of PEM fuel cell durability: Degradation mechanisms and mitigation strategies. *Journal of Power Sources, 184*(1), 104–119.
82. Pei, P., & Chen, H. (2014). Main factors affecting the lifetime of Proton Exchange Membrane fuel cells in vehicle applications: A review. *Applied Energy, 125*, 60–75.
83. Redondo-Iglesias, E., Venet, P., & Pelissier, S. (2018). Calendar and cycling ageing combination of batteries in electric vehicles. *Microelectronics Reliability, 88*, 1212–1215.
84. Lorenzo, C., Bouquain, D., Hibon, S., & Hissel, D. (2021). Synthesis of degradation mechanisms and of their impacts on degradation rates on proton-exchange membrane fuel cells and lithium-ion nickel–manganese–cobalt batteries in hybrid transport applications. *Reliability Engineering and System Safety, 212*, 107369.
85. Naseri, F., Karimi, S., Farjah, E., & Schaltz, E. (2022). Supercapacitor management system: A comprehensive review of modeling, estimation, balancing, and protection techniques. *Renewable and Sustainable Energy Reviews, 155*, 111913.
86. Shekhawat, M., & Bansal, H. O. (2023). An extensive review on hybrid electric vehicles powered by fuel cell-enabled hybrid energy storage system. *Environmental Science and Pollution Research, 30*(57), 119750–119771.
87. Ragab, A., Marei, M. I., & Mokhtar, M. (2023). Comprehensive study of fuel cell hybrid electric vehicles: Classification, topologies, and control system comparisons. *Applied Sciences, 13*(24), 13057.
88. Wang, B., Najjar, L., Xiong, N. N., & Chen, R. C. (2013). *Stochastic optimization: Theory and applications. Hindawi Publishing Corporation.*
89. Tang, X., et al. (2024). Reinforcement learning-based energy management for hybrid power systems: State-of-the-art survey, review, and perspectives. *Chinese Journal of Mechanical Engineering, 37*(1), 43.
90. Walker, J. M., Coraddu, A., Savio, S., & Oneto, L. (2023). Shallow and deep learning models for vessel motions forecasting during adverse weather conditions. *Modelling and Optimisation of Ship Energy Systems, 2023.*
91. Valchev, I., Coraddu, A., Oneto, L., Kalikatzarakis, M., Tiddens, W., & Geertsma, R. D. (2022). 'Artificial Intelligence-based short-term forecasting of vessel performance parameters', Institute of Marine Engineering. *Science and Technology*, 31.
92. Walker, J. M., Coraddu, A., & Oneto, L. (2024). *Power demand forecasting for a hybrid marine energy system with shallow and deep learning* (pp. 1–8). IEEE.

93. Bellman, R., & Kalaba, R. (1957). Dynamic programming and statistical communication theory. *Proceedings of the National Academy of Sciences, 43*(8), 749–751.
94. Powell, W. B. (2007). *Approximate dynamic programming: Solving the curses of dimensionality* (Vol. 703). John Wiley & Sons.
95. Birge, J. R., & Louveaux, F. (2011). *Introduction to stochastic programming*. Springer Science & Business Media.
96. Holland, J. H. (1992). Genetic algorithms. *Scientific American, 267*(1), 66–73.
97. Goldberg, D. E. (1994). Genetic and evolutionary algorithms come of age. *Communications of the ACM, 37*(3), 113–120.
98. Kennedy, J., & Eberhart, R. (1995). Particle swarm optimization. *IEEE*, 1942–1948.
99. Wolsey, L. A. (2020). *Integer programming*. John Wiley & Sons.
100. Bertsekas, D. P., Hager, W., & Mangasarian, O. (1999). *Nonlinear programming*. Athena Scientific.
101. Colson, B., Marcotte, P., & Savard, G. (2005). Bilevel programming: A survey. *4or, 3*, 87–107.
102. Deb, K. (2001). Nonlinear goal programming using multi-objective genetic algorithms. *Journal of the Operational Research Society, 52*(3), 291–302.
103. Zhang, T., Li, W., & Wang, R. (2023). Surrogated-assisted multimodal multi-objective optimization for hybrid renewable energy system. *Complex & Intelligent Systems, 9*(4), 4075–4087.
104. Hartley, E. N., & Maciejowski, J. M. (2014). Reconfigurable predictive control for redundantly actuated systems with parameterised input constraints. *Systems and Control Letters, 66*, 8–15.
105. Goodfellow, I., Bengio, Y., & Courville, A. (2016). *Deep learning*. MIT Press.
106. Aggarwal, C. C. (2018). *Neural networks and deep learning*. Springer.
107. Bishop, C. M., & Bishop, H. (2023). *Deep learning: Foundations and concepts*. Springer Nature.
108. Shalev-Shwartz, S., & Ben-David, S. (2014). *Understanding machine learning: From theory to algorithms*. Cambridge University Press.
109. Breiman, L. (2001). Random forests. *Machine Learning, 45*, 5–32.
110. Chen, T., & Guestrin, C. (2016). Xgboost: A scalable tree boosting system. In *Proceedings of the 22nd acm sigkdd international conference on knowledge discovery and data mining* (pp. 785–794).
111. Rasmussen, C. E. (2003). *Gaussian processes in machine learning*. Springer.
112. Van Houdt, G., Mosquera, C., & Nápoles, G. (2020). A review on the long short-term memory model. *Artificial Intelligence Review, 53*(8), 5929–5955.
113. Bai, S., Kolter, J. Z., & Koltun, V. (2018). An empirical evaluation of generic convolutional and recurrent networks for sequence modeling. *arXiv preprint arXiv:1803.01271*.
114. Vaswani, A. (2017). Attention is all you need. *Advances in Neural Information Processing Systems*.
115. Tang, X., Zhou, H., Wang, F., Wang, W., & Lin, X. (2022). Longevity-conscious energy management strategy of fuel cell hybrid electric vehicle based on deep reinforcement learning. *Energy, 238*, 121593.
116. Hu, H., Yuan, W.-W., Su, M., & Ou, K. (2023). Optimizing fuel economy and durability of hybrid fuel cell electric vehicles using deep reinforcement learning-based energy management systems. *Energy Conversion and Management, 291*, 117288.
117. Ris-Ala, R. (2023). *Fundamentals of reinforcement learning*. Springer.
118. Wen, Y., Su, Q., Shen, M., & Xiao, N. (2022). Improving the exploration efficiency of DQNs via the confidence bound methods. *Applied Intelligence, 52*(13), 15447–15461.
119. Tokic, M., & Palm, G. (2011). *Value-difference based exploration: Adaptive control between epsilon-greedy and softmax* (pp. 335–346). Springer.
120. Jang, B., Kim, M., Harerimana, G., & Kim, J. W. (2019). Q-learning algorithms: A comprehensive classification and applications. *IEEE Access, 7*, 133653–133667.

121. Kim, J., & Yang, I. (2020). Hamilton-Jacobi-Bellman equations for Q-learning in continuous time. *PMLR*, 739–748.
122. Jha, A. V., Appasani, B., Gupta, D. K., Ramavath, S., & Khan, M. S. (2023). Machine learning and deep learning approaches for energy management in Smart Grid 3.0. In *Smart Grid 3.0: Computational and communication technologies* (pp. 121–151). Springer.
123. Zheng, C., Li, W., Li, W., Xu, K., Peng, L., & Cha, S. W. (2022). A deep reinforcement learning-based energy management strategy for fuel cell hybrid buses. *International Journal of Precision Engineering and Manufacturing-Green Technology, 9*(3), 885–897.
124. Zhang, D., et al. (2024). An improved soft actor-critic-based energy management strategy of fuel cell hybrid vehicles with a nonlinear fuel cell degradation model. *International Journal of Precision Engineering and Manufacturing-Green Technology, 11*(1), 183–202.
125. Guo, X., et al. (2024). Energy management system for hybrid ship: Status and perspectives. *Ocean Engineering, 310*, 118638.
126. Xie, P., et al. (2021). Optimization-based power and energy management system in shipboard microgrid: A review. *IEEE Systems Journal, 16*(1), 578–590.
127. Jaurola, M., Hedin, A., Tikkanen, S., & Huhtala, K. (2019). Optimising design and power management in energy-efficient marine vessel power systems: A literature review. *Journal of Marine Engineering & Technology, 18*(2), 92–101.
128. Ge, Y., Zhang, J., Zhou, K., Zhu, J., & Wang, Y. (2023). Research on energy management for ship hybrid power system based on adaptive equivalent consumption minimization strategy. *Journal of Marine Science and Engineering, 11*(7), 1271.
129. Truong, H. V. A., Do, T. C., & Dang, T. D. (2024). Enhancing efficiency in hybrid marine vessels through a multi-layer optimization energy management system. *Journal of Marine Science and Engineering, 12*(8), 1295.
130. Balsamo, F., Capasso, C., Lauria, D., & Veneri, O. (2020). Optimal design and energy management of hybrid storage systems for marine propulsion applications. *Applied Energy, 278*, 115629.
131. Feng, X., Zainudin, E. B., Tseng, K. J., & Low, K. C. A. (2022). *On maritime electrification–electrification technologies, charging infrastructure and energy management strategies* (p. 012034). IOP Publishing.
132. Hofman, T., Steinbuch, M., Van Druten, R., & Serrarens, A. (2007). Rule-based energy management strategies for hybrid vehicles. *International Journal of Electric and Hybrid Vehicles, 1*(1), 71–94.
133. Hu, X., Zou, C., Tang, X., Liu, T., & Hu, L. (2020). Cost-optimal energy management of hybrid electric vehicles using fuel cell/battery health-aware predictive control. *IEEE Transactions on Power Electronics, 35*(1), 382–392.
134. Lü, X., et al. (2020). Energy management of hybrid electric vehicles: A review of energy optimization of fuel cell hybrid power system based on genetic algorithm. *Energy Conversion and Management, 205*, 112474.
135. Xie, P., Tan, S., Bazmohammadi, N., Guerrero, J. M., & Vasquez, J. C. (2022). A real-time power management strategy for hybrid electrical ships under highly fluctuated propulsion loads. *IEEE Systems Journal, 17*(1).
136. Trovão, J. P. F., & Pereirinha, P. J. G. (2015). Control scheme for hybridised electric vehicles with an online power follower management strategy. *IET Electrical Systems in Transportation, 5*(1), 12–23.
137. Bassam, A. M., Phillips, A. B., Turnock, S. R., & Wilson, P. A. (2016). An improved energy management strategy for a hybrid fuel cell/battery passenger vessel. *International Journal of Hydrogen Energy, 41*(47), 22453–22464. https://doi.org/10.1016/j.ijhydene.2016.08.049
138. Tritschler, P. J., Bacha, S., Rullière, E., & Husson, G. (2010). Energy management strategies for an embedded fuel cell system on agricultural vehicles. In *19th International Conference on Electrical Machines, ICEM 2010*. IEEE.

139. Yang, J., Zhang, T., Hong, J., Zhang, H., Zhao, Q., & Meng, Z. (2021). Research on driving control strategy and Fuzzy logic optimization of a novel mechatronics-electro-hydraulic power coupling electric vehicle. *Energy, 233*, 121221.
140. Sorrentino, M., Rizzo, G., & Arsie, I. (2011). Analysis of a rule-based control strategy for on-board energy management of series hybrid vehicles. *Control Engineering Practice, 19*(12), 1433–1441.
141. Liu, H., Gegov, A., & Cocea, M. (2016). Rule-based systems: A granular computing perspective. *Granular Computing, 1*.
142. Liu, Y., Liu, J., Zhang, Y., Wu, Y., Chen, Z., & Ye, M. (2020). Rule learning based energy management strategy of fuel cell hybrid vehicles considering multi-objective optimization. *Energy, 207*, 118212.
143. Akbulut, O., Cavus, M., Cengiz, M., Allahham, A., Giaouris, D., & Forshaw, M. (2024). Hybrid intelligent control system for adaptive microgrid optimization: Integration of rule-based control and deep learning techniques. *Energies (Basel), 17*(10), 2260.
144. Fayyazi, M., et al. (2023). Artificial intelligence/machine learning in energy management systems, control, and optimization of hydrogen fuel cell vehicles. *Sustainability, 15*(6), 5249.
145. Lee, S., Seon, J., Hwang, B., Kim, S., Sun, Y., & Kim, J. (2024). Recent trends and issues of energy management systems using machine learning. *Energies (Basel), 17*(3), 624.
146. Wolpert, D. H., & Macready, W. G. (1997). No free lunch theorems for optimization. *IEEE Transactions on Evolutionary Computation, 1*(1), 67–82. https://doi.org/10.1109/4235.585893
147. Xie, P., Asgharian, H., Vasquez, J. C., Guerrero, J., Araya, S. S., & Liso, V. (2023). Day-ahead energy management for hybrid electric vessel with different PEM fuel cell modular configurations. In *2022 the 3rd international conference on power, energy and electrical engineering* (Vol. 9, pp. 99–106). https://doi.org/10.1016/j.egyr.2022.12.091
148. Xie, P., Asgharian, H., Guerrero, J. M., Vasquez, J. C., Araya, S. S., & Liso, V. (2024). A two-layer energy management system for a hybrid electrical passenger ship with multi-PEM fuel cell stack. *International Journal of Hydrogen Energy, 50*, 1005–1019. https://doi.org/10.1016/j.ijhydene.2023.09.297
149. Han, J., Charpentier, J.-F., & Tang, T. (2014). An energy management system of a fuel cell/battery hybrid boat. *Energies (Basel), 7*(5), 2799–2820.
150. Rawlings, J. B., Mayne, D. Q., & Diehl, M. (2017). *Model predictive control: Theory, computation, and design* (2nd ed.). Nob Hill Publishing.
151. Peng, X., Chen, H., & Guan, C. (2023). Energy management optimization of fuel cell hybrid ship based on particle swarm optimization algorithm. *Energies (Basel), 16*(3), 1373.
152. Banaei, M., Boudjadar, J., & Khooban, M.-H. (2021). Stochastic model predictive energy management in hybrid emission-free modern maritime vessels. *IEEE Transactions on Industrial Informatics, 17*(8), 5430–5440. https://doi.org/10.1109/TII.2020.3027808
153. Serrao, L., Onori, S., & Rizzoni, G. (2011). A comparative analysis of energy management strategies for hybrid electric vehicles. *Journal of Dynamic Systems, Measurement and Control, Transactions of the ASME, 133*(3).
154. Jung, W., & Chang, D. (2023). Deep reinforcement learning-based energy Management for Liquid Hydrogen-Fueled Hybrid Electric Ship Propulsion System. *Journal of Marine Science and Engineering, 11*(10), 2007. https://doi.org/10.3390/jmse11102007
155. Wu, P., Partridge, J., & Bucknall, R. (2020). Cost-effective reinforcement learning energy management for plug-in hybrid fuel cell and battery ships. *Applied Energy, 275*, 115258.
156. Gan, M., Hou, H., Wu, X., Liu, B., Yang, Y., & Xie, C. (2022). Machine learning algorithm selection for real-time energy management of hybrid energy ship. *Energy Reports, 8*, 1096–1102.
157. Geng, P., Xu, X., & Tarasiuk, T. (2020). State of charge estimation method for Lithium-ion batteries in all-electric ships based on LSTM neural network. *Polish Maritime Research, 27*(3), 100–108. https://doi.org/10.2478/pomr-2020-0051

158. Mazzi, Y., Ben Sassi, H., Errahimi, F., & Es-Sbai, N. (2024). A soft actor-critic reinforcement learning framework for optimal energy management in electric vehicles with hybrid storage. *Journal of Energy Storage, 99*, 113344.
159. Wen, S., Zhang, C., Lan, H., Xu, Y., Tang, Y., & Huang, Y. (2019). A hybrid ensemble model for interval prediction of solar power output in ship onboard power systems. *IEEE Trans Sustain Energy, 12*(1), 14–24.
160. Kalikatzarakis, M., Geertsma, R. D., Boonen, E. J., Visser, K., & Negenborn, R. R. (2018). Ship energy management for hybrid propulsion and power supply with shore charging. *Control Engineering Practice, 76*, 133–154.
161. Zhu, D., Pritchard, E., Dadam, S. R., Kumar, V., & Xu, Y. (2022). Optimization of rule-based energy management strategies for hybrid vehicles using dynamic programming. *Combustion Engines, 184*(1), 3–10.
162. Fares, D., Chedid, R., Panik, F., Karaki, S., & Jabr, R. (2015). Dynamic programming technique for optimizing fuel cell hybrid vehicles. *International Journal of Hydrogen Energy, 40*(24), 7777–7790.
163. Song, K., Wang, X., Li, F., Sorrentino, M., & Zheng, B. (2020). Pontryagin's minimum principle-based real-time energy management strategy for fuel cell hybrid electric vehicle considering both fuel economy and power source durability. *Energy, 205*, 118064.
164. Jafarzadeh, S., & Schjølberg, I. (2018). Operational profiles of ships in Norwegian waters: An activity-based approach to assess the benefits of hybrid and electric propulsion. *Transp Res D Transp Environ, 65*, 500–523.

Green Hydrogen Production Chain Onboard Maritime Transport: A Case Study on Hydrogen Production Through Electrolysis Powered by the Waste Heat Recovery

Doha Elrhoul, Manuel Naveiro, and Manuel Romero Gómez

Nomenclature

A	Cell area	$\dot{n}$	Molar flow
ACM	Aspen custom modeller	NO_x	Nitrogen oxides
AE	Alkaline electrolysis	O_2	Oxygen
Bg	Flow permeability	P	Pressure
BOG	Boil-off gas	PEC	Photoelectrochemical
BOR	Boil-off rate	PEME	Proton exchange membrane electrolysis
CO_2	Carbon dioxide	P_i	Partial pressure
D^{eff}_{H2O}	Effective diffusion coefficient	PV	Photovoltaic
D_{H2-H2O}	Molecular diffusion	Q	Heat
D_{H2O-K}	Knudsen diffusion	R	Gas constant
d_i	Electrode thickness	Rad	Average pore radius
E	Rate of electric exergy input	SA	Scavenge air
E_{act}	Energy activation	SOEC	Solid oxide electrolysis
EG	Exhaust gas	SO_x	Sulphur oxides
E_{H2}	Standard chemical exergy of H_2	SRC	Steam Rankine cycle
E_{heat}	Rate of thermal exergy input	T	Temperature

(continued)

D. Elrhoul (✉)
University of A Coruna, Ferrol Industrial Campus, CITENI, Energy Engineering Research Group, University Institute of Maritime Studies, ETSNM, A Coruña, Spain
e-mail: doha.elrhoul@udc.es

M. Naveiro · M. Romero Gómez
Energy Engineering Research Group, University Institute of Maritime Studies, Nautical Sciences and Marine Engineering Department, ETSNM, University of A Coruña, A Coruña, Spain
e-mail: manuel.naveiro@udc.es; m.romero.gomez@udc.es

C. O. Colpan et al. (eds.), *Fuel Cell and Hydrogen Technologies in Maritime Transportation*, https://doi.org/10.1007/978-3-031-86110-9_7

A	Cell area	$\dot{n}$	Molar flow
F	Faraday constant	T_0	Environment reference temperature
FC	Fuel cell	V	Stack voltage
gH_2	Green H_2	V_0	Standard potential
GHG	Greenhouse gases	V_{act}	Activation overpotential
H_2	Hydrogen	V_{cell}	Cell voltage
H_2O	Water	V_{conc}	Concentration overpotential
HFO	Heavy fuel oil	V_{ohm}	Ohmic overpotential
HHV	Higher heating value	V_{tn}	Thermoneutral voltage
HTC	Thermochemical cycles	W	Work
I	Current	WHR	Waste heat recovery
ICE	Internal combustion engine	WI	Waste incinerator
IMO	International maritime organization	y_i	Molar fraction
J	Current density	ΔG	Gibbs energy
JW	Jacket water	ΔH	Energy consumption
L	Layer thickness	ΔQ	Thermal energy
LCA	Life cycle assessment	ε	Electrode porosity
LH_2	Liquefied H_2	η	Efficiency
LHV	Lower heating value	μ	Dynamic viscosity
LNG	Liquefied natural gas	ξ	Electrode tortuosity
$\dot{m}$	Mass flow	σ	Collision diameter
M_j	Molecular mass	Ω_D	Collision integral
N	Number of cells		

1 Introduction

Maritime transport is vital to the global economy due to its reliability, efficiency, flexibility and affordability over land and air transport [1, 2]. However, the main drawback of this means of transportation is its emissions, for instance, sulphur oxide (SO_x) emissions have already reached 13% in 2012 and nitrogen oxides (NO_x) 15% in the same year, while carbon dioxide (CO_2) registered 2.9% in 2018 and it is expected to rise to between 12% and 18% by 2050 [3–8]. The International Maritime Organization (IMO) is forcing the maritime industry to reduce and cut off their emissions part by applying several measures and restrictions regarding ships' energy efficiency and emissions reduction [2, 9, 10]. The maritime industry is adopting new technologies and efficient procedures to enhance its energy systems in response to the IMO regulations. However, the transition to carbon-free fuels is necessary to meet the emissions reduction objectives on time, as current activities seem insufficient and inefficient [11]. Among the existing marine fuels, liquefied natural gas (LNG) and heavy fuel oil (HFO) are the most commonly used, thanks to their maturity and prices [12], with LNG being the cleanest fuel compared to HFO

[13]. Solar and wind energy are also considered as energy source onboard ships dealing with greenhouse gas emissions (GHG); however, they are limited by their availability and ship's structure making it difficult to rely only on these renewables without additional energy support [6, 11, 14–16]. Therefore, a cost-effective hydrogen economy is required to achieve an emission-free maritime transport [17].

According to life cycle assessment studies (LCA), tackling environmental issues using hydrogen (H_2) energy seems to have a promising future [16, 18–21]. H_2 is distinguished by its properties such as being the simplest and lightest element, switching between low and high pressures, high energy density, and among the most abundant chemical elements on the Earth and in the universe [22–25].

Dutta [26] has reviewed the production chain of H_2 as an energy vector. In addition, the author has discussed the safety issues when dealing with H_2 in automobiles and refilling stations. Dutta concluded that creating and increasing public awareness and safety issues programs might be beneficial to accept H_2 as a fuel. Tashie-Lewis and Nnabuife [27] have conducted a technology review of the H_2 production chain in an H_2 economy, focusing mainly on transportation and power generation sectors. They concluded that the main barrier to adopting H_2 is the price competitiveness with fossil fuels that are still very cheap. Dincer and Acar [28] have focused their review on evaluating the H_2 production methods from both renewable and non-renewable sources for better sustainability. They concluded that although photonic-based H_2 production methods have the lowest energy and exergy efficiencies, their global warming potential (GWP) and acidification potential (AP) are negligible (almost zero). Among the 19 H_2 production methods, the cost of H_2 produced from fossil fuels is still the cheapest followed by coal gasification and plasma arc decomposition. In addition, H_2 production from electrical-based methods is cheap and more efficient. Abdalla et al. [29] reviewed the H_2 production, storage and transportation highlighting the key challenges to facilitate the adoption of H_2 as a clean energy alternative. Abdalla et al. concluded that H_2 is still mainly produced from non-renewable sources; hence, an increase in carbon dioxide (CO_2) emission. The H_2 storage and use are bounded by the tanks' volume, expensiveness and efficiency. Hoecke et al. [30] have discussed H_2 production and storage in the maritime sector focusing mainly on the storage methods that are still the bottleneck of the H_2 production chain. They concluded that so far, there has been no better storage method as each of them has its advantages and disadvantages. Osman et al. [21] have reviewed the H_2 production chain with a main focus on their environmental impact. Inal et al. [31] have reviewed some of the H_2 production methods towards decarbonization of the maritime sector but they focused more on the storage methods. Ishaq et al. [32] have focused their review on comparing the H_2 production methods as well as the main challenges facing H_2 adoption as an alternative fuel.

This chapter reviews the different green hydrogen (gH_2) production chain methods to be adopted onboard maritime transport. Section 2 presents the available onboard energy sources (2.1) as well as the possible H_2 production methods (2.2). Section 3 summarizes the storage methods. Section 4 is dedicated to the H_2 use onboard ships, such as in internal combustion engines (ICEs) or fuel cells (FCs). Finally, a case study of the H_2 production through solid oxide electrolysis (SOEC) powered by a steam Rankine cycle (SRC) is presented in Sect. 5 from stack modelling in Aspen Custom Modeler (ACM) to the H_2 production simulation in Aspen Hysys.

2 Green Hydrogen Production Methods Onboard Ships

Research studies such as [17, 22] have already reviewed the green H_2 (gH_2) production methods; hence, this section will outline only the possible applicable methods on board maritime transport together with the available onboard energy sources as summarized in Fig. 1.

2.1 Available Onboard Energy Sources

The available energy sources for the onboard H_2 production are divided into renewables (solar, wind energy) and waste heat recovery energy (WHR).

2.1.1 Solar Energy

Solar energy is one of the most environmentally friendly renewable sources to be adopted onboard ships [6]. According to Pan et al. [34], photovoltaic solar systems (PV) are used to generate electric energy onboard ships using batteries as an excess energy storage medium. The application of this energy generation technology is limited due to the low power generation (few kW). Therefore, PV is used only as a power auxiliary source or for small and medium-scale ships [34]. The use of solar

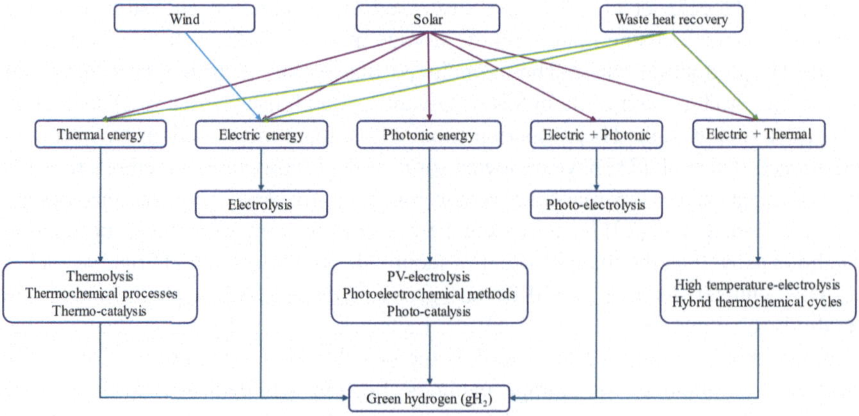

Fig. 1 H_2 production methods and their possible energy sources [33]

energy onboard ships is bounded by both solar irradiation and onboard space availability [15].

2.1.2 Wind Energy

According to Refs [34, 35], a possible electric generation onboard ships is through offshore wind systems. There are two main methods namely: wind-assisted generation (WASP) distinguished by connected sails to a computer for power assistance and control, and the traditional wind power generators through vertical (VAWT) and horizontal (HAWT) turbines [36, 37].

2.1.3 Waste Heat Recovery Energy

A marine propulsion engine wastes more than 50% of its energy through its different streams without any use [38–40]. As illustrated in Fig. 2, heat is wasted mainly from the exhaust gas stream (25.5%) followed by the air cooler (16.5%), the jacket water stream (5.2%), and a few amount through the lubricating oil (2.9%) and heat dissipated by radiation (0.6%) [40, 42]. Recovering the waste heat and use it for power generation improves the ships' efficiency while reducing costs and emissions without affecting the engine reliability [38, 39, 43]. Table 1 presents the main waste

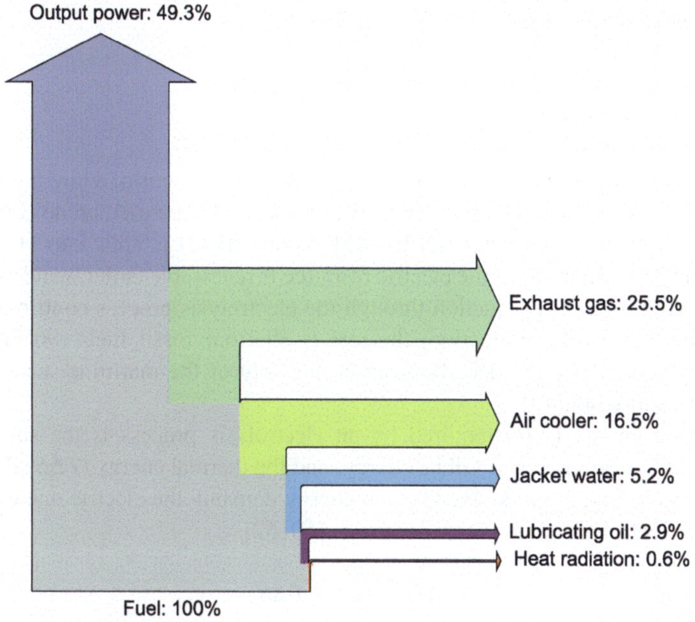

Fig. 2 Engine's waste heat sources [41]

Table 1 WH sources' temperature ranges from marine engines [33, 40, 42]

WH source	Temperature range (°C)
Engine cooling water (JW)	70–125
Scavenge air (SA)	100–160
Exhaust gas (EG)	200–500

Table 2 WHR application onboard ships [33, 39, 44]

WHR system application	Temperature range (°C)
Dehumidification	80–85
Absorption and adsorption refrigeration	120–140
Organic Rankine cycle	65
Steam Rankine cycle	530

heat sources of a marine engine with their temperature ranges varying from low to high. As summarized in Table 2, the main applications of the WHR sources are, depending on their temperature, dehumidification systems, refrigeration, as well as organic and steam Rankine cycles.

2.2 *Production Methods*

The energy used for the H_2 production onboard ships is divided into single (electric, thermal or photonic) and hybrid (a combination of electrical + thermal or electric + photonic) extraction forms.

2.2.1 Electric Energy

The electric energy is used to drive the water-splitting reactions of a low-temperature electrolysis process, namely alkaline (AE), proton exchange membrane (PEME) or direct electrolysis of seawater (DES) [45]. Water (H_2O) is split into H_2 and O_2 without any emissions since the electric resource is renewable (solar, wind or WHR) [46–48]. Although H_2 production through the electrolysis process contributes only 4% of the worldwide production, the rest is all from fossil fuels (96%), it is a promising process for the decarbonization not only of the maritime sector but all the other sectors [48, 49].

The total energy (ΔH) required by an electrolysis process is the sum of the electric (ΔG), also known as Gibbs energy, and the thermal energy ($T\Delta S$) demands. As depicted in Fig. 3, unlike the thermal energy demand, the electric one decreases when the electrolysis temperature increases [50–52].

$$\Delta H = \Delta G + T\Delta S \quad (1)$$

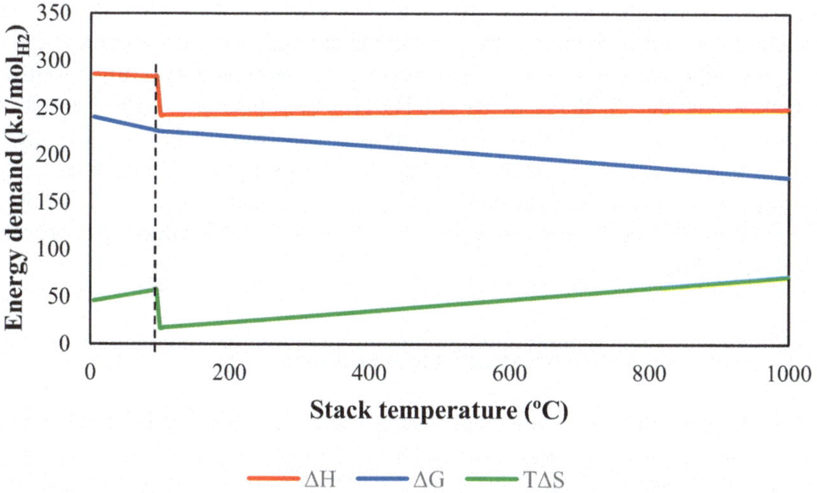

Fig. 3 Variation of total energy demand ΔH, electric ΔG and thermal energy $T\Delta S$ as a function of an ideal electrolysis stack temperature

2.2.2 Thermal Energy

H_2 can be produced through thermolysis, thermo-catalysis and thermochemical splitting processes using thermal energy. Thermolysis requires very high temperatures exceeding 2500 K to conduct the water-splitting reactions; hence, steam is dissociated without any additional reaction [17, 53–55]. H_2 and O_2 are effectively separated through a palladium membrane after being cooled down to temperatures between 1500 K and 2000 K, this latter occurs in a few milliseconds [56, 57]. The high H_2 production cost, materials' toxicity and corrosiveness are the main disadvantages of this process [48, 57]. Producing H_2 through thermo-catalysis has no application yet and the process requires high temperatures that could be supplied by heat recovery, or concentrated solar heat [17]. Thermochemical processes are divided into two categories; hydrogen sulphide (H_2S) is split into H_2 and sulphur (S) under high temperatures supplied by the same means of energy of thermo-catalysis. The second category is water splitting into H_2 and O_2 under closed chemical cycles [17, 26, 57].

2.2.3 Photonic Energy

There exist three different processes taking advantage of the photonic energy to produce H_2, namely PV-electrolysis, photoelectrochemical methods and photocatalysis. The PV-electrolysis consists of supplying an H_2 production electrolysis process with energy generated through PV panels. Although this process has a higher energy efficiency among other photonic-based H_2 production processes, it is still less efficient than others and 20 times more expensive than fossil fuels [17, 28, 57]. The photoelectrochemical (PEC) methods are still under development due to efficiency improvement. PECs are distinguished by their compact configuration and the ability to produce both electricity and H_2. However, the same as PV-electrolysis, producing H_2 from fossil fuels is still cheaper than from PEC [28]. The photocatalysis process is less expensive than PV-electrolysis and PEC, in addition, it has the ability to convert directly the photons to H_2 without electricity [53, 57]. However, this process is limited by the recombination of exited electrons affecting the process performance [46].

2.2.4 Extraction of Electrical and Thermal Energy

The combination of both thermal and electric energy results in the production of H_2 through high-temperature electrolysis such as SOEC or via hybrid thermochemical cycles (HTCs). SOEC is characterized by its high energy efficiency compared to other electrolysis types (AE and PEME), thanks to their high operating temperatures (600–1000 °C) [45, 48, 58]. It is a promising technology for onboard H_2 production if high temperature sources are available [45, 58, 59]. As depicted in Fig. 3, in the case of high temperature electrolysis, the thermal energy substitutes the electric energy during the water-splitting reactions, hence, minimizing the H_2 production cost [50, 51, 60]. HTCs also operate at high temperatures of no more than 1000 °C leading to a high energy efficiency [28]. They are considered to be cheaper than pure thermochemical cycles and fossil fuels [28, 61].

2.2.5 Extraction of Electrical and Photonic Energy

H_2 is produced through photo-electrolysis using both photonic (a semiconductor) and electric (water electrolysis) energy [17, 29]. This process is the least efficient and the most expensive among other H_2 production processes [29, 48].

3 Green Hydrogen Storage Methods Onboard Ships

The storage of H_2 is challenging due to its properties such as flammability risk when mixed with other substances and low volumetric density requiring huge tanks and rising its cost [26, 29, 62–64]. H_2 is stored in a gaseous, solid or liquid state [65].

3.1 Gaseous Storage

Storing H_2 as a gas requires passing through a compression stage due to its low volumetric density [23]. Compressed H_2 is a simple and mature technology widely used in many applications such as in fuel-cell vehicles and refuelling stations [29, 53, 66–68]. The compression pressure is a key parameter for the H_2 storage [69]. There are mainly five types (I, II, III, IV and V) of H_2 storage tanks distinguished by their pressure resistance and weight [29, 65, 70]. Types III and IV have high pressure resistance (350–700 bar) making them one of the best options to be adopted onboard [23, 71]. Type I is 70% heavier than type IV; however, the latter is more expensive [53, 65]. Type V is characterized by its high pressure (estimated to be>700 bar) and its full composite composition [30].

3.2 Liquid Storage

Unlike gaseous storage, liquefied H_2 (LH_2) storage has the advantage of being more efficient, lighter, has a higher energy storage, enabling the H_2 to reach a high purity level as a liquid, and low storage price [29, 65, 68, 71]. H_2 liquefaction is a similar process to that of NG with different critical refrigeration temperatures (−253 °C and − 160 °C, respectively), higher insulation, and more safety precautions [66]. The main drawbacks of the LH_2 storage are its requirement of expensive super-insulated tanks tolerating the low temperatures, and short shelf-life [23, 29, 53, 72]. Usually, the cost of LH_2 storage is 30% higher than the gaseous storage due to the additional liquefication stage; however, this can be reduced in case of storing huge H_2 quantities [71].

A combination of liquefied (−253 °C) and compressed (>300 bar) H_2 results in a cryo-compressed H_2 reducing the energy losses due to liquefied H_2 and increasing the system's storage density and safety [29, 68, 73, 74].

3.3 Solid Storage

H_2 could be stored in metal hydrides (MHs), chemical hydrides (CHs), metal-organic frameworks (MOFs), glasses or zeolites, among others. Heating MHs or reducing their external pressure result in H_2 release and storage at a reasonable pressure and temperature avoiding hazards [65, 69, 75]. This technology is cheap compared to LH_2 and decreases the storage system's volume making it a good H_2 fuel carrier [76]. Using CHs for H_2 production is impractical as they require high energy activation and temperature [67, 72]. Compared to other materials, MOFs have the ability to provide larger H_2 storage areas and are characterized by a high porosity making them an attractive H_2 storage technology [29, 53, 67]. Both zeolites and glasses are used for H_2 storage under temperatures from 200 °C to 400 °C [69, 77]. These technologies require large storage areas, have low-density storage making them expensive and impractical onboard ships [69, 77].

Ammonia is used as a medium for H_2 storage, thanks to its higher energy density, safety, easy long-distance transportation, cheap price and being the only carbon-free H_2 carrier [12, 66, 71, 76, 78]. However, extra infrastructure is needed for the onboard application [12]. Table 3 presents some of the storage applications in the maritime sector.

Table 3 Hydrogen storage applications in ships (reproduced from [33, 68])

Category	Name	H_2 storage method
Yachts and sport boats	Cobalt 233 Zet	3 H_2 containers of 350 bar
	Yacht no.1[1]	6 kg H_2 tanks at 300 bar
	Riviera 600	High-pressure containers
	Energy observer[2]	8 tanks of carbon fibre and aluminium (62 kg of H_2)
	Yacht XV 1	CH_2 or MH with light composite containers
Pusher	RiverCell-Elektra	Six containers of 500 bar holding 740 kg CH_2
Underwater vehicles (AUVs)	URASHIMA	MH
	URASHIMA second generation	
	IDEF	Pressurized H_2 gas bottle of 300 bar and 50 L
Fishing vessels	N/A	One LH_2 container of 450 L
Passenger vessels	Xperience NX	Four exchangeable H_2 gas containers with 200 bar 30 L H_2 gas in each
	Elding	High-pressure (350 bar) containers
	Hydroville	12 H_2 tanks of 205 L at 200 bar
	Hydra	MH
	Hydroxy 3000	Pressurized H_2 gas bottle of 200 bar and 76 L
	Ross Barlow	Fixed H_2 containers and solid H_2 store
	Tuckerboot	Exchangeable H_2 pressure containers
	Nemo H_2	40 kg H_2 in 8 cylinders at 350 bar
	ALSTERWASSER	12 containers of 350 bar 50 kg H_2
	SF-breeze	1200 kg LH_2 container
Submarines	Class 212 A and 214	Lightweight outer hull and pressure hull

[1] The first yacht powered with H_2 (2003)
[2] The first emission-free ship with H_2 auto-production from seawater (2017)

Table 4 European and international H_2_related standards [33, 79]

Standard/Code	Brief description	H_2 storage method
ISO/TR 15916	– Risks and safety guidelines related to H_2 use (temperature, materials, H_2 embrittlement, etc.)	Gaseous Liquid
PED. EN 1252–1:1998	– Materials used for storage containers – Irrelevant to cryogenic tanks for LNG neither to cast materials and unalloyed materials	Compressed gas
EN 1797:2001	– Compatibility between the gas and material used	
EN 13648 part 1, 2 and 3	– Extreme pressure safety and protection	
ISO 15399 Gaseous Hydrogen—Cylinders and tubes for stationary storage	– Covers volumes of 10,000 l and pressures of 110 MPa	
EIGA code of practice IGC 15/06	– Tanks filling – H_2 compression – End-storage installation – H_2 purification	
IGC and IGF codes	– LH_2 storage on board vessels	Liquid

3.4 International and European H_2 Storage Standards

Many standards and codes, summarized in Table 4, are used to manage the H_2 storage including safety measures and material use.

4 Green Hydrogen Consumption Methods Onboard Ships

H_2 is used to power the propulsion system of ships such as in fuel cells (FCs), burned in internal combustion engines (ICEs) or in gas turbines (GTs) [30, 80, 81].

4.1 Fuel Cells Systems

H_2 is used as the main fuel of FCs, similar to electrolysis, FCs vary between low, medium and high temperature technologies [30, 80, 82]. Low temperature (50 to 150 °C) FCs include alkaline (AFC) and proton exchange membrane (PEMFC), phosphoric acid (PAFC) are considered as medium temperature (around 200 °C) FCs, while high temperature (up to 1000 °C) comprise direct carbon (DCFC), molten carbonate (MCFC) and solid oxide (SOFC) FCs [33, 80, 82]. Table 5 presents some

Table 5 FC applications for ships [33, 79, 82–86]

Project	FC-type	Vessel-type	FC power rating	R&D partners	Fuel
FCSHIP	MCFC SOFC PEMFC	Passenger ferry	Up to 1 MW	DNV-GL	Hydrocarbons Hydrocarbons Hydrogen
Hydra	AFC	NA	6.9 kW	NA	Metal hydride
Hydrocell Oy			30 kW		
Zemship Alsterwasser	PEMFC	Passenger ferry	96 kW	Proton motor	Hydrogen
Nemo H_2		Passenger ferry	60 kW	Alewijnse	
MARANDA		Research vessel	Two 82.5 kW (165 kW)	ABB and VTT	
Raicho project for next-generation water transportation		Research vessel	30 kW	Tokyo University of Marine Science and Technology and Toshiba	
Hybrid power lab at NTNU		Laboratory Scale demo	Two 30 kW (60 kW)	ABB, SINTEF Ocean, NTNU	
Elding		NA	10 kW	NA	
Hornblower hybrid			32 kW		
Hydrogenesis			12 kW		
SF-BREEZE			120 kW		
Cobalt 233 Zet			50 kW		
MF Vågen	HT-PEMFC		12 kW		
RiverCell ELEKTRA			3 X 100 kW		

FC applications in the maritime sector. FCs are a mature technology for the road sector; however, their output power (few MWs) is limiting their application only for small ships, while for large ships they can only be used as auxiliary systems [80, 87, 88].

4.2 *Combustion Propulsion System*

Most ships rely on ICEs as their propulsion system [87, 89]. However, they are powered mainly by fossil fuels resulting in GHG emissions; hence, one solution is replacing fossil fuels with clean and green fuels such as H_2 [80, 89, 90]. Feeding ICEs with H_2 does not require significant modifications of the propulsion system and

ICEs are able to produce directly mechanical power [91]. Therefore, less adaptation cost compared with FCs, for instance which need additional equipment [80, 91]. GTs are mainly used as auxiliary power systems for cruise liners operating under Brayton cycles [80]. It is recommended to blend H_2 with NG to fully take advantage of the H_2 usage in both GTs and ICEs, this is mainly because of the H_2 combustion properties [80].

5 Case Study: Green H_2 Production Onboard a Ship Through a SOEC Powered by an SRC

As stated earlier, more than 50% of the energy produced by the ship's engine is wasted ineffectively. Recovering this later through a WHR system for the onboard H_2 production for its further use is advantageous both economically and efficiently. Producing H_2 onboard ships through electrolysis is a promising alternative to the one derived from fossil fuel methods (e.g. steam methane reforming). Software programs such as AspenTech tools (viz. ACM, Aspen Plus and Aspen Hysys) are used for the simulation of the H_2 production plant. For instance, authors such as Mónica et al. [92] have modelled an AE stack for H_2 production using ACM. They have suggested a model according to experimental data. The examples file of the AspenTech V12.1 [93] includes a modified model of the one suggested by [92] adding potassium hydroxide (KOH) as a component in the simulation process rather than just estimating it with water input, and the consideration of the cooling loop. Zaccara et al. [94] compared the H_2 production through PEME, SOEC and biomass processes using Aspen Plus. Im-orb et al. [95] developed a SOEC model in Aspen Plus for the H_2 production investigating the effect of some key parameters on the electrolysis performances, mainly the temperature and current density, among others. Compared to Aspen Plus, Aspen Hysys is more user-friendly, especially for beginners; therefore, this section serves as a guide for the H_2 production through a SOEC stack onboard maritime transport using both ACM and Aspen Hysys. First, the required electrochemical equations are determined according to the different existing research studies such as [36, 95–103] then used for the SOEC stack modelling in ACM. The created stack model is exported to Aspen Hysys for results validation, then for H_2 production simulation using the power recovered and supplied by an SRC.

5.1 *Electrochemical Modelling Equations*

Both electric energy and heat are needed to split H_2O into H_2 and O_2 in this electrolysis process.

The electric energy (W) is determined as the product of the stack voltage (V) by the stack current (I) as follows:

$$W = V \cdot I \tag{2}$$

$$V = V_{\text{cell}} \cdot N \tag{3}$$

$$I = J \cdot A \tag{4}$$

where V_{cell} is the cell voltage, N is the number of stack cells, J is the current density and A is the cell area.

Additional heat ($\dot{Q}$) is needed in the process of water splitting through the high-temperature water electrolysis, calculated by the following equations:

$$\dot{Q} = (V_{\text{cell}} - V_{\text{tn}}) \cdot I \cdot N \tag{5}$$

$$V_{\text{tn}} = \frac{\Delta H}{2F} \tag{6}$$

where V_{tn} is the thermoneutral voltage calculated as a function of the stack energy consumption (ΔH), and F is the Faraday constant.

According to Eq. 5, when $V_{\text{tn}} > V_{\text{cell}}$, the process is called endothermic which means it requires additional heat to achieve its energy needs. While when $V_{\text{tn}} < V_{\text{cell}}$ the process is exothermic generating extra heat to the surroundings. In the case when $V_{\text{tn}} = V_{\text{cell}}$, the electrolysis process does not exchange any heat with the surroundings (adiabatic). For accurate results, the V_{tn} equation is correlated to a third-degree polynomial equation as a function of the stack's temperature using the Engineering Equation solver (EES) as shown in Fig. 4 and Eq. 7:

$$V_{\text{tn}} = 1.25028255 + 6.23391235 \cdot 10^{-5} \cdot T - 2.32482542 \cdot 10^{-8} \cdot T^2 + 2.53527575 \cdot 10^{-12} \cdot T^3 \tag{7}$$

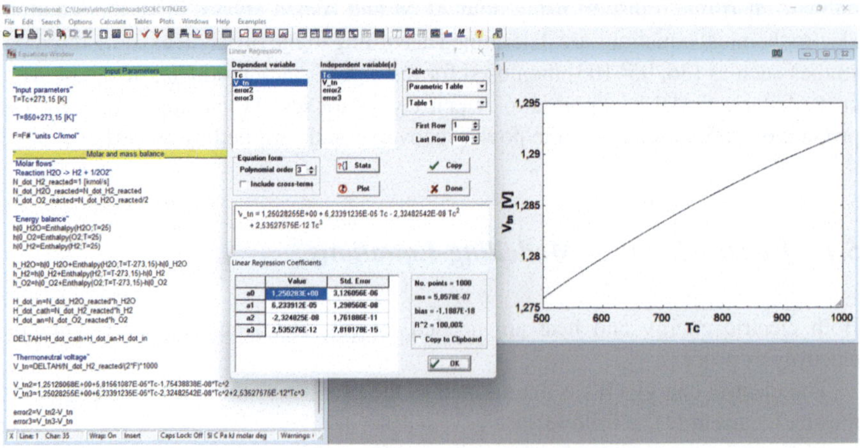

Fig. 4 V_{tn} correlation as a function of the stack temperature using EES

The energy and exergy efficiencies of the stack are determined by the following equations:

$$\eta_{\mathrm{LHV}} = 100 \cdot \frac{\dot{m}_{H_{2\mathrm{pro}}} \cdot LHV_{H_2}}{W + Q_{h-\mathrm{in}}} \tag{8}$$

$$Q_{h-\mathrm{in}} = |Q| + Q_{\mathrm{losses}} + Q_{\mathrm{heat}} \tag{9}$$

$$Q_{\mathrm{losses}} = 10\% \cdot |Q| \tag{10}$$

$$\eta_{\mathrm{exergy}} = \frac{\dot{n}_{H_{2\mathrm{pro}}} \cdot E_{H_2}}{E + E_{\mathrm{heat}}} \tag{11}$$

$$E = W \tag{12}$$

$$E_{\mathrm{heat}} = Q_{h-in}\left(1 - \frac{T_0}{T}\right) \tag{13}$$

where $LHVH2$ and $\mathrm{HHV}_{\mathrm{H2}}$ are the lower and higher heating values of H_2, $\dot{m}_{H_{2\mathrm{pro}}}$ and $\dot{n}_{H_{2\mathrm{prod}}}$ are the mass and molar flow rates of the H_2 produced during the electrolysis process, respectively. Q_{losses} are the SOEC heat losses estimated as 10% of the stack thermal heat, $Q_{\mathrm{h\text{-}in}}$ is the total heat needed by the SOEC during the H_2O splitting reactions, while Q_{heat} is any extra heat added to the SOEC. EH_2 is the standard chemical exergy of H_2, E and E_{heat} are the rates of the electric and thermal exergy inputs. While T_0 and T are the reference environment and stack temperatures.

The **overall cell voltage (V_{cell})** is calculated as the sum of the irreversible voltage (V_{rev}), ohmic overpotentials (V_{ohmic}), activation overpotentials (V_{act}) and the concentration overpotentials (V_{conc}):

$$V_{\mathrm{cell}} = V_{\mathrm{rev}} + V_{\mathrm{ohmic}} + V_{\mathrm{act}} + V_{\mathrm{conc}} \tag{14}$$

$$V_{\mathrm{rev}} = V_0 + \frac{RT}{2F} \ln\left(\frac{P^0_{H2}\left(P^0_{O2}\right)^{1/2}}{P^0_{H2O}}\right) \tag{15}$$

$$V_0 = 1.253 - 2.4516 \cdot 10^{-4} T \tag{16}$$

$$P_j^0 = y_j \cdot P, \qquad (j = H_2O, H_2, O_2) \tag{17}$$

where $T(K)$ is the stack temperature, V_0 is the standard potential, P_{i}^0 are the partial pressures of species j and y_{i} are their corresponding input molar fractions.

The ohmic overpotential (V_{ohm}) is calculated by Eq. 18 as a function of the electrolyte's layer thickness (L), the stack's temperature (T) and the current density (J):

$$V_{\mathrm{ohm}} = 2.99 \cdot 10^{-5} \cdot \exp\left(\frac{10300}{T}\right) \cdot J \cdot L \tag{18}$$

The total activation overpotential (V_{act}) of the cathode and anode electrodes is calculated as outlined in the following equation:

$$V_{act} = V_{act,a} + V_{act,c} \tag{19}$$

$$V_{act,i} = \frac{RT}{F} \cdot \ln\left[\frac{J}{2J_{0,i}} + \sqrt{\left(\frac{J}{2J_{0,i}}\right)^2 + 1}\right], \qquad (i = a, c) \tag{20}$$

$$J_{0,i} = J_{exp,i} \cdot exp\left(\frac{-E_{act,i}}{RT}\right) \tag{21}$$

where $E_{act,i}$ is the activation energy, $J_{exp,i}$ is the pre-exponential factor and $J_{0,i}$ is the exchange current density at anode and cathode electrodes.

The total concentration overpotential (V_{conc}) combining the cathode and anode electrodes is determined as detailed in the following equations:

$$V_{conc} = V_{conc,a} + V_{conc,c} \tag{22}$$

$$V_{conc,a} = \frac{RT}{4F} \cdot \ln\left(\frac{\sqrt{\left(P^0_{O_2}\right)^2 + \frac{R \cdot T \cdot J \cdot \mu \cdot d_a}{2F \cdot B_g}}}{P^0_{O2}}\right) \tag{23}$$

$$V_{conc,c} = \frac{RT}{2F} \cdot \ln\left(\frac{1 + \frac{J \cdot R \cdot T \cdot d_c}{2F \cdot D^{eff}_{H_2O} \cdot P^0_{H_2}}}{1 - \frac{J \cdot R \cdot T \cdot d_c}{2F \cdot D^{eff}_{H_2O} \cdot P^0_{H_2O}}}\right) \tag{24}$$

$$B_g = \frac{\varepsilon^3}{72 \cdot \xi(1-\varepsilon)^2} \cdot (2\text{rad})^2 \tag{25}$$

$$\begin{aligned}\mu = & -1.692 + 889.75\left(\frac{T}{1000}\right) - 892.79\left(\frac{T}{1000}\right)^2 + 905.98\left(\frac{T}{1000}\right)^3 \\ & - 598.36\left(\frac{T}{1000}\right)^4 + 221.64\left(\frac{T}{1000}\right)^5 - 34.75\left(\frac{T}{1000}\right)^6\end{aligned} \tag{26}$$

$$\frac{1}{D^{eff}_{H_2O}} = \frac{\xi}{\varepsilon}\left(\frac{1}{D_{H_2-H_2O}} + \frac{1}{D_{H_2O-k}}\right) \tag{27}$$

$$D_{H_2O-k} = \frac{2}{3}\text{rad}\sqrt{\frac{8R \cdot T}{\pi M_{H_2O}}} \tag{28}$$

$$D_{H_2-H_2O} = 0.00133\left(\frac{1}{M_{H_2}} + \frac{1}{M_{H_2O}}\right)^{1/2} \frac{T^{3/2}}{P \cdot (\sigma_{H_2-H_2O})^2 \cdot \Omega_D} \tag{29}$$

$$\Omega_D = \frac{1.06036}{T^{*0.1561}} + \frac{0.193}{\exp(0.47635T^*)} + \frac{1.03587}{\exp(1.52996T^*)} + \frac{1.76474}{\exp(3.89411T^*)} \tag{30}$$

$$T^* = \frac{T}{\frac{\varepsilon_{H_2-H_2O}}{k}} \tag{31}$$

$$\frac{\varepsilon_{H_2-H_2O}}{k} = \sqrt{\frac{\varepsilon_{H_2}}{k} \cdot \frac{\varepsilon_{H_2O}}{k}} \tag{32}$$

$$\sigma_{H_2-H_2O} = \frac{\sigma_{H_2} + \sigma_{H_2O}}{2} \tag{33}$$

where d_i is the electrode thickness, Bg, μ, ξ, ε and rad are the flow permeability, dynamic viscosity, electrode tortuosity, electrode porosity and the average pore radius, in this order. Note that the partial pressures P_i^0 are in Pascal. $D^{eff}{}_{H2O}$, $D_{H2O\text{-}K}$ and $D_{H2\text{-}H2O}$ are the effective diffusion coefficient, Knudsen and the molecular diffusions. σ_j and M_j are the collision diameter and molecular mass of species j ($j = H_2O$, H_2). Ω_D and T^* are the dimensionless diffusion collision integral and temperature.

The molar flow rates of the reactant (H_2O) and products (H_2, O_2) of the water-splitting process, considering the reactions that occurred at both the anode (a) and cathode (c) electrodes, are defined by the following equations:

$$\dot{n}_{H_2\text{pro}} = \frac{1}{2F} \cdot I \cdot N \tag{34}$$

$$\dot{n}_{O_2\text{pro}} = \frac{1}{2} \cdot \dot{n}_{H_2\text{pro}} \tag{35}$$

$$\dot{n}_{H_2O_\text{react}} = \dot{n}_{H_2\text{pro}} \tag{36}$$

$$\dot{n}_{O_{2a}} = \dot{n}_{O_2\text{pro}} \tag{37}$$

$$\dot{n}_{H_{2c}} = \dot{n}_{H_2\text{pro}} \tag{38}$$

$$\dot{n}_{H_2O_c} = \dot{n}_{H_2O_\text{in}} - \dot{n}_{N_{H_2O_\text{react}}} \tag{39}$$

Table 6 summarizes the parameters adopted in this study for stack modelling.

5.2 SOEC Stack Modelling and Simulation

First, the SOEC stack is modelled in the ACM using the aforementioned equations. Then, after compiling and running the model without any warning or errors, it is exported to the Aspen Hysys for simulations.

Table 6 SOEC electrochemical modelling parameters [96, 104]

Parameter	Value
L	$12.5 \cdot 10^{-6}$ m
da	$17.5 \cdot 10^{-6}$ m
dc	$12.5 \cdot 10^{-6}$ m
ε	0.48
ξ	5.4
rad	$1.385 \cdot 10^{-6}$ m
ε_{H2O}/k	809.1 K
ε_{H2}/k	59.7 K
σ_{H2O}	2.641 m
σ_{H2}	2.827 m
$J_{exp,a}$	$2.051 \cdot 10^{9}$ A/m²
$J_{exp,c}$	$1.344 \cdot 10^{10}$ A/m²
$E_{act,a}$	$1.2 \cdot 10^{5}$ J/mol
$E_{act,c}$	$1 \cdot 10^{5}$ J/mol

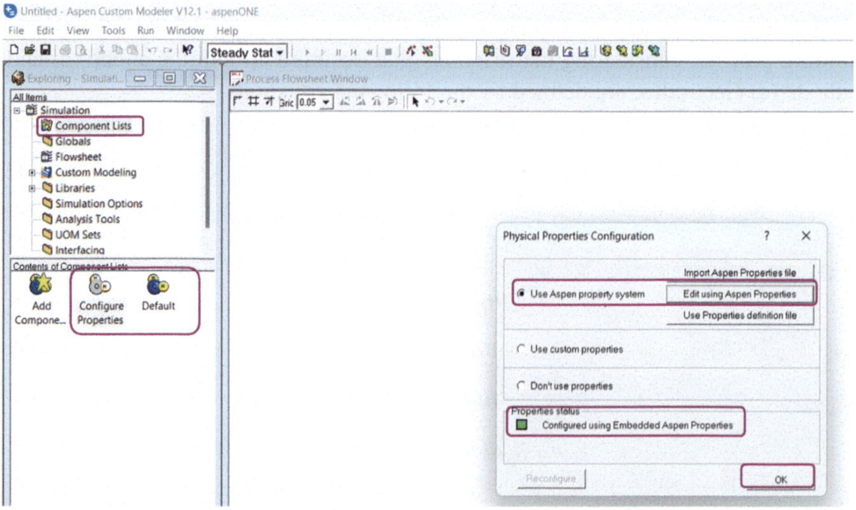

Fig. 5 Component list creation in ACM

5.2.1 Stack Modelling in ACM

The electrolysis stack is modelled according to the following steps:

(a) Open ACM and create the component list, in this example: water (H_2O, hydrogen (H_2) and oxygen (O_2),

- Click on the **component list** in the simulation part of the exploring panel as shown in Fig. 5.

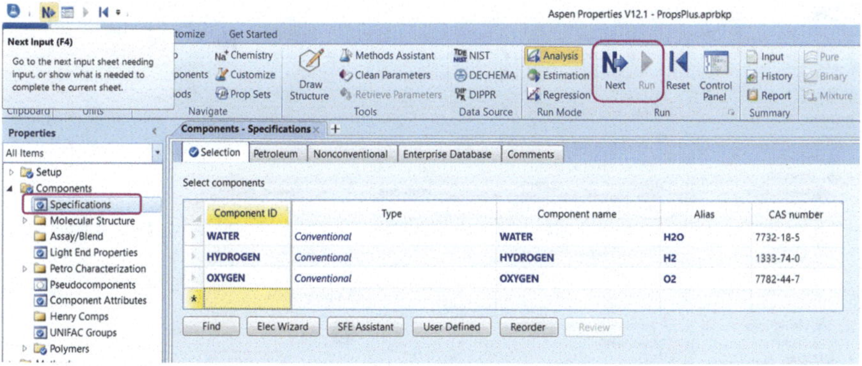

Fig. 6 Thermodynamic configuration of the component list through Aspen properties

- Click on **configure properties**, then **edit using Aspen properties** (Fig. 5). This will launch the **Aspen properties** for thermodynamic estimations (Fig. 6).
- In Aspen properties (Fig. 6), start by selecting the model **component** list, click on **next** to show if any additional step is needed to complete the sheet, then **run** to go next step. Choose the appropriate calculation **method** (Fig. 7) from the available **base methods**, then click on **next** input. If all is correct, a message will show **the available results** (Fig. 7) on the left bottom side of the window. **Save changes** and go back to ACM,
- In ACM, a **green rectangle** (Fig. 5) appears showing that the configuration is successfully done, click on **OK**.
- Click on **default** (Fig. 8) to build the component list by moving the component from left to right side, then click on **OK**.

(b) Add the model from the **custom modelling** item and choose a name (Fig. 9), in this example "SOEC_STACK", be careful that only underscore (_) is accepted, neither hyphen (–) nor spaces are used for the model's name.

(c) A window will launch where to introduce the electrochemical modelling equations (Fig. 9). Note that in order to successfully export and use the model in Aspen Hysys, the component list should be defined as full names in the script and not symbols. For instance, "hydrogen" and not "H_2".

5.2.2 Stack Modelling Script

The electrochemical modelling equations of the SOEC stack are introduced in the model window (Fig. 9) in the following manner:

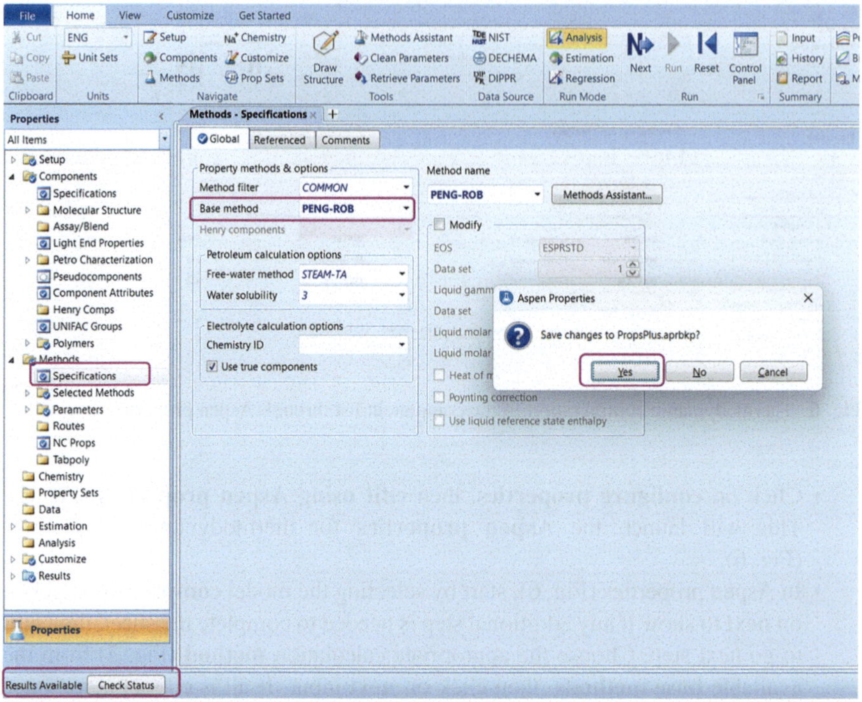

Fig. 7 Calculation method specification

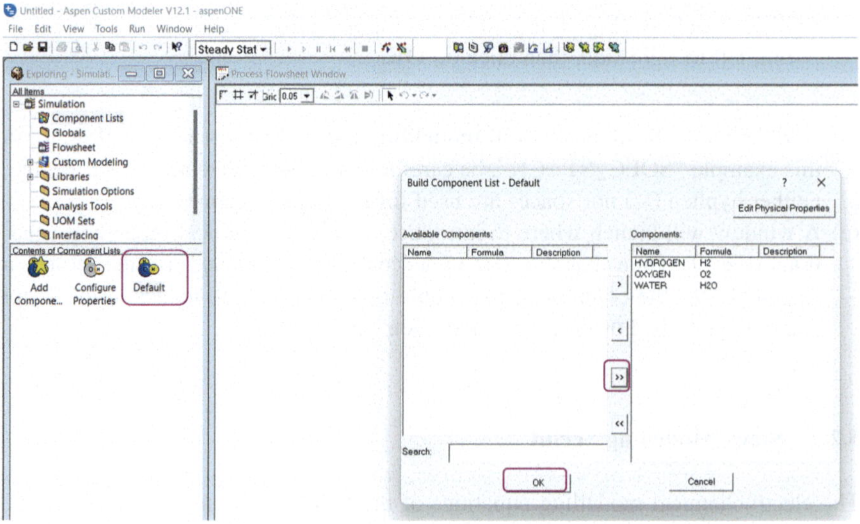

Fig. 8 Build component list

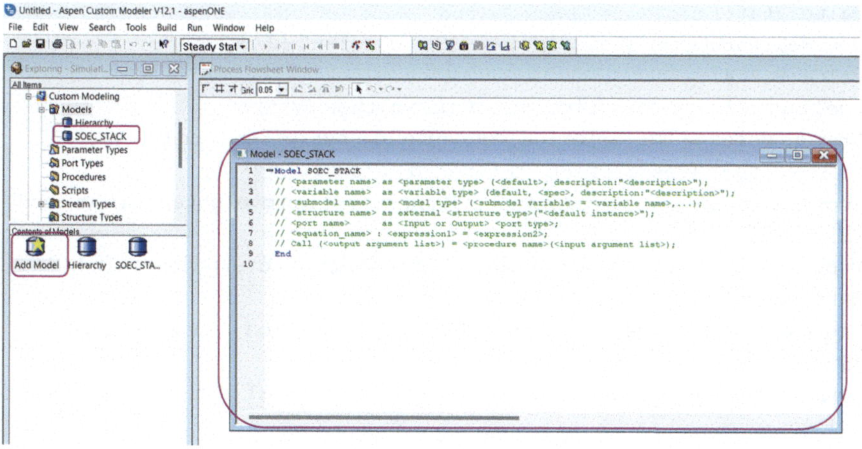

Fig. 9 Electrochemical modelling script

```
Model SOEC_STACK
Parameter Comp Uses StringParameter
Valid as StringSet;
End
CompH2 as Comp;
CompO2 as Comp;
CompH2O as Comp;
CompH2.Valid : ComponentList;
CompO2.Valid : ComponentList;
CompH2O.Valid : ComponentList;
CompH2 : "Hydrogen";
CompO2 : "Oxygen";
CompH2O : "Water";
T as temperature (Description: "SOEC operating temperature (°C)", fixed,
800);
TK as RealVariable (Description: "SOEC operating temperature (K)");
TK = T+273.15;
P as pressure (Description: "SOEC operating pressure (bar)", fixed,
1.01325);
N as RealVariable (Description: "Number of SOEC cells ", fixed, 130000);
A as area (Description: "Active area (m²)", fixed, 0.04);
I as current (Description: "Cell current (A)");
J as RealVariable (Description: "Cell current density (A/m²)",
useUOMof: "I/A", fixed, 8000); // useUOMof to set the unit of J
I = J*A; // Eq.3
z as RealParameter (Description: "Number of moles", value:2);
F as RealParameter (Description: "Faraday number (C/mol)",
```

```
value:96486);
R as RealParameter (Description: "Gas constant (J/mol.K)",
value:8.314);
Vcell as voltage (Description: "Cell voltage (V)");
Vrev as voltage (Description: "Reversible overpotential (V)");
V0 as voltage (Description: "Standard potential (V)");
Vconc_a as voltage (Description: "Concentration overpotential at anode
(V)");
Vconc_c as voltage (Description: "Concentration overpotential at
cathode (V)");
Vconc as voltage (Description: "SOEC concentration overpotential
(V)");
Vohm as voltage (Description: "Ohmic overpotential of the electrolyte
(V)");
Vact_a as voltage (Description: "Activation overpotential at anode
(V)");
Vact_c as voltage (Description: "Activation overpotential at cathode
(V)");
Vact as voltage (Description: "SOEC activation overpotential (V)");
PO2 as pressure (Description: "Partial pressure of O2 (bar)");
PH2 as pressure (Description: "Partial pressure of H2 (bar)");
PH2O as pressure (Description: "Partial pressure of H2O (bar)");
yO2 as realvariable (Description: "Input molar fraction of O2", fixed,
1); // 100% of O2
yH2 as realvariable (Description: "Input molar fraction of H2");
yH2O as realvariable (Description: "Input molar fraction of H2O", fixed,
0.8); // 80% of steam
yH2 = 1 - yH2O; // total=100%
PO2 = yO2*P; // Eq.18
PH2 = yH2*P; // Eq.18
PH2O = yH2O*P; // Eq.18
//Reversible voltage calculation
V0 = 1.253 - 2.4516e-4*TK; // Eq.17
Vrev = V0 + (R*TK/(2*F))*loge(PH2*PO2^0.5/PH2O); // Eq.16
//Ohmic overpotential calculation
L as length (Description: "Electrolyte layer thickness (m)", fixed,
12.5e-6); //m
Vohm = 2.99E-5*J*L*exp(10300/TK); // Eq.19
//Activation overpotential calculation
J0_a as realvariable (Description: "Exchange current density of anode
(A/m²)", useUOMof: "I/A");
J0_c as realvariable (Description: "Exchange current density of cathode
(A/m²)", useUOMof: "I/A");
Ja as realvariable;
Jc as realvariable;
```

```
Jexp_a as realvariable (Description: "Pre-exponential factor of anode
(A/m²)", useUOMof: "I/A", fixed, 2.051e9);
Jexp_c as realvariable (Description: "Pre-exponential factor of
cathode (A/m²)", useUOMof: "I/A", fixed, 1.344e10);
Eacta as realvariable (Description: "Activation energy of anode
(kJ/mol)", fixed, 1.2e5); // J/mol
Eactc as realvariable (Description: "Activation energy of cathode
(kJ/mol)", fixed, 1e5); // J/mol
J0_a = Jexp_a*exp(-Eacta/(R*TK)); // Eq.22
J0_c = Jexp_c*exp(-Eactc/(R*TK)); // Eq.22
Ja = J/(2*J0_a); // used for equation simplification
Jc = J/(2*J0_c); // used for equation simplification
Vact_a = (R*TK/F)*loge(Ja + sqrt(Ja^2 + 1)); // Eq.21
Vact_c = (R*TK/F)*loge(Jc + sqrt(Jc^2 + 1)); // Eq.21
Vact = Vact_a + Vact_c; // Eq.20
//Concentration overpotentials calculation
aO2 as realvariable;
nu as realvariable (Description: "Dynamic viscosity");
Bg as realvariable (Description: "Flow permeability");
epsilon as RealParameter (Description: "Electrode porosity",
value:0.3);
ksi as RealParameter (Description: "Tortuosity", value:5);
rad as RealParameter (Description: "Average pore radius (m)",
value:2.77*1e-6/2);
dan as length (Description: "Anode thickness (m)", fixed, 17.5*1e-6);
dca as length (Description: "Cathode thickness (m)", fixed, 12.5*1e-6);
nu = (-1.692+(889.75*(TK/1000))-(892.79*(TK/1000)^2)+(905.98*
(TK/1000)^3)-(598.36*(TK/1000)^4)+(221.64*(TK/1000)^5)-(34.75*
(TK/1000)^6))*1e-7; // Eq.27
Bg = (epsilon^3*(2*rad)^2)/(72*ksi*(1-epsilon)^2); // Eq.26
aO2 = J*R*TK*nu*dan/(2*F*Bg);
Vconc_a = (R*TK/(4*F))*loge(sqrt((PO2*1e5)^2 + aO2)/(PO2*1e5)); //
Eq.24
aH2 as realvariable; // used for equation simplification
aH2O as realvariable; // used for equation simplification
Deff_H2O as realvariable (Description: "Effective diffusion
coefficient");
D_H2O_K as realvariable (Description: "Knudsen diffusion");
D_H2_H2O as realvariable (Description: "Molecular diffusion");
omegad as realvariable (Description: "Dimensionless diffusion
collision integral");
tau as realvariable;
tau = TK/sqrt(59.7*809.1); // 59.7 and 809.1 are epsilon/k of H2 and H2O
respectively// Eq.32
sigm as realvariable (Description: "Collision diameter");
```

```
sigm= (2.827 + 2.641)/2; //2.872 and 2.641 are sigm_H2 and sigm_H2O
respectively // Eq.34
M_H2_H2O as realvariable;
MW(componentlist) as molweight; //kg/kmol
call(MW) = pMolWeights() componentlist;
omegad = (1.06036/tau^0.1561)+(0.193/exp(0.47635*tau))+(1.03587/
exp(1.52996*tau))+(1.76474/exp(3.89411*tau)); // Eq.31
D_H2_H2O*1e4 = 0.00133*sqrt((1/MW("Hydrogen"))+(1/MW("Water")))*
(TK^(3/2))/(P*sigm^2*omegad); //P in bar// Eq.30
D_H2O_k = (2/3)*rad*sqrt(8*R*TK/((22/7)*(MW("Water")/1000))); //22/
7 = pi // Eq.29
DH2H2O as realvariable;
DH2OK as realvariable;
DEffH2O as realvariable;
DH2H2O = 1/D_H2_H2O; //used for equation simplification
DH2OK = 1/D_H2O_K; //used for equation simplification
DEffH2O = (ksi/epsilon)*(DH2H2O + DH2OK); // Eq.28
1/DEffH2O = Deff_H2O; // Eq.28
aH2 = J*R*TK*dca/(2*F*Deff_H2O*PH2*1e5);
aH2O = J*R*TK*dca/(2*F*Deff_H2O*PH2O*1e5);
Vconc_c = (R*TK/(2*F))*loge((1 + aH2)/(1 - aH2O)); // Eq.25
Vconc = Vconc_a + Vconc_c; // Eq.23
//Cell Voltage calculation
Vcell = Vrev + Vohm + Vact + Vconc; // Eq.15
/*--------------------------------------------------------*/
Power as power_electric (free);
Welec as input WorkPort;
Power*1000 = N*Vcell*I; //kW// Eq.1 & Eq.2
Welec.W = Power; //kW
H2_prod as flow_mol (Description: "Hydrogen production rate at the
cathode (kmol/hr)");
H2_prod = 3.6*(I*N)/(z*F); // 3.6 to convert from (mol/s) to (kmol/hr)
// Eq.35
H2O_cons as flow_mol (Description: "Water consumed in the process (kmol/
hr)");
H2O_cons = H2_prod; // Eq.37
O2_prod as flow_mol (Description: "Oxygen production rate at the anode
(kmol/hr)");
O2_prod = H2_prod/2; // Eq.36
//Inlet flows calculation
flow_in (componentlist) as flow_mol;
Inlet as input MoleFractionPort;
flow_in ("Water") = Inlet.z("Water")*Inlet.F;
flow_in("Hydrogen") = Inlet.z("Hydrogen")*Inlet.F;
flow_in("Oxygen") = Inlet.z("Oxygen")*Inlet.F;
```

```
//Sweep gas calculation
Inlet_O2 as input MoleFractionPort;
flowin_O2swga(componentlist) as flow_mol;
flowin_O2swga("Water") = Inlet_O2.z("Water")*Inlet_O2.F;
flowin_O2swga("Hydrogen") = Inlet_O2.z("Hydrogen")*Inlet_O2.F;
flowin_O2swga("Oxygen") = Inlet_O2.z("Oxygen")*Inlet_O2.F;
//Cathode flows calculation
H2_cat as flow_mol (Description:"H2 at cathode (kmol/hr)");
H2_cat = H2_prod + flow_in("Hydrogen");
O2_cat as flow_mol (Description:"O2 at cathode (kmol/hr)");
O2_cat = 0;
H2O_cat as flow_mol (Description:"H2O at cathode (kmol/hr)");
H2O_cat = (flow_in("Water") - H2O_cons); // Eq.40
//Anode flows calculation
H2_an as flow_mol (Description: "H2 at cathode (kmol/hr)");
H2_an = 0 + flowin_O2swga("Hydrogen");
O2_an as flow_mol (Description:"O2 at cathode (kmol/hr)");
O2_an = H2_prod/2 + flow_in("Oxygen") + flowin_O2swga("Oxygen");
H2O_an as flow_mol (Description:"H2O at cathode (kmol/hr)");
H2O_an = 0 + flowin_O2swga("Water");
//Cathode out
Cathode as output MoleFractionPort;
Cathode.F = H2O_cat + O2_cat + H2_cat;
Cathode.z("Oxygen") = O2_cat/Cathode.F;
Cathode.z("Hydrogen") = H2_cat/Cathode.F;
Cathode.z("Water") = H2O_cat/Cathode.F;
Cathode.T=T;
Cathode.P=P;
Cathodedens as dens_mol;
Cathode.V*Cathodedens = 1;
Call(Cathode.h)=pEnth_Mol(T,P,Cathode.z);
Call(Cathodedens) = pDens_Mol_liq(T, P, Cathode.z);
//Anode out
Anode as output MoleFractionPort;
Anode.F = H2O_an + O2_an + H2_an;
Anode.z("Oxygen") = O2_an/Anode.F;
Anode.z("Hydrogen") = H2_an/Anode.F;
Anode.z("Water") = H2O_an/Anode.F;
Anode.T=T;
Anode.P=P;
Anodedens as dens_mol;
Anode.V*Anodedens = 1;
Call(Anode.h)=pEnth_Mol(T,P,Anode.z);
Call(Anodedens) = pDens_Mol_liq(T, P, Anode.z);
// Heat
```

```
Qsoec as realvariable (Description: "Heat required/generated by the
SOEC (kW)");
Qloss as realvariable (Description: "Heat losses by the SOEC (kW)");
Qexcess as realvariable (Description: "Excess heat by the SOEC (kW)");
H2prod as flow_mass;
H2prod = H2_prod*MW("Hydrogen"); //convert molar flows to mass flows
O2prod as flow_mass;
O2prod = O2_prod*MW("Oxygen"); //convert molar flows to mass flows
//Thermoneutral voltage calculation  (Vtn = deltaH/(z*F))
Vtn as voltage (Description: "Thermoneutral voltage (V)");
Vtn=1.25028255+6.23391235E-05*T-2.32482542E-08*T^2+2.53527575E-
12*T^3; // Eq.7
Qsoec = N*I*(Vcell - Vtn)*0.001;//kW// Eq.4
Qloss = 0.1*Qsoec; // Eq.11
Qexcess = Qsoec - Qloss;
Q_soec as output HeatPort;
Q_loss as output HeatPort;
Q_excess as output HeatPort;
Q_soec.Q = Qsoec*3600*1e-6;
Q_loss.Q = Qloss*3600*1e-6;
Q_excess.Q = Qexcess*3600*1e-6;
//Energy and Exergy efficiencies
EnerEff_HHV as realvariable;
HHV_H2 as realvariable (fixed, 142);
EnerEff_LHV as realvariable;
LHV_H2 as realvariable (fixed, 120);
ExerEff as realvariable (Description: "Exergy efficiency (%)");
E_H2 as RealVariable (Description: "Exergy content of H2 (kJ/kmol)",
value:238490, spec: Fixed);
E_elec as RealVariable (Description: "Rate of electric exergy input
(kW)");
E_stack as RealVariable (Description: "Rate of thermal exergy input for
stack (kW)");
T0 as temperature (Description: "Temperature of reference environment
(°C)", Fixed, 25);
E_elec = Welec.W; //kW// Eq.13
E_stack = (Qsoec + Qloss)*(1-(T0+273)/TK); // Eq.14
if (Q_soec.Q >=0) then //heat generated (exothermic)
EnerEff_HHV = 100*(HHV_H2 * H2prod * (1000/3600))/(Welec.W); // Eq.9
EnerEff_LHV = 100*(LHV_H2 * H2prod * (1000/3600))/(Welec.W); // Eq.8
ExerEff = (100*E_H2*H2_prod)/(3600*(E_elec)); // Eq.12
else //heat required (endothermic)
EnerEff_HHV = 100*(HHV_H2 * H2prod * (1000/3600))/(Welec.W + abs (Qsoec
+ Qloss));
EnerEff_LHV = 100*(LHV_H2 * H2prod * (1000/3600))/(Welec.W + abs (Qsoec
```

```
+ Qloss));
ExerEff = (100*E_H2*H2_prod)/(3600*(E_elec + abs (E_stack)));
endif
End
```

(a) Click right on the script window (or F8) to **compile** the model then **run** it. Errors and warnings are shown in the **simulation messages** panel.
(b) If a **red triangle** (Fig. 10) shows at the bottom of the window, that means the model is whether under-specified (pointing down) which means that there are fewer fixed variables "free", or over-specified (pointing up), more fixed variables.
(c) Place the model in the **process flowsheet** (Fig. 10), the model could be personalized by **adding an icon.** Double-click on and change the appropriate variable between fixed or free according to the model specifications. In this example, the inlet flows, and their compositions are considered as fixed variables; therefore, the specification should be changed from "free" to "fixed" (Fig. 11).
(d) If no warnings or errors are detected and the model runs, then it is ready to be exported and used in Aspen Hysys (Fig. 11).
(e) **Create script**, **save** the model and **accept to install** it.

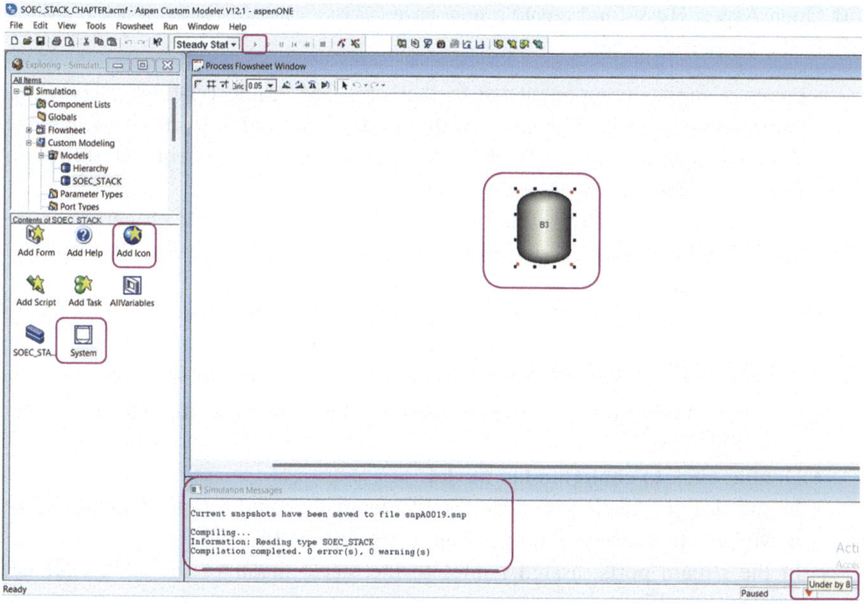

Fig. 10 Stack model created

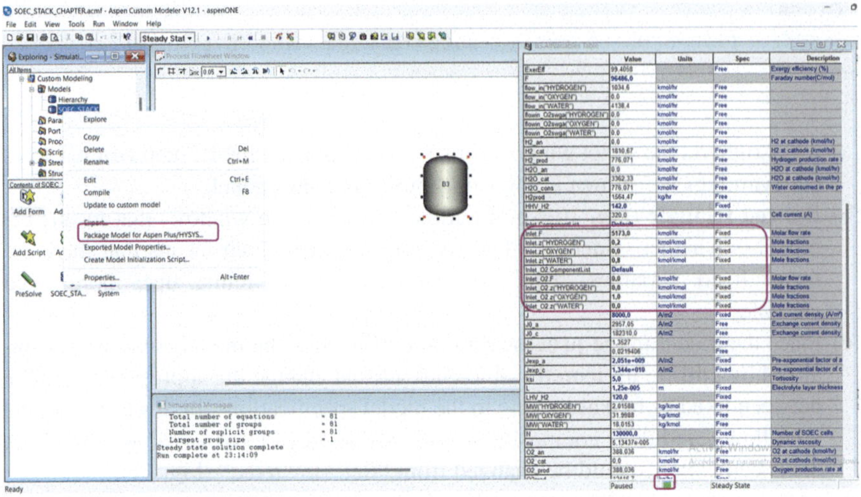

Fig. 11 Stack ready for exportation

5.2.3 SOEC Simulation in Aspen Hysys

(a) Open Aspen Hysys and create a new case.
(b) In the properties panel:

- create a similar component list to that of ACM. Choose between Hysys or Aspen properties. In this case, as the model is imported from ACM, choose Aspen properties (Fig. 12), then add the studied components as shown in Fig. 13 (water, hydrogen and oxygen).
- Add the Aspen properties fluid package by choosing the same properties as the component list and selecting the appropriate calculation method. It should be similar to the one used in ACM (Fig. 14).

(c) Move to the simulation panel:

- Move the **ACM** from the **external model** of the **model palette** to the flowsheet case (Fig. 15).
- Double-click to configure the model,
- In the design panel, select the SOEC from the model type, then initialize it. Make sure the right fluid package is used (Fig. 16).
- In the stream ports, assign names to the stack streams (e.g. inlet, cathode, anode) (Fig. 17).
- Enter the problem conditions in the worksheet (temperature, pressure, composition, flow) as shown in Fig. 18. Always check the message panel to solve any existing simulation issues.

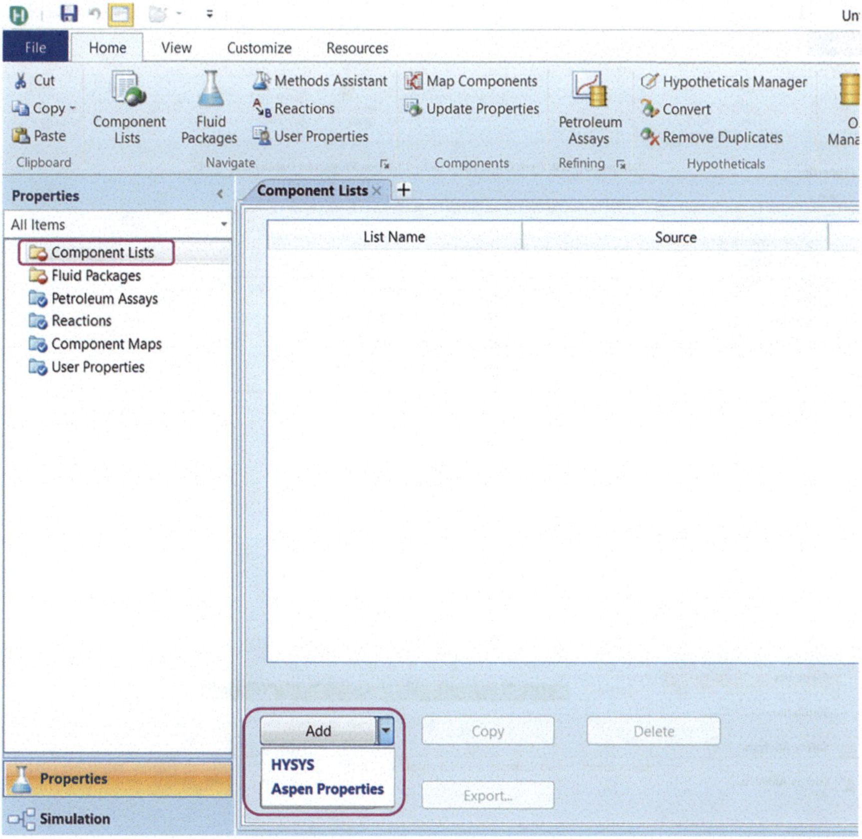

Fig. 12 Properties selection

- Check the variables, for additional information and results. Make sure the units are correct, otherwise change them.
- The stack is run and ready.

5.3 SOEC Results Validation with Existing Findings

The SOEC stack modelling is validated with the finding results obtained by Refs [96, 97]. The simulation results summarized in Table 7 (extracted from Figs. 19 and 20) as well as the cell voltage–current density curve illustrated in Fig. 21 are consistent with the experimental data and findings of Refs [96, 97].

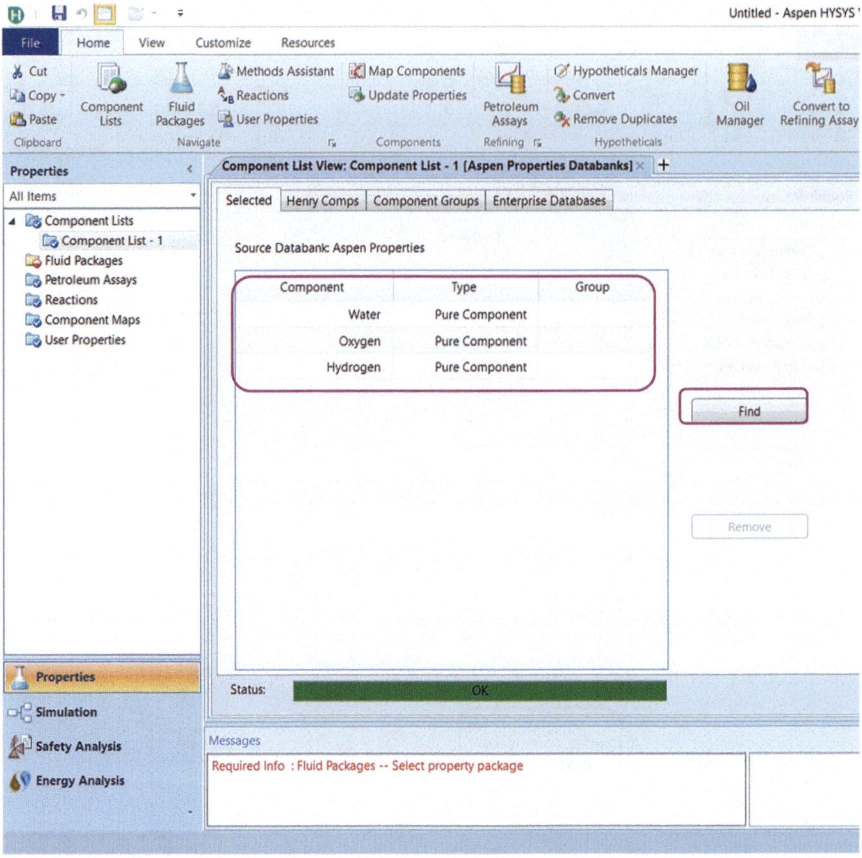

Fig. 13 Component list selection

5.4 H_2 Production Onboard a Ship Through an SRC

This section studied an example of the H_2 production onboard a ship through SOEC powered by an SRC. The ships model adopted is an LNG carrier with a capacity of a 170 K m^3 and a dual fuel diesel electric propulsion with four-stroke engines. The ship's data is summarized in Table 8.

5.4.1 WHR System Description

Figure 22 illustrates the process diagram of the SRC adopted in this study. Saturated liquid water (1) leaves the condenser (HE1) at low pressure and is then pumped to the atmospheric pressure (2). The water is preheated (3) by the jacket water (HE2)

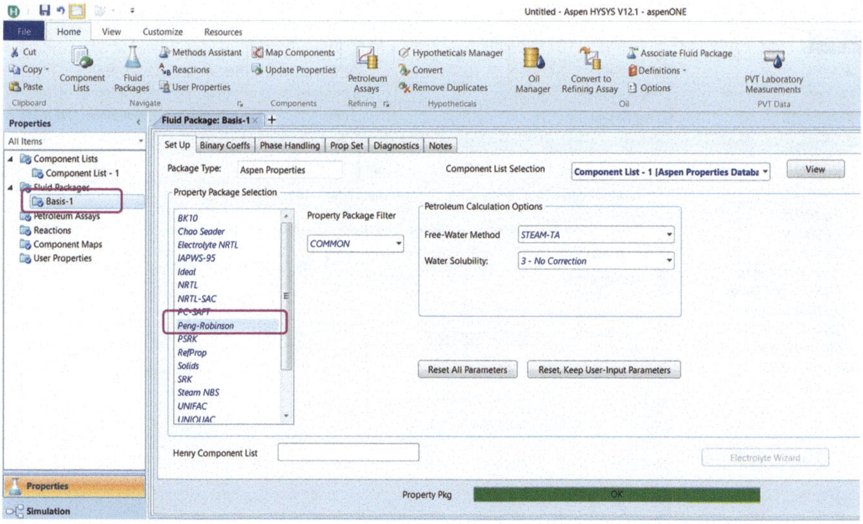

Fig. 14 Fluid package and calculation method selection

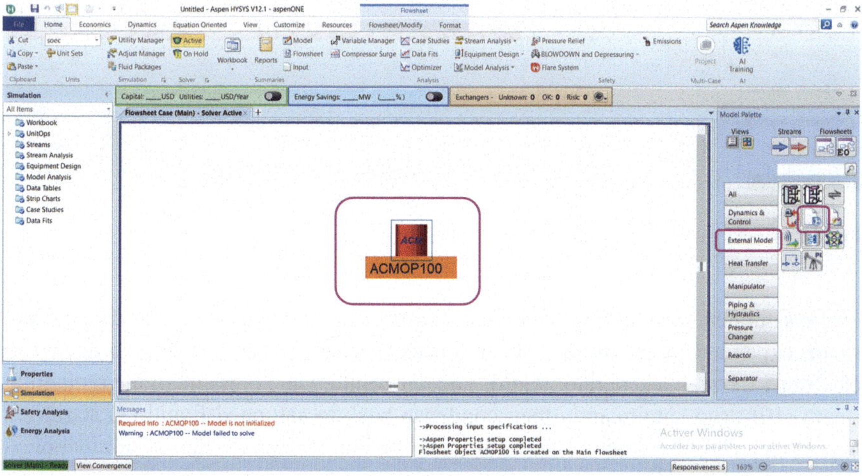

Fig. 15 Model exportation from ACM—part 1

heat, mixed (4) with the ship's steam consumption (12), then pumped to a higher pressure of 10 bar (5) following an isentropic compression process. The scavenge air (HE3) heat is used for further water preheating (6). This latter is converted to saturated steam (7) via an evaporator (HE4) and then superheated (8) in HE5 using the exhaust gases' heat (17–19) ensuring that the steam (9) entering the turbine (T) is dry and at a higher temperature. Hence, improving the cycle's efficiency. Part

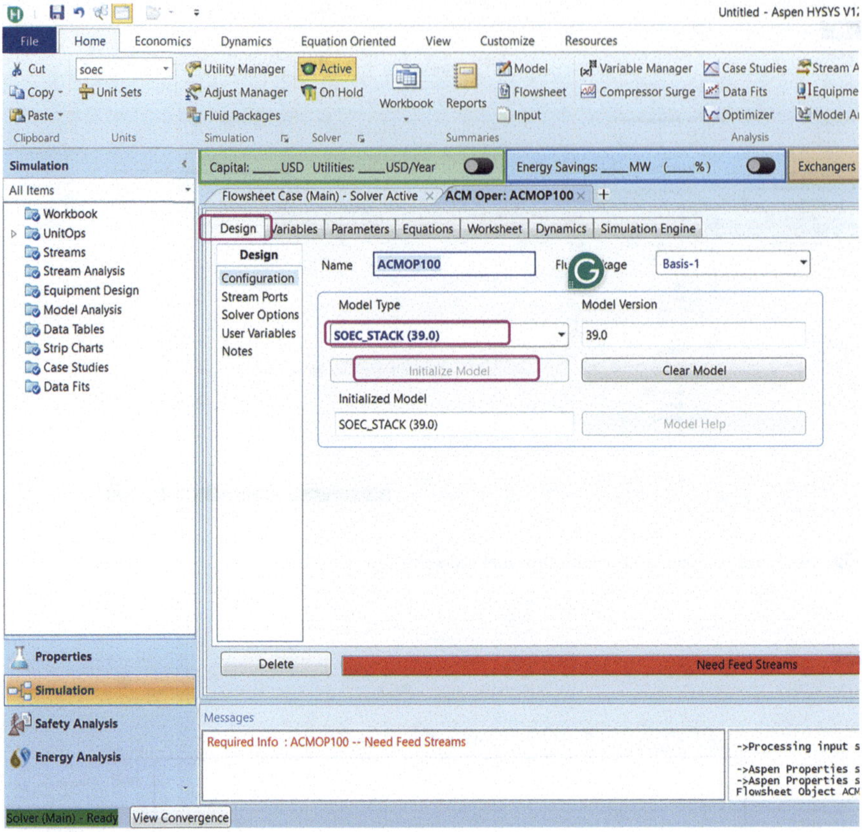

Fig. 16 Model exportation from ACM—part 2

of the saturated steam (11) is transferred to the auxiliary services. The superheated steam expanded in the turbine (T) produces mechanical work and is converted to electricity through a generator (Gen). The temperature and pressure of the remaining steam (10) drop significantly then condense back to saturated liquid (1) starting a new cycle. The flowsheet of the overall simulation process for the H_2 production is presented in Fig. 23.

The molar composition of the exhaust gases (17) is similar to the one of the boil-off gas (BOG) and it is determined by the following equations:

$$\dot{m}_{\text{BOG}} = \dot{m}_{\text{BOG}_\text{N}} - \frac{\rho_{\text{BOG}}}{\rho_{\text{LNG}}}(\dot{m}_{\text{LNG}} + \dot{m}_{\text{BOG}_\text{N}}) \tag{40}$$

$$\dot{m}_{\text{BOG}_\text{N}} = BOR \cdot V_{\text{tank}} \cdot \rho_{\text{LNG}} \tag{41}$$

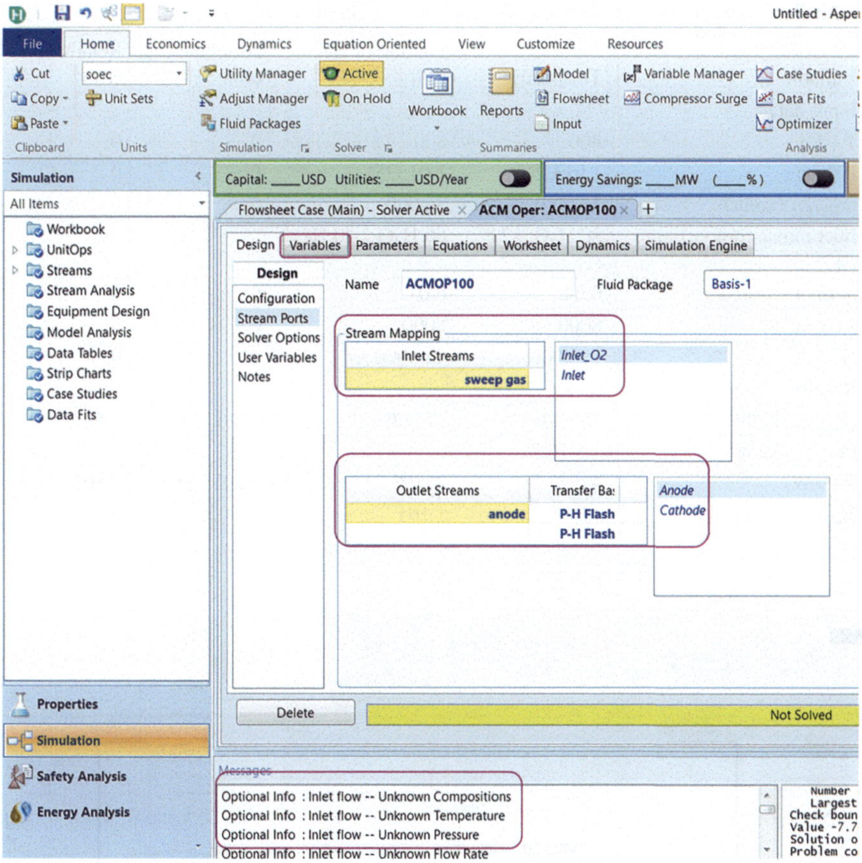

Fig. 17 Streams assignation

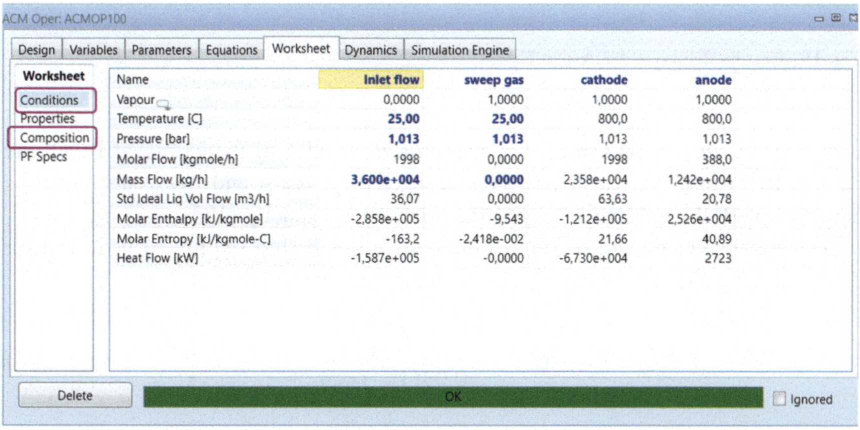

Fig. 18 Problem conditions

Table 7 Comparison of simulation results with the finding of Ref [97]

Variable	This study	Finding of Ref [96]	Absolute error	Relative error
Input data				
T (°C)	800	800	0	0
P (bar)	1.01325	1.01325	0	0
$\dot{m}_{H2O\text{-}in}$ (kg/hr)	36,000	36,000	0	0
Inlet molar composition (%)	80 H_2O—20 H_2	80 H_2O—20 H_2	–	–
A (m^2)	0.04	0.04	0	0
J (A /m^2)	8000	8000	0	0
N	130,000	130,000	0	0
Results				
$\dot{m}_{H2\text{-}pro}$ (kg/hr)	1564.47	1551.6	12.87	0.83%
$\dot{m}_{O2\text{-}pro}$ (kg/hr)	12416.7	–	–	–
W (kW)	44752.8	45808.51	−1055.71	2.30%
V_{cell} (V)	1.076	1.101	−0.025	−2.29%

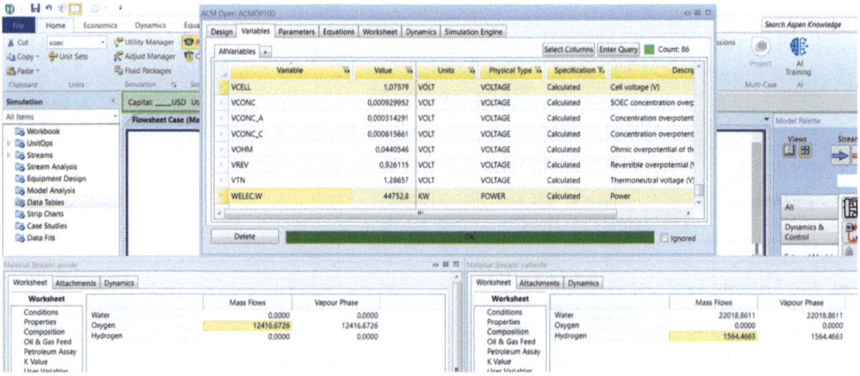

Fig. 19 Results displayed by Aspen Hysys

where $\dot{m}_{BOG}$, $\dot{m}_{BOG_N}$ and $\dot{m}_{LNG}$ are the mass flow rates of the extracted BOG, the natural BOG and the extracted LNG, respectively. ρ_{BOG} and ρ_{LNG} are the mass densities of BOG and LNG. V_{tank} is the total cargo capacity [105].

The freshwater generated ($\dot{m}_{fr.w}$) onboard a given ship requires a heat transfer determined by Eq. 43, where $t_{fr.\ w}$ is the tolerance (15%) [106, 107].

$$Q_{fr.w} = \frac{\dot{m}_{fr.w}}{0.03(1 - t_{fr.w})} \quad (42)$$

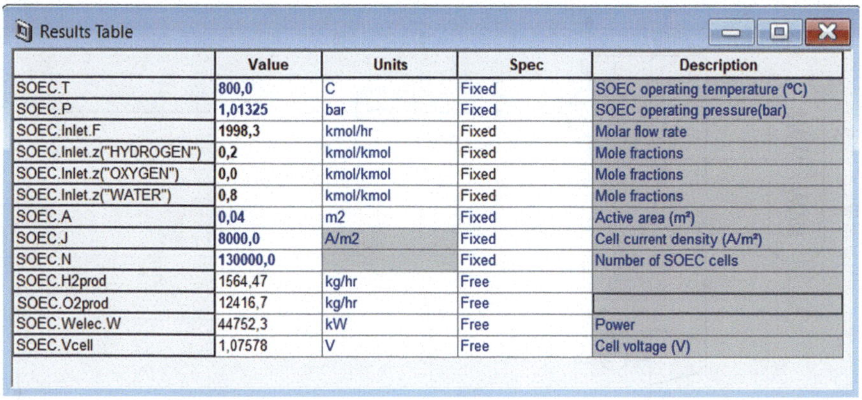

Results Table

	Value	Units	Spec	Description
SOEC.T	800,0	C	Fixed	SOEC operating temperature (°C)
SOEC.P	1,01325	bar	Fixed	SOEC operating pressure(bar)
SOEC.Inlet.F	1998,3	kmol/hr	Fixed	Molar flow rate
SOEC.Inlet.z("HYDROGEN")	0,2	kmol/kmol	Fixed	Mole fractions
SOEC.Inlet.z("OXYGEN")	0,0	kmol/kmol	Fixed	Mole fractions
SOEC.Inlet.z("WATER")	0,8	kmol/kmol	Fixed	Mole fractions
SOEC.A	0,04	m2	Fixed	Active area (m²)
SOEC.J	8000,0	A/m2	Fixed	Cell current density (A/m²)
SOEC.N	130000,0		Fixed	Number of SOEC cells
SOEC.H2prod	1564,47	kg/hr	Free	
SOEC.O2prod	12416,7	kg/hr	Free	
SOEC.Welec.W	44752,3	kW	Free	Power
SOEC.Vcell	1,07578	V	Free	Cell voltage (V)

Fig. 20 Results displayed by ACM

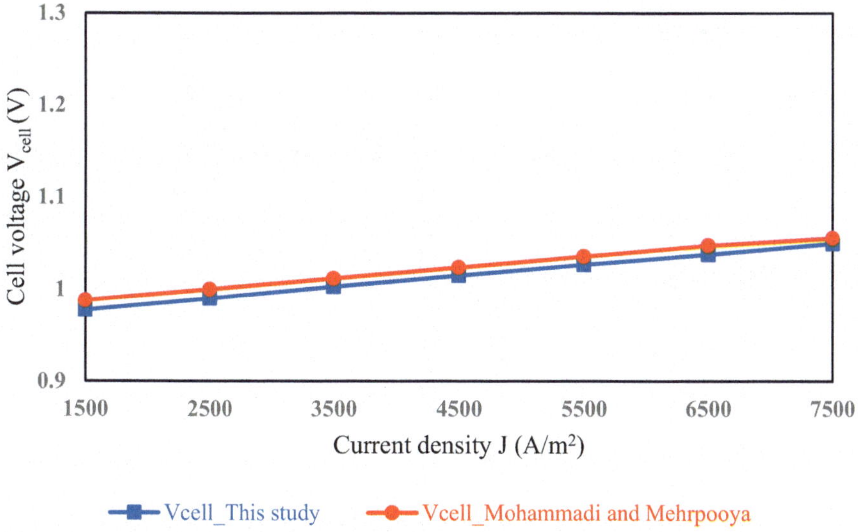

Fig. 21 Cell voltage: Current density validation curve with results funding of [97]

Table 8 Main characteristics of the ship model

Ship's type	LNG carrier
Ship's capacity	170 K m^3
Freshwater generation ($\dot{m}_{fr.w}$)	20,000 kg/day
Propulsion system	Dual fuel diesel electric
Engine type	MAN 51/60DF
Engine power	1050 kW/cyl
Engine cylinders' configuration	3 × 12 + 1 × 6
Boil-off rate (BOR)	0.10%
Total steam consumption	1999.2 kg/hr
Auxiliary power demand	2240 kW

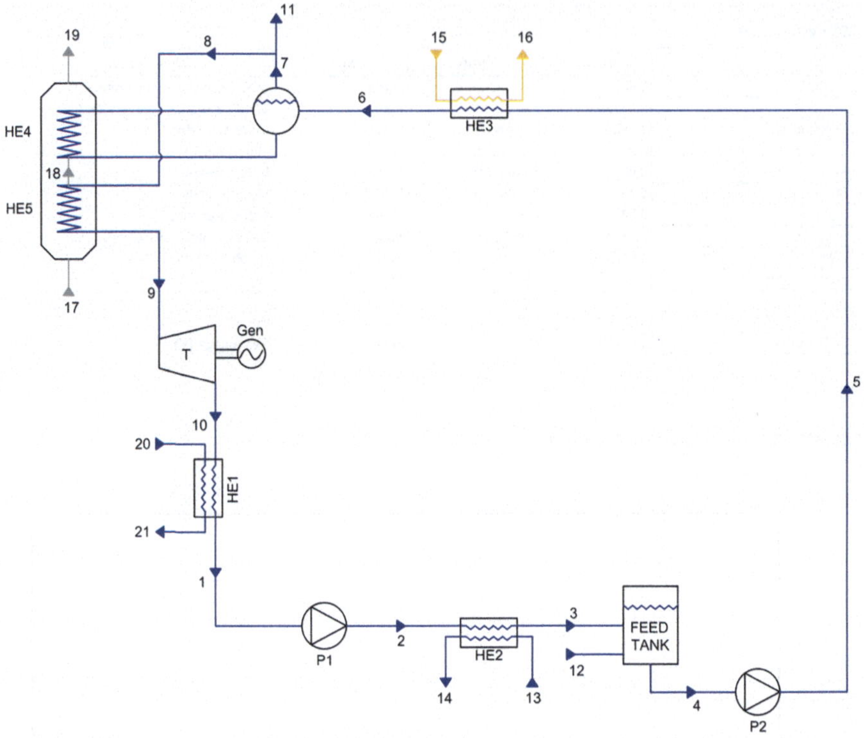

Fig. 22 Waste heat recovery system through an SRC

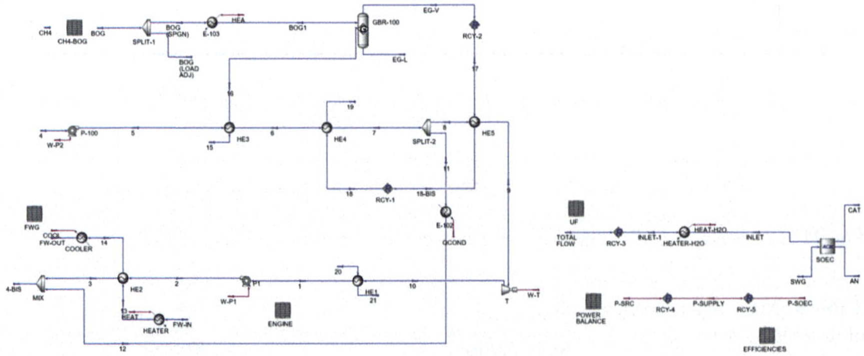

Fig. 23 Flowsheet of the H_2 production through an SOEC powered by an SRC

5.4.2 Results and Discussions

The simulation results are summarized in this section. Tables 9 and 10 present the stream properties and parameters of the SRC and the WHR sources, while Table 11

Table 9 Stream properties and conditions of the SRC

	Temperature (°C)	Pressure (bar)	Mass Flow (kg/hr)	Mass Enthalpy (kJ/kg)	Vapour Fraction
1	40.00	0.073	14,550	−15,823	0.000
2	40.02	1.013	14,550	−15,823	0.000
3	85.00	1.013	14,550	−15,628	0.000
4	85.65	1.013	16,550	−15,625	0.000
4-BIS	85.60	1.013	16,550	−15,625	0.000
5	85.76	10.000	16,550	−15,624	0.000
6	149.94	10.000	16,550	−15,338	0.000
7	180.04	10.000	16,550	−13,159	1.000
8	180.04	10.000	14,550	−13,159	1.000
9	362.13	10.000	14,550	−12,784	1.000
10	40.00	0.073	14,550	−13,503	0.956
11	180.04	10.000	1999.2	−13,159	1.000
12	90.00	1.013	1999.2	−15,606	0.000
13	90.00	1.013	178,029	−15,606	0.000
14	86.35	1.013	178,029	−15,622	0.000
15	154.94	1.013	181,154	132.989	1.000
16	129.76	1.013	181,154	106.884	1.000
17	387.13	1.013	186,125	−961.030	1.000
18	361.56	1.013	186,125	−990.304	1.000
18-BIS	361.52	1.013	186,125	−990.350	1.000
19	188.78	1.013	186,125	−1184.0	1.000
20	25.00	1.013	782,666	−15,888	0.000
21	35.00	1.013	782,666	−15,845	0.000
FW-IN	75.00	1.013	178,029	−15,672	0.000
FW-OUT	82.72	1.013	178,029	−15,638	0.000

Table 10 WHRS parameters as a function of the ship's load

WHR source	Temperature (°C)	Pressure (bar)	Mass flow (kg/hr)	Heat flow (kW)
Jacket water	90	1.01325	178,029	788.2
Scavenge air	154.9	1.01325	181,154	1314
Exhaust gas	387.1	1.01325	186,125	1516
Total WH recovered				3618.2

presents the results of the H_2 production by the SOEC. For calculation purposes, the ship load is set at 70% and the flow of the sweep gas is neglected.

According to the results, an excess energy of around 1.42 MW is available onboard the ship without any use. This power will be supplied to a SOEC for the gH_2 production and its use when needed, for instance, as a clean fuel. The process can produce around 57.179 kg/hr. of gH_2.

Table 11 Simulation results of the SOEC gH_2 production

Variable	Unit	Value
Power	kW	1417.63
V_{cell}	V	0.9324
J	A/m^2	292.386
Temperature	°C	800
Pressure	Bar	1.01325
$\dot{m}_{in}$	Kg/hr	646.4
Inlet molar composition	%	90% H_2O + 10% H_2
$\dot{m}_{H2\text{-}pro}$	Kg/hr	57.179
$\dot{m}_{O2\text{-}pro}$	Kg/hr	453.808
η_{LHV}	%	69.24
η_{SRC}	%	21.26

6 Conclusions

This chapter first summarizes and reviews the gH_2 production chain (production, storage and consumption) methods adopted onboard maritime transport. Electrolysis is a promising alternative to fossil fuel (e.g. steam methane reforming) for clean and green H_2 fuel production. Therefore, this chapter also presents a case study of gH_2 production through a SOEC powered by an SRC. The main conclusions are outlined as follows:

- Among the different available energy source onboard ships, taking advantage of the engine's waste heat is promising as solar and wind energy are bounded by the meteorological conditions and the ship's structure. Recovering the heat wasted reduces fuel consumption; hence, costs as well as emissions, and improves the ship's efficiency. Producing gH_2 through electrolysis is an emission-free process with O_2 as the only byproduct.
- The H_2 storage is still challenging, due mainly to the H_2 properties, as each method has its advantages and disadvantages. Compressed H_2 storage is widely used, thanks to the technology maturity and simplicity. However, the huge compressors' requirement makes this technology more expensive. Although storing H_2 in a liquefied state is around 30% cheaper than the gaseous state, this technology has a short shelf-life and requires more safety precautions and super-insulated tanks to deal with the low H_2 storage temperatures. Solid H_2 storage seems to be a promising technology, thanks to its safety and cheap price compared to both gaseous and liquefied H_2 storage, as well as the resulting H_2 purity. On the other part, using ammonia as an H_2 storage medium is also promising, thanks to its overall properties such as safety (easy smell detection), high energy density, and is the only carbon-free H_2 carrier.
- ICEs are widely adopted as the propulsion system of maritime transport, while FCs are mainly adopted by small-scale ships or used as auxiliary systems in the case of large-scale ships. Unlike adopting a new propulsion system for the gH_2

fuel use, injecting H_2 into ICE requires only some engine adaptation, therefore, reducing retrofitting costs.
- According to the case study, the SRC recovers around 3.62 MW of heat that would be lost to the environment without any effective use. The system produces an electric energy of around 1.42 MW supplied to a SOEC which produces an amount of 57.179 kg/hr. of gH_2. The SRC is 21.26% efficient, while the SOEC achieves an efficiency of almost 70%.

To meet the IMO regulations and restrictions towards emissions reductions and the decarbonization of the maritime sector, special and serious attention should be given to the adoption of clean and green fuels such as H_2.

References

1. UNCTAD. (2017). *Review of maritime transport*. UNCTAD. https://unctad.org/system/files/official-document/rmt2017_en.pdf.
2. IEA. (2020, 3rd January). *International shipping*. IEA. https://www.iea.org/reports/international-shipping
3. E. Commission. (7th June). *Reducing emissions from the shipping sector*. https://ec.europa.eu/clima/eu-action/transport-emissions/reducing-emissions-shipping-sector_en
4. I. M. O. (IMO). (2021). *Fourth IMO GHG Study 2020 Full Report*. I. M. O. (IMO). https://wwwcdn.imo.org/localresources/en/OurWork/Environment/Documents/Fourth%20IMO%20GHG%20Study%202020%20-%20Full%20report%20and%20annexes.pdf
5. Yang, R., & Zhang, W. (2019). Research on modeling and simulation of new energy ship power system. In *2019 IEEE 3rd conference on energy internet and energy system integration (EI2)* (pp. 1288–1292). IEEE. https://doi.org/10.1109/EI247390.2019.9062149
6. Yuan, Y., Wang, J., Yan, X., Shen, B., & Long, T. (2020). A review of multi-energy hybrid power system for ships. *Renewable Sustainable Energy Reviews, 132*, 110081. https://doi.org/10.1016/j.rser.2020.110081
7. Fan, Y. V., Perry, S., Klemeš, J. J., & Lee, C. T. (2018). A review on air emissions assessment: Transportation. *Journal of Cleaner Production, 194*, 673–684. https://doi.org/10.1016/j.jclepro.2018.05.151
8. Inal, O. B., & Deniz, C. (2020). Assessment of fuel cell types for ships: Based on multi-criteria decision analysis. *Journal of Cleaner Production, 265*, 121734. https://doi.org/10.1016/j.jclepro.2020.121734
9. Geertsma, R., Negenborn, R., Visser, K., & Hopman, J. (2017). Design and control of hybrid power and propulsion systems for smart ships: A review of developments. *Applied Energy, 194*, 30–54. https://doi.org/10.1016/j.apenergy.2017.02.060
10. Anon. (2018). Initial IMO strategy on reduction of GHG emissions from ships. *Resolution MEPC, 304*(72), 1.
11. Nuchturee, C., Li, T., & Xia, H. (2020). Energy efficiency of integrated electric propulsion for ships–a review. *Renewable and Sustainable Energy Reviews, 134*, 110145. https://doi.org/10.1016/j.rser.2020.110145
12. McKinlay, C. J., Turnock, S. R., & Hudson, D. A. (2020). A comparison of hydrogen and ammonia for future long distance shipping fuels. In *Presented at the LNG/LPG and alternative fuels, London, UK, 29th – 30th January 2020*. https://eprints.soton.ac.uk/437555/
13. Zhuo, R., & Wang, H. (2022). Decarbonising shipping and the role of LNG: International law and policy trends. In D. S. Olawuyi & E. G. Pereira (Eds.), *The Palgrave handbook of natural*

gas and global energy transitions (pp. 319–343). Springer. https://doi.org/10.1007/978-3-030-91566-7_13
14. Diab, F., Lan, H., & Ali, S. (2016). Novel comparison study between the hybrid renewable energy systems on land and on ship. *Renewable and Sustainable Energy Reviews, 63*, 452–463. https://doi.org/10.1016/j.rser.2016.05.053
15. Nasirudin, A., Chao, R.-M., & Utama, I. K. A. P. (2017). "solar powered boat design optimization," presented at the 10th international conference on marine technology. *MARTEC, 2016*, 260.
16. Yilmaz, F., Ozturk, M., & Selbas, R. (2020). Design and thermodynamic modeling of a renewable energy based plant for hydrogen production and compression. *International Journal of Hydrogen Energy, 45*(49), 26126–26137. https://doi.org/10.1016/j.ijhydene.2019.12.133
17. Dincer, I. (2012). Green methods for hydrogen production. *International Journal of Hydrogen Energy, 37*(2), 1954–1971. https://doi.org/10.1016/j.ijhydene.2011.03.173
18. Aydin, M. I., & Dincer, I. (2022). An assessment study on various clean hydrogen production methods. *Energy, 245*, 123090. https://doi.org/10.1016/j.energy.2021.123090
19. Abuşoğlu, A., Özahi, E., Kutlar, A. İ., & Demir, S. (2017). Exergy analyses of green hydrogen production methods from biogas-based electricity and sewage sludge. *International Journal of Hydrogen Energy, 42*(16), 10986–10996. https://doi.org/10.1016/j.ijhydene.2017.02.144
20. Chen, Z. S., & Lam, J. S. L. (2022). Life cycle assessment of diesel and hydrogen power systems in tugboats. *Transportation Research Part D: Transport and Environment, 103*, 103192. https://doi.org/10.1016/j.trd.2022.103192
21. Osman, A. I., et al. (2022). Hydrogen production, storage, utilisation and environmental impacts: A review. *Environmental Chemistry Letters, 20*(1), 153–188. https://doi.org/10.1007/s10311-021-01322-8
22. Dincer, I., & Zamfirescu, C. (2017). *Sustainable hydrogen production* (p. 492). Elsevier. https://doi.org/10.1016/C2014-0-00658-2
23. Bethoux, O. (2020). Hydrogen fuel cell road vehicles and their infrastructure: An option towards an environmentally friendly energy transition. *Energies, 13*(22), 6132. https://doi.org/10.3390/en13226132
24. Sarafraz, M. M., Christo, F. C., & Safaei, M. R. (2022). Potential of plasmonic microreactor for Photothermal hydrogen-enriched fuel production from biomethane. *International Journal of Hydrogen Energy, 47*, 26355. https://doi.org/10.1016/j.ijhydene.2022.01.078
25. Ozturk, M., & Dincer, I. (2020). An integrated system for clean hydrogen production from municipal solid wastes. *International Journal of Hydrogen Energy, 46*, 6251. https://doi.org/10.1016/j.ijhydene.2020.11.145
26. Dutta, S. (2014). A review on production, storage of hydrogen and its utilization as an energy resource. *Journal of Industrial Engineering Chemistry, 20*(4), 1148–1156. https://doi.org/10.1016/j.jiec.2013.07.037
27. Tashie-Lewis, B. C., & Nnabuife, S. G. (2021). Hydrogen production, distribution, storage and power conversion in a hydrogen economy—A technology review. *Chemical Engineering Journal Advances, 8*, 100172. https://doi.org/10.1016/j.ceja.2021.100172
28. Dincer, I., & Acar, C. (2015). Review and evaluation of hydrogen production methods for better sustainability. *International Journal of Hydrogen Energy, 40*(34), 11094–11111. https://doi.org/10.1016/j.ijhydene.2014.12.035
29. Abdalla, A. M., Hossain, S., Nisfindy, O. B., Azad, A. T., Dawood, M., & Azad, A. K. (2018). Hydrogen production, storage, transportation and key challenges with applications: A review. *Energy Conversion Management, 165*, 602–627. https://doi.org/10.1016/j.enconman.2018.03.088
30. Hoecke, L. V., Laffineur, L., Perreault, P., Verbrugge, S. W., & Lenaerts, S. (2021). Challenges in the use of hydrogen for maritime applications. *Energy & Environmental Science, 14*(2), 815–843. https://doi.org/10.1039/D0EE01545H

31. Inal, O. B., Zincir, B., & Dere, C. (2022). Hydrogen as maritime transportation fuel: A pathway for decarbonization. In A. K. Agarwal & H. Valera (Eds.), *Greener and scalable E-fuels for decarbonization of transport* (pp. 67–110). Springer. https://doi.org/10.1007/978-981-16-8344-2_4
32. Ishaq, H., Dincer, I., & Crawford, C. (2022). A review on hydrogen production and utilization: Challenges and opportunities. *International Journal of Hydrogen Energy, 47*(62), 26238–26264. https://doi.org/10.1016/j.ijhydene.2021.11.149
33. Elrhoul, D., Gómez, M. R., & Naveiro, M. (2023). Review of green hydrogen technologies application in maritime transport. *International Journal of Green Energy, 20*, 1–26. https://doi.org/10.1080/15435075.2023.2194384
34. Pan, P., Sun, Y., Yuan, C., Yan, X., & Tang, X. (2021). Research progress on ship power systems integrated with new energy sources: A review. *Renewable and Sustainable Energy Reviews, 144*, 111048. https://doi.org/10.1016/j.rser.2021.111048
35. Talluri, L., Nalianda, D. K., Kyprianidis, K. G., Nikolaidis, T., & Pilidis, P. (2016). Techno economic and environmental assessment of wind assisted marine propulsion systems. *Ocean Engineering, 121*, 301–311. https://doi.org/10.1016/j.oceaneng.2016.05.047
36. Nejadian, M. M., Ahmadi, P., & Houshfar, E. (2023). Comparative optimization study of three novel integrated hydrogen production systems with SOEC, PEM, and alkaline electrolyzer. *Fuel, 336*, 126835. https://doi.org/10.1016/j.fuel.2022.126835
37. Pope, K., Dincer, I., & Naterer, G. F. (2010). Energy and exergy efficiency comparison of horizontal and vertical axis wind turbines. *Renewable Energy, 35*(9), 2102–2113. https://doi.org/10.1016/j.renene.2010.02.013
38. Konur, O., Colpan, C. O., & Saatcioglu, O. Y. (2022). A comprehensive review on organic Rankine cycle systems used as waste heat recovery technologies for marine applications. *Energy Sources, Part A: Recovery, Utilization, and Environmental Effects, 44*(2), 4083–4122. https://doi.org/10.1080/15567036.2022.2072981
39. Senary, K., Tawfik, A., Hegazy, E., & Ali, A. (2016). Development of a waste heat recovery system onboard LNG carrier to meet IMO regulations. *Alexandria Engineering Journal, 55*(3), 1951–1960. https://doi.org/10.1016/j.aej.2016.07.027
40. Singh, D. V., & Pedersen, E. (2016). A review of waste heat recovery technologies for maritime applications. *Energy Conversion and Management, 111*, 315–328. https://doi.org/10.1016/j.enconman.2015.12.073
41. MAN, "Waste Heat Recovery System (WHRS) for Reduction of Fuel Consumption, Emission and EEDI," 2014. https://www.biofuels.co.jp/waste-heat-recovery-system.pdf
42. Mondejar, M. E., Andreasen, J. G., Pierobon, L., Larsen, U., Thern, M., & Haglind, F. (2018). A review of the use of organic Rankine cycle power systems for maritime applications. *Renewable and Sustainable Energy Reviews, 91*, 126–151. https://doi.org/10.1016/j.rser.2018.03.074
43. Olaniyi, E. O., & Prause, G. (2020). Investment analysis of waste heat recovery system installations on ships. *Engines, 8*(10), 811. https://doi.org/10.3390/jmse8100811
44. Shu, G., Liang, Y., Wei, H., Tian, H., Zhao, J., & Liu, L. (2013). A review of waste heat recovery on two-stroke IC engine aboard ships. *Renewable and Sustainable Energy Reviews, 19*, 385–401. https://doi.org/10.1016/j.rser.2012.11.034
45. d'Amore-Domenech, R., Santiago, Ó., & Leo, T. J. (2020). Multicriteria analysis of seawater electrolysis technologies for green hydrogen production at sea. *Renewable Sustainable Energy Reviews, 133*, 110166. https://doi.org/10.1016/j.rser.2020.110166
46. Aydin, M. I., Karaca, A. E., Qureshy, A. M., & Dincer, I. (2021). A comparative review on clean hydrogen production from wastewaters. *Journal of Environmental Management, 279*, 111793. https://doi.org/10.1016/j.jenvman.2020.111793
47. Dincer, I., & Zamfirescu, C. (2016). Hydrogen production by electrical energy. In *Sustainable hydrogen production* (pp. 99–161). Elsevier. https://doi.org/10.1016/B978-0-12-801563-6.00003-0

48. Nikolaidis, P., & Poullikkas, A. (2017). A comparative overview of hydrogen production processes. *Renewable Sustainable Energy Reviews, 67*, 597–611. https://doi.org/10.1016/j.rser.2016.09.044
49. Nicita, A., Maggio, G., Andaloro, A., & Squadrito, G. (2020). Green hydrogen as feedstock: Financial analysis of a photovoltaic-powered electrolysis plant. *International Journal of Hydrogen Energy, 45*(20), 11395–11408. https://doi.org/10.1016/j.ijhydene.2020.02.062
50. Priest, C. M., et al. (2023). Challenges in practical button cell testing for hydrogen production from high temperature electrolysis of water (in English). *Original Research, 11*, 2023. https://doi.org/10.3389/fenrg.2023.1278203
51. Bespalko, S., & Mizeraczyk, J. (2022). Overview of the hydrogen production by plasma-driven solution electrolysis. *Energies, 15*(20), 7508. https://doi.org/10.3390/en15207508
52. Buttler, A., & Spliethoff, H. (2018). Current status of water electrolysis for energy storage, grid balancing and sector coupling via power-to-gas and power-to-liquids: A review. *Renewable Sustainable Energy Reviews, 82*, 2440–2454. https://doi.org/10.1016/j.rser.2017.09.003
53. Abdin, Z., Zafaranloo, A., Rafiee, A., Mérida, W., Lipiński, W., & Khalilpour, K. R. (2020). Hydrogen as an energy vector. *Renewable and Sustainable Energy Reviews, 120*, 109620. https://doi.org/10.1016/j.rser.2019.109620
54. Acar, C., & Dincer, I. (2015). Impact assessment and efficiency evaluation of hydrogen production methods. *International Journal of Energy Research, 39*(13), 1757–1768. https://doi.org/10.1002/er.3302
55. Roeb, M., Agrafiotis, C., & Sattler, C. (2015). Hydrogen production via thermochemical water splitting. In *Compendium of Hydrogen Energy* (pp. 319–347). Elsevier. https://doi.org/10.1016/B978-1-78242-361-4.00011-X
56. Baykara, S. (2004). Experimental solar water thermolysis. *International Journal of Hydrogen Energy, 29*(14), 1459–1469. https://doi.org/10.1016/j.ijhydene.2004.02.011
57. Acar, C., & Dincer, I. (2018). 3.1 hydrogen production. In *Comprehensive energy systems* (Vol. 3, pp. 1–40). Elsevier. https://doi.org/10.1016/B978-0-12-809597-3.00304-7
58. Dincer, I., & Acar, C. (2018). Smart energy solutions with hydrogen options. *International Journal of Hydrogen Energy, 43*(18), 8579–8599. https://doi.org/10.1016/j.ijhydene.2018.03.120
59. Dincer, I., & Zamfirescu, C. (2016). Hydrogen production by thermal energy. In *Sustainable hydrogen production* (pp. 163–308). Elsevier. https://doi.org/10.1016/B978-0-12-801563-6.00004-2
60. Sapountzi, F. M., Gracia, J. M., Weststrate, C. J., Fredriksson, H. O. A., & Niemantsverdriet, J. W. (2017). Electrocatalysts for the generation of hydrogen, oxygen and synthesis gas. *Progress in Energy and Combustion Science, 58*, 1–35. https://doi.org/10.1016/j.pecs.2016.09.001
61. El-Emam, R. S., & Özcan, H. (2019). Comprehensive review on the techno-economics of sustainable large-scale clean hydrogen production. *Journal of Cleaner Production, 220*, 593–609. https://doi.org/10.1016/j.jclepro.2019.01.309
62. Atilhan, S., Park, S., El-Halwagi, M. M., Atilhan, M., Moore, M., & Nielsen, R. B. (2021). Green hydrogen as an alternative fuel for the shipping industry. *Current Opinion in Chemical Engineering, 31*, 100668. https://doi.org/10.1016/j.coche.2020.100668
63. GL D. (2019) "*Comparison of alternative marine fuels*," 2019–0567, Rev. 4, 2019-09-25. https://sea-lng.org/wp-content/uploads/2020/04/Alternative-Marine-Fuels-Study_final_report_25.09.19.pdf.
64. Gondal, I. A. (2019). Offshore renewable energy resources and their potential in a green hydrogen supply chain through power-to-gas. *Sustainable Energy & Fuels, 3*(6), 1468–1489. https://doi.org/10.1039/C8SE00544C
65. Qubeissi, M. A., El-Kharouf, A., Qubeissi, M. A., El-Kharouf, A., & Soyhan, H. S. (Eds.). (2020). *Renewable energy: Resources, challenges and applications*. IntechOpen. https://doi.org/10.5772/intechopen.81765

66. Lenivova V. (2020). "*The potential role of power-to-gas in the Russian-European energy system.*" master of science in environmental and resources management master thesis Faculty of Environment and Natural Sciences, Brandenburgische Technische Universität, Cottbus-Senftenberg.
67. Krishna, R., et al. (2012). Hydrogen storage for energy application. In *Hydrogen storage*. IntechOpen. https://doi.org/10.5772/51238
68. Wang, Z., Wang, Y., Afshan, S., & Hjalmarsson, J. (2021). A review of metallic tanks for H2 storage with a view to application in future green shipping. *International Journal of Hydrogen Energy, 46*(9), 6151–6179. https://doi.org/10.1016/j.ijhydene.2020.11.168
69. Fernández, I. A., Gomez, M. R., Gómez, J. R., & López-González, L. M. (2020). Generation of H2 on board LNG vessels for consumption in the propulsion system. *Polish Maritime Research, 27*(1), 83–95. https://doi.org/10.2478/pomr-2020-0009
70. Barthelemy, H., Weber, M., & Barbier, F. (2017). Hydrogen storage: Recent improvements and industrial perspectives. *International Journal of Hydrogen Energy, 42*(11), 7254–7262. https://doi.org/10.1016/j.ijhydene.2016.03.178
71. IAEA. (2013). Hydrogen production using nuclear energy. In *IAEA nuclear energy series* (Vol. NP-T-4.2). International Atomic Energy Agency (IAEA). https://www-pub.iaea.org/MTCD/Publications/PDF/Pub1577_web.pdf
72. Dincer, I., & Zamfirescu, C. (2016). Hydrogen and its production. In I. Dincer & C. Zamfirescu (Eds.), *Sustainable hydrogen production* (pp. 65–97). Elsevier. https://doi.org/10.1016/B978-0-12-801563-6.00002-9
73. Moreno-Blanco, J., Petitpas, G., Espinosa-Loza, F., Elizalde-Blancas, F., Martinez-Frias, J., & Aceves, S. M. (2019). The storage performance of automotive cryo-compressed hydrogen vessels. *International Journal of Hydrogen Energy, 44*(31), 16841–16851. https://doi.org/10.1016/j.ijhydene.2019.04.189
74. Moreno-Blanco, J., Petitpas, G., Espinosa-Loza, F., Elizalde-Blancas, F., Martinez-Frias, J., & Aceves, S. M. (2019). The fill density of automotive cryo-compressed hydrogen vessels. *International Journal of Hydrogen Energy, 44*(2), 1010–1020. https://doi.org/10.1016/j.ijhydene.2018.10.227
75. Gkanas, E., & Khzouz, M. (2018). Metal hydride hydrogen compression systems–materials, applications and numerical analysis. In *Hydrogen storage technologies* (pp. 1–37). Wiley. https://doi.org/10.1002/9781119460572.ch1
76. Dvoynikov, M., Buslaev, G., Kunshin, A., Sidorov, D., Kraslawski, A., & Budovskaya, M. (2021). New concepts of hydrogen production and storage in Arctic region. *Resources, 10*(1), 3. https://doi.org/10.3390/resources10010003
77. Faye, O., Szpunar, J. A., Szpunar, B., & Beye, A. C. (2017). Hydrogen adsorption and storage on palladium—Functionalized graphene with NH-dopant: A first principles calculation. *Applied Surface Science, 392*, 362–374. https://doi.org/10.1016/j.apsusc.2016.09.032
78. Jiang, L., & Fu, X. (2021). An ammonia–hydrogen energy roadmap for carbon neutrality: Opportunity and challenges in China. *Engineering, 7*(12), 1688–1691. https://doi.org/10.1016/j.eng.2021.11.004
79. Tronstad, T., Åstrand, H. H., Haugom, G. P., & Langfeldt, L. (2017). *Study on the use of fuel cells in shipping*. European Maritime Safety Agency. http://www.dieselduck.info/machine/01%20prime%20movers/2016%20EMSA%20Study%20on%20the%20use%20of%20Fuel%20Cells%20in%20Shipping.pdf
80. Ammar, N. R., & Alshammari, N. (2018). Overview of the green hydrogen applications in marine power plants onboard ships. *International Journal of Multidisciplinary and Current Research, 6*(01), 1. https://doi.org/10.14741/ijmcr.v6i01.10912
81. Sarafraz, M. M., Safaei, M. R., Goodarzi, M., & Arjomandi, M. (2019). Reforming of methanol with steam in a micro-reactor with cu–SiO2 porous catalyst. *International Journal of Hydrogen Energy, 44*(36), 19628–19639. https://doi.org/10.1016/j.ijhydene.2019.05.215
82. Shakeri, N., Zadeh, M., & Nielsen, J. B. (2020). Hydrogen fuel cells for ship electric propulsion: Moving toward greener ships. *IEEE Electrification Magazine, 8*, 27–43. https://

ieeexplore.ieee.org/stamp/stamp.jsp?arnumber=9110636&casa_token=OY1zppGqo3kAAAAA:I1rmGz5gn9s9goHFyGJB1ddmbZ4W5P607O1Xwkmt3O1usqPDFW7g19nP2UyEgdI1GYsmm3w11m4
83. Xing, H., Stuart, C., Spence, S., & Chen, H. (2021). Fuel cell power systems for maritime applications: Progress and perspectives. *Sustainability, 13*(3), 1213. https://doi.org/10.3390/su13031213
84. McConnell, V. P. (2010, 2010, 2010). Now, voyager? The increasing marine use of fuel cells. *Fuel Cells Bulletin*, (5), 12–17. https://doi.org/10.1016/S1464-2859(10)70166-8
85. Biert, L. V., Godjevac, M., Visser, K., & Aravind, P. V. (2016). A review of fuel cell systems for maritime applications. *Journal of Power Sources, 327*, 345–364. https://doi.org/10.1016/j.jpowsour.2016.07.007
86. Markowski, J., & Pielecha, I. (2019). The potential of fuel cells as a drive source of maritime transport. *IOP Conference Series: Earth and Environmental Science, 214*, 012019. https://doi.org/10.1088/1755-1315/214/1/012019
87. Seddiek, I. S., Elgohary, M. M., & Ammar, N. R. (2015). The hydrogen-fuelled internal combustion engines for marine applications with a case study. *Brodogradnja: Teorija i praksa brodogradnje i pomorske tehnike, 66*(1), 23–38.
88. Biert L. V., Hart M. P. T., and Mrozewski K. (2021). *"Public final report: Inventory of the application of fuel cells in the MARitime sector (FCMAR)."* https://www.koersenvaart.nl/files/MIIP%20007-2020%20FCMAR%2003022021.pdf.
89. Prussi, M., Scarlat, N., Acciaro, M., & Kosmas, V. (2021). Potential and limiting factors in the use of alternative fuels in the European maritime sector. *Journal of Cleaner Production, 291*, 125849. https://doi.org/10.1016/j.jclepro.2021.125849
90. Viana, M., et al. (2020). Estimated health impacts from maritime transport in the Mediterranean region and benefits from the use of cleaner fuels. *Environment International, 138*, 105670. https://doi.org/10.1016/j.envint.2020.105670
91. Korberg, A. D., Brynolf, S., Grahn, M., & Skov, I. R. (2021). Techno-economic assessment of advanced fuels and propulsion systems in future fossil-free ships. *Renewable and Sustainable Energy Reviews, 142*, 110861. https://doi.org/10.1016/j.rser.2021.110861
92. Delgado M. S. (2019). *"Desarrollo y validación de un modelo para la simulación de sistemas de electrólisis alcalina para la producción de hidrógeno a partir de energías renovables."* Doctoral, Energía y Combustibles, E.T.S.I. de Minas y Energía (UPM).
93. AspenTech. (2021). *"Integrated Custom Model of Alkaline Electrolysis System for H_2 Production"*.
94. Zaccara, A., et al. (2020). Renewable hydrogen production processes for the off-gas valorization in integrated steelworks through hydrogen intensified methane and methanol syntheses. *Metals, 10*(11), 1535. https://doi.org/10.3390/met10111535
95. Im-orb, K., Visitdumrongkul, N., Saebea, D., Patcharavorachot, Y., & Arpornwichanop, A. (2018). Flowsheet-based model and exergy analysis of solid oxide electrolysis cells for clean hydrogen production. *Journal of Cleaner Production, 170*, 1–13. https://doi.org/10.1016/j.jclepro.2017.09.127
96. Wang, F., Wang, L., Zhang, H., Xia, L., Miao, H., & Yuan, J. (2021). Design and optimization of hydrogen production by solid oxide electrolyzer with marine engine waste heat recovery and ORC cycle. *Energy Conversion and Management, 229*, 113775. https://doi.org/10.1016/j.enconman.2020.113775
97. Mohammadi, A., & Mehrpooya, M. (2018). Techno-economic analysis of hydrogen production by solid oxide electrolyzer coupled with dish collector. *Energy Conversion and Management, 173*, 167–178. https://doi.org/10.1016/j.enconman.2018.07.073
98. Ni, M., Leung, M. K. H., & Leung, D. Y. C. (2007). Parametric study of solid oxide steam electrolyzer for hydrogen production. *International Journal of Hydrogen Energy, 32*(13), 2305–2313. https://doi.org/10.1016/j.ijhydene.2007.03.001
99. Visitdumrongkul, N., Tippawan, P., Authayanun, S., Assabumrungrat, S., & Arpornwichanop, A. (2016). Enhanced performance of solid oxide electrolysis cells by integration with a partial

oxidation reactor: Energy and exergy analyses. *Energy Conversion and Management, 129*, 189–199. https://doi.org/10.1016/j.enconman.2016.10.023

100. Wang, F., et al. (2021). Thermodynamic analysis of solid oxide electrolyzer integration with engine waste heat recovery for hydrogen production. *Case Studies in Thermal Engineering, 27*, 101240. https://doi.org/10.1016/j.csite.2021.101240
101. Zhao, Y., et al. (2021). System level heat integration and efficiency analysis of hydrogen production process based on solid oxide electrolysis cells. *International Journal of Hydrogen Energy, 46*(77), 38163–38174. https://doi.org/10.1016/j.ijhydene.2021.09.105
102. AlZahrani A. A. (2017) "*Investigation of a novel high temperature solid oxide Electrolyzer for solar hydrogen production*," Doctor of Philosophy, University of Ontario Institute of Technology.
103. Luo, Y., Wu, X.-y., Shi, Y., Ghoniem, A. F., & Cai, N. (2018). Exergy analysis of an integrated solid oxide electrolysis cell-methanation reactor for renewable energy storage. *Applied Energy, 215*, 371–383. https://doi.org/10.1016/j.apenergy.2018.02.022
104. Zhang, S., Zhang, N., Smith, R., & Wang, W. (2022). A zero carbon route to the supply of high-temperature heat through the integration of solid oxide electrolysis cells and H2–O2 combustion. *Renewable and Sustainable Energy Reviews, 167*, 112816. https://doi.org/10.1016/j.rser.2022.112816
105. GTT. https://gtt.fr/technologies/markiii-systems
106. Man, "*MAN B&W G70ME-C10.5-GI. Project Guide Electronically Controlled Dual Fuel Two-stroke Engines*," 2023, https://man-es.com/applications/projectguides/2stroke/content/printed/G70ME-C10_5-GI.pdf.
107. Naveiro, M. (2018). *Sistemas De Recuperación De Calor Residual En Buques*. Universidade Da CoruñA.

Examining the Effect of Port Injection (PI) and Direct Injection (DI) of Hydrogen on Engine Performance and Emissions in Diesel Engines

Kubilay Bayramoğlu and Tolga Bayramoğlu

1 Introduction

Energy is recognised as a significant factor in determining the economic development of any society and as one of the important factors indicating the quality of life. In most countries, energy policy is well exemplified by the country's growing energy shortage and the increasing emphasis on searching for alternative energy sources [1]. A range of variables, including rising living standards, an expanding global population and the accelerating industrialisation process, are causing a large increase in the worldwide demand for energy [2]. Most of the energy sources used today are fossil fuels, which cause various problems, such as climate change and global warming [3]. CO_2 emissions from fossil fuels are one of the leading environmental problems [4]. Therefore, the use of more environmentally friendly energy sources is being mandated by national and international regulations [5]. In line with the regulations, it is envisaged that renewable energy sources or more environmentally friendly fuels will play an active role instead of marine diesel engines, especially in transport [6]. The only chemical without carbon atoms is the H_2 molecule, which provides a clean combustion in terms of greenhouse emissions [7]. Hydrogen is widely recognised as a high-quality fuel for diesel engines due to its low ignition delay and immediate flame speed. It may control the ignition of fuel and the process

K. Bayramoğlu
Department of Mechanical Engineering, Zonguldak Bulent Ecevit University, Zonguldak, Turkey
e-mail: kubilay.bayramoglu@beun.edu.tr

T. Bayramoğlu (✉)
Program of Motor Vehicles and Transportation Technologies, Zonguldak Bulent Ecevit University, Zonguldak, Turkey
e-mail: t.bayramoglu@beun.edu.tr

C. O. Colpan et al. (eds.), *Fuel Cell and Hydrogen Technologies in Maritime Transportation*, https://doi.org/10.1007/978-3-031-86110-9_8

of combustion for internal combustion engines using the low-burn technique [8]. These properties prove hydrogen's high potential as a clean energy source.

There are various techniques for internal combustion engines utilising hydrogen. These techniques have been discussed in different studies. Babayev et al. [9] have tested with a non-premixed combustion model using a hydrogen engine that uses pilot H_2 fuel to ignite a direct injection (DI) compression ignition (CI) hydrogen engine. Wang et al. [10] discussed and simulated the feasibility of adding ammonia and hydrogen with air from the intake manifold. Du et al. [11] carried out combustion performance and emissions in diesel engines by hydrogen addition at different ratios into the combustion chamber by direct injection method. According to the data obtained, as the hydrogen addition rate increases, increased mean effective pressure, enhanced combustion speed and better thermal efficiency can be observed. At an excess air coefficient of 1.5, a significant reduction in CO and NO_X and an increase in effective thermal efficiency were observed. Kim et al. [12] experimentally tested the impact of hydrogen to improve the combustion instability of gasoline engines. They found that as the addition of hydrogen rate increases, thermal efficiency rises, combustion is enhanced and all emissions aside from NO_X are decreased. By computational fluid dynamics (CFD) simulation. Zhu and Duan [13] investigated in-cylinder combustion progression and NO production in hydrogen-fuelled internal combustion engines. It was observed that rapid turbulent combustion occurs and OH concentration decreases rapidly, while NO mass increases at peak temperatures. Zhu and Duan [13] conducted emission and engine performance using pure hydrogen in internal combustion engines. It was observed that CO and total hydrocarbon emissions decreased while NOx increased. Jiang et al. [14] researched the engine performance and emission characteristics of hydrogen and ammonia combinations at variable equivalence ratios and intake air temperatures. They numerically investigated combustion times, combustion temperature and NOx evolution mechanisms. Manzoor et al. [15] numerically investigated the combustion modes of hydrogen with different spark timings, equivalence ratios and compression ratios. Akar and Özener [16] conducted the mixing behaviour of hydrogen in the cylinder in a direct injection hydrogen internal combustion engine. Chen et al. [17] have tested how different hydrogen direct injection techniques affected the combustion process and hydrogen–air mixture. They found that the gaseous fuel concentration gradually stratifies from the ignition centre towards the periphery region, according to the optimal mixture stratification law. Liang et al. [18] investigated the effect of combined split injection techniques and various valve timings on hydrogen direct injection engine combustion, efficiency and emissions. Musy et al. [19] investigated engine performance and emissions in hydrogen internal combustion engines using port injection (PI) and direct injection methods. Ramsay et al. [20] conducted a comprehensive CFD analysis to explore the efficiency of a dual-fuel used hydrogen-diesel fuel in the diesel engine using a new combination technique where most diesel and combustion parameters injection occurs in a certain volume. Singh et al. [21] examined thorough research on combustion parameters throughout the whole operational map to improve the performance of turbocharged direct injection hydrogen internal combustion engines. Boretti [22] studied different hydrogen injection modes to ensure that hydrogen injected through the intake port does not block the intake port injection of the intake air. While some studies have investigated how hydrogen burns in the combustion chamber, many studies have assessed how hydrogen affects emissions and combustion efficiency. According to the literature search, the most popular approach usually involves injecting hydrogen into the intake manifold rather than the combustion chamber directly [22].

Literature studies have generally dealt with the combustion processes of hydrogen in internal combustion engines either as a mixture of different fuels or by sending it from the intake manifold to the combustion chamber. However, the widespread use of hydrogen in recent years also includes the processes of pilot injection or injection directly to the combustion zone with diesel fuel alone. In a novel approach, this study numerically evaluated the effect of direct injection and port injection processes on engine combustion performance and harmful exhaust products.

In this study, the combined use of diesel fuel and hydrogen fuel was investigated. The hydrogen additive is injected with the diesel fuel through an injector, either directly or through the intake port with intake air. The variation of combustion chamber pressure and combustion temperature, which are engine performance indicators, with crank angle (CA) was obtained for both injection types. CO and NO_x emissions are also presented comparatively for both port injection (PI) and direct injection (DI) conditions. Thus, this study will make important contributions to the literature in terms of selecting hydrogen injection strategies with future engine technologies.

2 Methodology

Hydrogen addition in diesel engines takes place in two different ways. These are the direct injection of hydrogen through an injector and the port injection of hydrogen from the intake air [23]. Port injection is a standard fuel injection system for internal combustion engines. During the intake stroke, fuel is injected into the intake manifold. Modifying traditional port injection of internal combustion engines to hydrogen involves a relatively simple modification of the injection system. In port injection systems, the engine displaces some of the combustion air, limiting the power density of the engine and eliminating undesirable conditions such as knocking. The port injection also increases the work required during the compression stroke compared to direct injection. These factors generally reduce power output and efficiency in hydrogen port injection engines [24].

One potential strategy to boost engine performance in internal combustion engines is to use an injector to inject hydrogen straight into the combustion chamber along with pilot diesel fuel [25]. This will prevent the port injection arrangement from firing back because fuel injection happens after the intake valves have closed all the way. In addition, pre-ignition problems may be somewhat prevented by limiting the amount of time the hydrogen blend is exposed to hot regions. If

Table 1 Specifications of the diesel engine

Specification	Units	Value and specification
Engine type	–	Single cylinder-caterpillar 3400 Heavy-duty engine
Bore × stroke	mm	137 × 165.1
Number of nozzles	–	6 × 0.23
Inlet valve closing (IVC)	CA	190.3
Exhaust valve opening (EVO)	CA	505.3

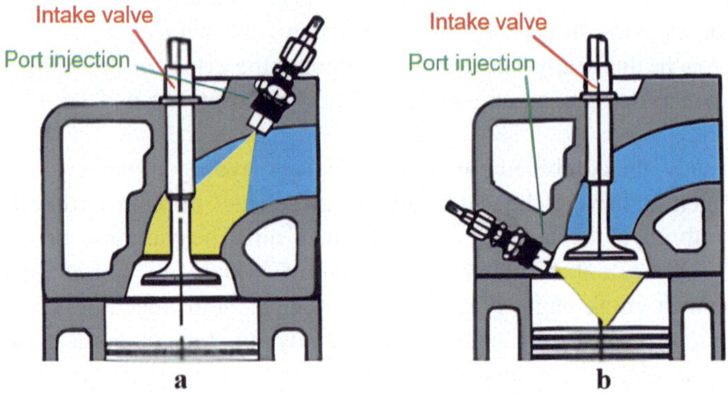

Fig. 1 Schematic representation of hydrogen injection methods in a diesel engine: (**a**) port injection and (**b**) direct injection

injection occurs after the intake ports are closed, the volumetric efficiency decreases with port injection which is caused by the air movement of hydrogen, as previously mentioned, and becomes insignificant. Hydrogen pressure for injection (≥100 bar) is needed to counteract the elevated in-cylinder pressure given that injecting fuel later in the stroke of compression. To support high-load operation, increased pressure of injection may also boost fuel mass flow rate compared to standard low-pressure port injection, which can supply more fuel energy with the equivalent injection period [26, 27]. Therefore, port injection and direct injection have different advantages over each other. In the study, the combustion model was carried out in a 3D diesel engine combustion chamber. Diesel engine specifications are given in Table 1. The selection of the specified motor is based on the experimental and numerical analysis of the motor, which is widely available in the literature. Thus, the numerical studies are verified with numerical and experimental studies in the literature. Thus, new numerical studies are verified with the given test engine enabling the parametrical derivation of new studies. Basic engine specifications are expressed as bore stroke dimensions in mm, inlet valve closing (IVC) and exhaust valve opening (EVO) moments in crank angle (CA). IVC and EVO are essential parameters in the study because combustion analyses are performed for a closed volume where all ports are closed.

The combustion process was executed for a four-stroke diesel engine with a single-cylinder diesel engine. Figure 1a shows the port hydrogen injection process with an intake valve, and Fig. 1b shows the direct hydrogen injection process. Direct hydrogen injection is provided by a different injector in addition to diesel injection, which is the pilot fuel. Port injection provides hydrogen injection with intake air with an injector added from the intake manifold.

To achieve a deeper comprehension of the combustion process in diesel engines and the emissions released from the exhaust gas in numerical studies, some software is needed to analyse the flow and chemical reactions [28]. In combustion processes, to correctly predict the combustion reactions of the reacting fuels, the combustion reactions to be carried out under the appropriate ambient conditions of the selected fuel must be defined in the system. Considering all these conditions, the reduction reactions of appropriate chemical reactions should be verified by experimental data with the help of appropriate software. In numerical analyses of internal combustion engines, utilising a suitable chemical reduction approach is an extremely important factor in creating an efficient combustion model of the chemical fuel used as fuel. Although the broad-based reduction reactions used in numerical analyses of diesel engines and the computational performance of computers accelerate the solution, the size of the chemical reactions is also very critical. The chemical reaction magnitude is increased by the quantity of fuel components and the significant amount of carbon atoms in fuel types. In combustion reactions, various fuel types with varying types and numbers of reactions employ distinct reduction reactions. Structures that have been decreased to simpler mechanisms of reaction with fewer species and reactions are employed in place of intricate reduction processes in the reactions of combustion. In this section, the differences in the same combustion for different reduction reactions are analysed [5]. In the combustion analysis, the chemical kinetic mechanism was provided by Chemkin software, and the n-heptane reduction mechanism was used [29]. A chemical reduction mechanism containing a total of 35 species and 173 reactions was used [5, 13]. The combustion chamber flow conditions are turbulent, and the RNG k-epsilon turbulence model is used [13]. Using the appropriate turbulence model is crucial for engineering flow and combustion simulations. Especially in flow problems, k-ω, RNG k-ε and standard k-ε turbulence models are employed as stated in the correct estimation of solving the Navier–Stokes equations with a Reynolds average [30]. The RNG k-ε turbulence model derived from the k-ε turbulence model is used. The RNG k-ε model is more effective than the k-ε model despite having many similarities. These differences include the inclusion of a source term for the interactions between mean shear values and turbulence development, precise estimation of vortex in the combustion zone, and the availability of differential and mathematical equations for the Prandtl number [31].

In the given data, combustion chamber temperature limit conditions and intake air conditions for port injection are given. Within the scope of the study, the amount of hydrogen supplied to the system is the same for both conditions. All other combustion conditions and boundary conditions were taken as exact, and numerical analyses were carried out [14]. Boundary conditions in the engine geometry designed in CFD applications are parameters that will affect the analysis, such as inlet, outlet, moving

Table 2 Initial and boundary conditions of the computational model

Boundary and initial conditions	Units	Value and conditions
Temperature of cylinder head	K	400
Temperature of cylinder	K	400
Temperature of piston	K	400
Diesel starts of injection (SOI)	CA	340
Hydrogen starts of injection	CA	340
Injection duration	CA	20
Fuel injection model	–	Gaz jet model
Initial air-hydrogen temperature	K	298.15
Initial air-hydrogen pressure	Bar	1.2
Initial turbulence-kinetic energy	Cm/s^2	10,000

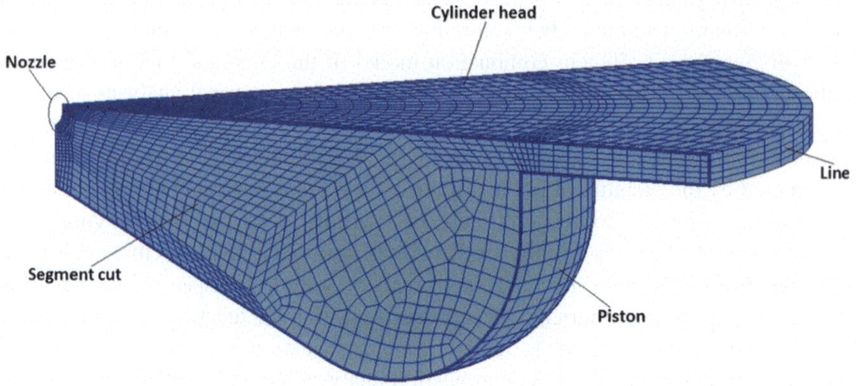

Fig. 2 Boundary conditions and mesh structure of the computational model

piston region, cylinder head temperature, piston temperature and port temperatures, engine speed, spray advance and fuel quantity. In determining the boundary conditions, the engine analysed was determined with the data taken from the catalogue and experimental studies. Table 2 shows the computational model boundary and initial conditions.

Approximately 400 k structural mesh is applied to the numerical model, and the numerical model boundary conditions are given in Fig. 2. The numerical model mainly consists of piston, cylinder head, line and nozzle boundary conditions.

Diesel engine combustion analyses were carried out by computational fluid dynamics method with Ansys Forte software [32]. The combustion phenomena in diesel engines are defined as chemical reactions that occur because of the combination of fuel injected into the combustion chamber under suitable conditions with enough air. The combustion analysis that was carried out inside the parameters of the research, was conducted by modelling the 60° sector model of the 3-D model using the catalogue and experimental data of a diesel engine with a single cylinder and direct injection. Experimental and numerical analysis results were compared, as

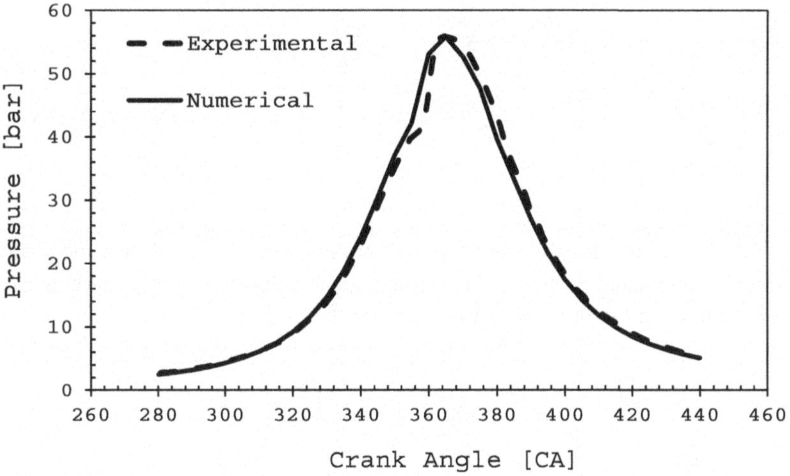

Fig. 3 Validation of the computational model

shown in Fig. 3. It was determined that the relationship between the test results derived from the experimental analysis [28] and the numerical study data was in agreement.

2.1 Governing Equations

Turbulent flows are the most complex problems to be modelled because they involve different parameter variables. To minimise the Navier–Stokes equations' computational period and obtain the approximate solution, the calculations are separated into time-averaged and unsteady parts [33]. Eq. (1) shows the standardised expressions used in the approximate solution of the Reynolds-averaged Navier–Stokes (RANS) equations.

$$\phi = \underline{\phi} + \phi' \tag{1}$$

where ϕ is any dependent variable, $\underline{\phi}$ is expressed as the sum of its mean value and ϕ' the sum of the values in the variable state. The conservation of fundamental mass, momentum and energy expressions are given by Eqs. (2), (3) and (4), respectively [34].

$$\frac{\partial \rho}{\partial t} + \frac{\partial \rho u_j}{\partial x_j} = \mathrm{S} \tag{2}$$

$$\frac{\partial \rho u_i}{\partial t}+\frac{\partial \rho u_i u_j}{\partial x_j}=-\frac{\partial P}{\partial j}+\frac{\partial \sigma_{ij}}{\partial x_j}+S \tag{3}$$

$$\frac{\partial \rho e}{\partial t}+\frac{\partial u_j \rho e}{\partial x_j}=-P\frac{\partial u_j}{\partial x_j}+\sigma_{ij}\frac{\partial u_i}{\partial x_j}+\frac{\partial}{\partial x_j}\left(K\frac{\partial T}{\partial x_j}\right)+\frac{\partial}{\partial x_j}\left(\rho D\sum\nolimits_m h_m\frac{\partial \lambda_m}{\partial x_j}\right)+S \tag{4}$$

The RNG k-ε model is a turbulence model that is more correct than the standard k-ε model, which can be corrected using statistical techniques. The equations for the RNG k-ε model are expressed as Eqs. (5) and (6) [35].

$$\frac{\partial \rho}{\partial t}k+\frac{\partial u_i \rho}{\partial i}k=\frac{\partial}{\partial x_j}\left[\left(\mu+\frac{\mu_t}{\sigma_k}\right)\frac{\partial k}{\partial x_j}\right]+P_k+\rho\epsilon \tag{5}$$

$$\frac{\partial \rho}{\partial t}\epsilon+\frac{\partial u_i \rho}{\partial i}\epsilon=\frac{\partial}{\partial x_j}\left[\left(\mu+\frac{\mu_t}{\sigma_\epsilon}\right)\frac{\partial \epsilon}{\partial x_j}\right]+C_{1\epsilon}\frac{\epsilon}{k}P_k+C_{2\epsilon}{}^{*}\rho\frac{\epsilon^2}{k} \tag{6}$$

In the given equations, C1ε, and C2ε are the model turbulence model constants expressed. Also, k is the kinetic energy of turbulence, and ϵ is the dissipation rate of this energy.

3 Results and Discussion

Two distinct methods exist by which hydrogen fuel for diesel engines is delivered to the combustion chamber. The first of these is spraying hydrogen fuel with different injectors with diesel, which is the pilot fuel, through an injector. An alternative approach is to send intake air and hydrogen fuel from the intake manifold together. The findings of the investigation objectives included examining how two distinct strategies for injecting hydrogen impacted emissions and engine combustion performance. The combustion chamber pressure value is a critical factor in determining engine performance. Engine power and indicated mean effective pressure are determined by measuring the combustion pressure in the diesel combustion zone. The combustion chamber calculated pressure is displayed in Fig. 4. The pressure of the combustion chamber is higher for the port injection condition. This is due to the more homogeneous distribution of air and hydrogen in the diesel engine combustion chamber in the port injection condition. Considering the combustion processes, the ignition delay between when the fuel is sprayed and when the pressure increases suddenly shows similar characteristics for both injection types. This situation is similar to the combustion processes taken from the literature [36, 37]. Better combustion is facilitated by the finer atomisation of the fuel, which decreases ignition delay as the crank angle increases and improves the combustion chamber pressure during combustion profiles [38].

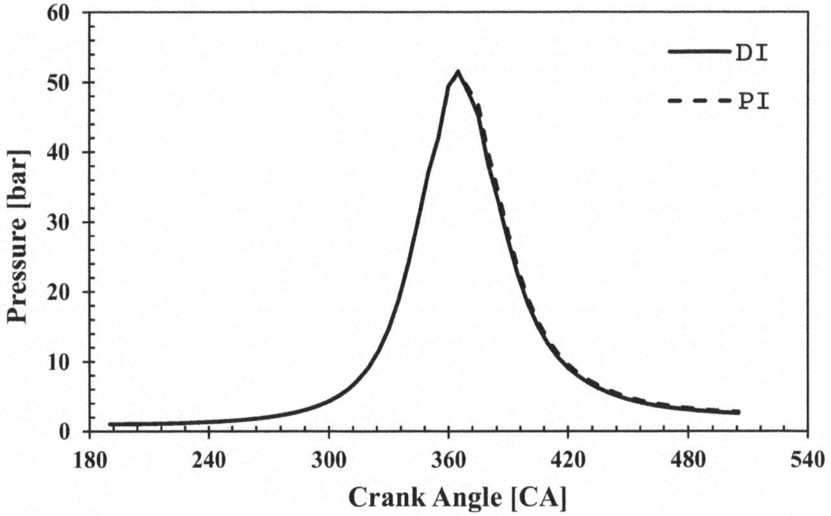

Fig. 4 The effect of crank angle on the combustion chamber

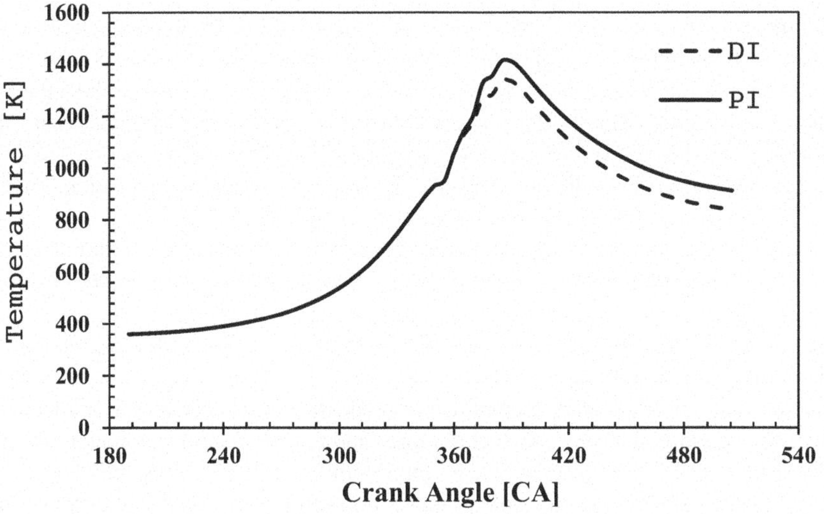

Fig. 5 Variation of combustion temperature against crank angle

The combustion chamber temperature change versus crank angle is shown in Fig. 5. The combustion chamber temperature is higher under port injection conditions in proportion to the pressure. Increased combustion efficiency is another result of the efficient air–fuel combination. While the maximum combustion chamber temperature is about 1500 K for port injection conditions, it is about 1400 K for direct injection conditions. In internal combustion engines, variable combustion

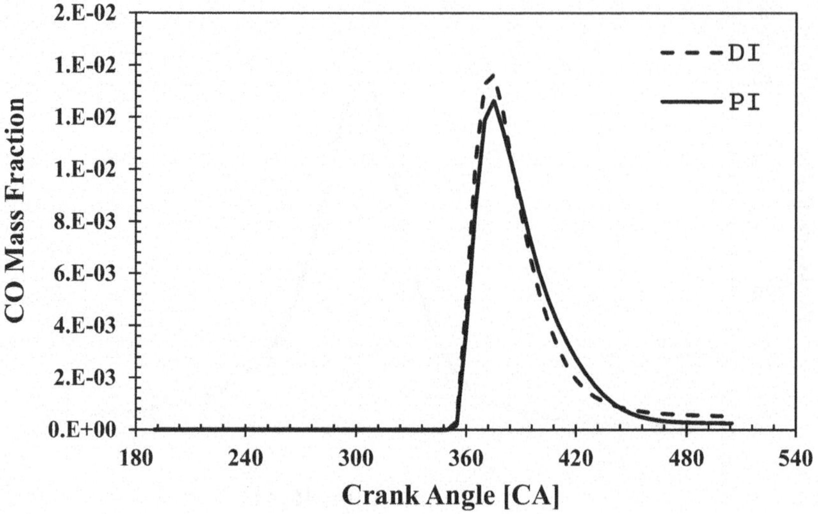

Fig. 6 The change of CO mole fraction with crank angle

parameters affect the combustion process. One of the critical parameters is the adiabatic flame temperature. For the process of combustion that occurs adiabatically without any shaft action, the temperature of the flue gas is referred to as the adiabatic flame temperature. This is the highest temperature that may be attained with the specific reactants. Lower temperatures are caused by dissociation, partial combustion and heat transfer. When the fuel and oxidiser are combined in a stoichiometric mixture, that is, when the proportions are correct, and all of the fuel and oxidiser are consumed, the highest adiabatic flame temperature is reached. To regulate the adiabatic flame temperature, the amount of extra air can be specifically designed into the system [39, 40]. Under port injection conditions, hydrogen is introduced into the system in a mixture with air, so the air–fuel mixture is more effective than under direct injection conditions.

Emissions in diesel engine combustion processes vary according to combustion chamber pressure, temperature, fuel mixtures and type. CO emissions decrease with increasing combustion efficiency. The minimum level of CO emissions is proportional to the effective air–fuel blend and high performance of diesel combustion. Figure 6 shows the change in carbon monoxide pollutants within the combustion zone against the crank angle. It is seen that CO emissions are higher for direct injection conditions than port injection. This is due to the higher combustion efficiency for port injection. Basically, the formation of CO is a result of insufficient combustion. Incomplete combustion occurs when there is insufficient air supply, so only around one-half as much oxygen is attached to the carbon and carbon monoxide is formed. Unlike carbon dioxide, carbon monoxide cannot exist by nature in the atmosphere. It is produced when diesel fuel is not combusted completely. Carbon monoxide is produced in the combustion chamber by low oxygen levels and low

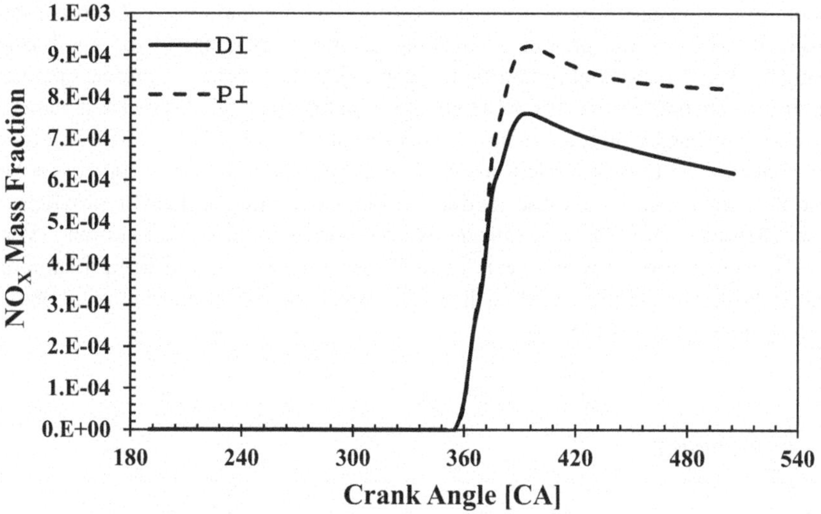

Fig. 7 Variation of NO_x emissions of diesel engine against crank angle

temperatures [41]. Also, in cases of insufficient combustion, some of the CO_2 components are converted into CO [42]. Therefore, in parallel with temperature and combustion conditions, combustion efficiency for port injection (PI) conditions is higher than for direct injection (DI).

NO_x emissions are formed in three different ways: thermal NO, prompt NO and fuel NO. Thermal NO is formed in proportion to the temperature of the combustion chamber. Prompt NO is formed in the combustion process and fuel NO is formed depending on the nitrogen elements in the structure of the fuel. Figure 7 illustrates how the crank angle affects the NO levels in the combustion chamber of a diesel engine. It is evident that with port injection conditions of use, the amount of NO in the combustion chamber is higher than with direct injection conditions. The higher temperature for the port injection condition can be shown as the main reason for this situation. In numerical studies of internal combustion engines, chemical reduction mechanisms are the main parameter for NO_X and other emissions. In combustion processes, flames spread through mixtures of different species capable of exploding quickly enough. A detailed and accurate understanding of chemical kinetic phenomena makes the focus on explosion reactions possible. Therefore, recent studies have focused on models that accurately predict the processes taking place in the combustion chamber. In this study, the exact chemical reduction mechanism was used for both types of injection. Most complex impurities from combustion reactions are released in stationary reactions, usually at low temperatures and immediately after the combustion step [43]. After determining the appropriate chemical reduction mechanisms, the combustion temperature, which is the main parameter directly affecting NO_X formation, is determined. Nitrogen oxides (NO_X) are chemical compounds formed in the combustion chamber because of the inability to decompose substances such as N_2 and O_2 into atomic particles and sub-particles during the combustion process, depending on the temperature of the combustion chamber. The process that produces nitrogen oxides is affected by engine operating parameters such as air excess coefficient, ignition timing, engine speed and manifold pressure. Nitrogen oxides are mainly composed of NO and NO_2 [44]. The Zeldovich mechanism used to describe nitrous oxide emissions in combustion processes best describes the formation of these chemical compounds. This mechanism provides the basic chemical steps for most nitrous oxide products for engine conditions. Much work has been done in recent years on acid formation for nitric acid formation and degradation. The original Zeldovich mechanism expresses important reactions for nitrous oxide products [45, 46].

4 Conclusions

The gradual decrease in fossil fuel resources in the world, national and international regulations to reduce the amount of harmful gases emitted into the atmosphere from internal combustion engines and the greenhouse effect of harmful gases emitted into the atmosphere have made it necessary to use alternative fuels to the fuels used in diesel engines, to change the engine geometry and spray parameters. One of the fuels most likely to be used in the coming regulations is hydrogen. However, hydrogen cannot be used alone in internal combustion engines today due to its high ignition and energy potential, as it causes knocking in engines. Therefore, different techniques exist for using it as an add-on to conventional fuels. The main ones are direct injection and port injection. This research looked at the consequences of combining diesel and hydrogen fuel. The combustion chamber was supplied with hydrogen fuel through both port and direct injection. According to the evaluated data results, the following conclusions can be given:

- It has been observed that hydrogen supplied via port injection mixes homogeneously with air at the entrance to the combustion region and accordingly, the measured data of temperature and pressure in the diesel engine combustion chamber are greater than direct injection. According to these data, combustion is more efficient in port injection.
- Even though it is well recognised that increased combustion efficiency reduces CO emissions, the data collected showed that using port injection results in lower CO emissions.
- It was determined that NOx emissions increase with temperature and that port injection configuration produces more NOx emissions than direct injection.

These results showed that using port injection results in a more uniform hydrogen–air mixture, which causes more efficient combustion. Additionally, it was discovered that depending on temperature, it results in a drop in CO emissions but a rise in the emission of NO_X.

References

1. Nguyen, V. N., Ganesan, N., Ashok, B., et al. (2024). Eco-friendly perspective of hydrogen fuel addition to diesel engine: An inclusive review of low-temperature combustion concepts. *International Journal of Hydrogen Energy, 104*, 344. https://doi.org/10.1016/j.ijhydene.2024.05.454
2. Tian, E., Lv, G., & Li, Z. (2024). Evaluation of emission of the hydrogen-enriched diesel engine through machine learning. *Energy, 132303*, 132303. https://doi.org/10.1016/j.energy.2024.132303
3. Bayramoğlu, K., Yılmaz, S., & Kaya, K. D. (2019). Numerical investigation of valve lifts effects on performance and emissions in diesel engine. *Int J Glob Warm, 18*, 287–303. https://doi.org/10.1504/IJGW.2019.101088
4. El-Adawy, M., Nemitallah, M. A., & Abdelhafez, A. (2024). Towards sustainable hydrogen and ammonia internal combustion engines: Challenges and opportunities. *Fuel, 364*, 131090. https://doi.org/10.1016/j.fuel.2024.131090
5. Bayramoğlu, K., & Nuran, M. (2020). Analyzing the effect of fuel injection timing and injection duration on performance and emissions in diesel engines. *J ETA Marit Sci, 8*, 38–52. https://doi.org/10.5505/jems.2020.99705
6. Zhang, X., Lin, Z., Hao, H., et al. (2024). Environment-economic comparison of potential alternative fuel heavy-duty trucks in China. *Transportation Research Part D Transport and Environment, 131*, 104206. https://doi.org/10.1016/j.trd.2024.104206
7. Stępień, Z. (2021). A comprehensive overview of hydrogen-fueled internal combustion engines: Achievements and future challenges. *Energies, 14*, 6504. https://doi.org/10.3390/en14206504
8. Long, Y., Li, G., Zhang, Z., et al. (2022). Hydrogen-rich gas generation via the exhaust gas-fuel reformer for the marine LNG engine. *International Journal of Hydrogen Energy, 47*, 14674–14686. https://doi.org/10.1016/j.ijhydene.2022.02.188
9. Babayev, R., Andersson, A., Dalmau, A. S., et al. (2021). Computational characterization of hydrogen direct injection and nonpremixed combustion in a compression-ignition engine. *International Journal of Hydrogen Energy, 46*, 18678–18696. https://doi.org/10.1016/j.ijhydene.2021.02.223
10. Wang, Y., Zhou, X., & Liu, L. (2021). Theoretical investigation of the combustion performance of ammonia/hydrogen mixtures on a marine diesel engine. *International Journal of Hydrogen Energy, 46*, 14805–14812. https://doi.org/10.1016/j.ijhydene.2021.01.233
11. Du, Y., Yu, X., Wang, J., et al. (2016). Research on combustion and emission characteristics of a lean burn gasoline engine with hydrogen direct-injection. *International Journal of Hydrogen Energy, 41*, 3240–3248. https://doi.org/10.1016/j.ijhydene.2015.12.025
12. Kim, J., Chun, K. M., Song, S., et al. (2018). Hydrogen effects on the combustion stability, performance and emissions of a turbo gasoline direct injection engine in various air/fuel ratios. *Applied Energy, 228*, 1353–1361. https://doi.org/10.1016/j.apenergy.2018.06.129
13. Zhu, H., & Duan, J. (2019). Research on emission characteristics of hydrogen fuel internal combustion engine based on more detailed mechanism. *International Journal of Hydrogen Energy, 44*, 5592–5598. https://doi.org/10.1016/j.ijhydene.2018.08.044
14. Jiang, M., Sun, W., Guo, L., et al. (2024). Numerical study of the combustion process and NOx evolution mechanism of ammonia-hydrogen compound fuel engine under different intake conditions. *Journal of the Energy Institute, 116*, 101759. https://doi.org/10.1016/j.joei.2024.101759
15. Manzoor, M. U., Dou, X., Yosri, M. R., et al. (2024). Capturing different modes of hydrogen combustion in a spark-ignition engine using numerical simulations. *Fuel, 375*, 132343. https://doi.org/10.1016/j.fuel.2024.132343
16. Akar, F., & Özener, O. (2024). Numerical investigation of in-cylinder gas motion dynamics in a heavy-duty direct injection hydrogen internal combustion engine. *International Journal of Hydrogen Energy, 86*, 730–741. https://doi.org/10.1016/j.ijhydene.2024.08.338

17. Chen, W., Lu, C., Zuo, Q., et al. (2024). Combustion characteristics analysis and performance evaluation of a hydrogen engine under direct injection plus lean burn mode. *Journal of Cleaner Production, 470*, 143323. https://doi.org/10.1016/j.jclepro.2024.143323
18. Liang, Z., Xie, F., Li, Q., et al. (2024). Co-optimization and prediction of high-efficiency combustion and zero-carbon emission at part load in the hydrogen direct injection engine based on VVT, split injection and ANN. *Energy, 308*, 133038. https://doi.org/10.1016/j.energy.2024.133038
19. Musy, F., Ortiz, R., Ortiz, I., & Ortiz, A. (2024). Hydrogen-fuelled internal combustion engines: Direct injection versus port-fuel injection. *International Journal of Hydrogen Energy*. https://doi.org/10.1016/j.ijhydene.2024.07.136
20. Ramsay, C. J., Dinesh, K. K. J. R., Fairney, W., & Vaughan, N. (2020). A numerical study on the effects of constant volume combustion phase on performance and emissions characteristics of a diesel-hydrogen dual-fuel engine. *International Journal of Hydrogen Energy, 45*, 32598–32618. https://doi.org/10.1016/j.ijhydene.2020.09.021
21. Singh, K., Dwivedi, G., Verma, T. N., & Shukla, A. K. (2024). Energy, exergy, emissions and sustainability assessment of hydrogen supplemented diesel dual fuel turbocharged common rail direct injection diesel engine. *International Journal of Hydrogen Energy, 104*, 378. https://doi.org/10.1016/j.ijhydene.2024.05.080
22. Boretti, A. (2020). Hydrogen internal combustion engines to 2030. *International Journal of Hydrogen Energy, 45*, 23692–23703. https://doi.org/10.1016/j.ijhydene.2020.06.022
23. Gong, C., Li, Z., Yi, L., & Liu, F. (2019). Comparative study on combustion and emissions between methanol port-injection engine and methanol direct-injection engine with H2-enriched port-injection under lean-burn conditions. *Energy Conversion and Management, 200*, 112096. https://doi.org/10.1016/j.enconman.2019.112096
24. Yip, H. L., Srna, A., Yuen, A. C. Y., et al. (2019). A review of hydrogen direct injection for internal combustion engines: Towards carbon-free combustion. *Applied Sciences, 9*, 4842. https://doi.org/10.3390/app9224842
25. Mohammadi, A., Shioji, M., Nakai, Y., et al. (2007). Performance and combustion characteristics of a direct injection SI hydrogen engine. *International Journal of Hydrogen Energy, 32*, 296–304. https://doi.org/10.1016/j.ijhydene.2006.06.005
26. Dimitriou, P., & Tsujimura, T. (2017). A review of hydrogen as a compression ignition engine fuel. *International Journal of Hydrogen Energy, 42*, 24470–24486. https://doi.org/10.1016/j.ijhydene.2017.07.232
27. Matthias, N. S., Wallner, T., & Scarcelli, R. (2012). A hydrogen direct injection engine concept that exceeds U.S. DOE light-duty efficiency targets. *SAE International Journal of Engines, 5*, 838–849. https://doi.org/10.4271/2012-01-0653
28. Yousefi, A., Birouk, M., & Guo, H. (2017). An experimental and numerical study of the effect of diesel injection timing on natural gas/diesel dual-fuel combustion at low load. *Fuel, 203*, 642–657. https://doi.org/10.1016/J.FUEL.2017.05.009
29. ANSYS: San Diego. (2017). *ANSYS chemkin-pro theory manual*. ANSYS.
30. Papageorgakis, G. C., & Assanis, D. N. (1999). Comparison of linear and nonlinear rng-based k-epsilon models for incompressible turbulent flows. *Numer Heat Transf Part B Fundam, 35*, 1–22. https://doi.org/10.1080/104077999275983
31. Bayramoğlu, K. (2018). *HAD (Hesaplamalı Akışkanlar Dinamiği) Yöntemiyle Gemi Dizel Motorlarinda Yanma Analizi*. Dokuz Eylul Üniversitesi.
32. ANSYS_Inc. (2015). *ANSYS Forte 17.2 Theory Guide* (pp. 1–110). ANSYS_Inc.
33. Versteeg, H. K., & Malalasekera, W. (2007). *An introduction to computational fluid dynamics: The finite volume method*. Pearson Education Limited.
34. Zhang, Z., Hu, J., Yang, D., et al. (2024). A comprehensive assessment over the environmental impact and combustion efficiency of using ammonia/ hydrogen/diesel blends in a diesel engine. *Energy, 303*, 131955. https://doi.org/10.1016/j.energy.2024.131955
35. Ansys. (2015). *ANSYS fluent theory guide*. ANSYS 162 Doc.

36. Liu, H., Zhang, X., Zhang, Z., et al. (2023). Effects of 2-ethylhexyl nitrate (EHN) on combustion and emissions on a compression ignition engine fueling high-pressure direct-injection pure methanol fuel. *Fuel, 341*, 127684. https://doi.org/10.1016/j.fuel.2023.127684
37. Biswas, S., & Banerjee, R. (2023). In-cylinder pressure profiling for CNG/ethanol enriched biodiesel combustion. *Fuel, 343*, 127970. https://doi.org/10.1016/j.fuel.2023.127970
38. Kumar, M., Bhowmik, S., & Paul, A. (2022). Effect of pilot fuel injection pressure and injection timing on combustion, performance and emission of hydrogen-biodiesel dual fuel engine. *International Journal of Hydrogen Energy, 47*, 29554–29567. https://doi.org/10.1016/j.ijhydene.2022.06.260
39. McAllister, S., Chen, J.-Y., & Fernandez-Pello, A. C. (2011). *Fundamentals of combustion processes*. Springer.
40. Bayramoğlu, K., & Yılmaz, S. (2021). Emission and performance estimation in hydrogen injection strategies on diesel engines. *International Journal of Hydrogen Energy, 46*, 29732–29744.
41. Jin, X., Zhou, Y., & Zheng, S. (2018). Numerical investigation of different effects of carbon dioxide properties and carbon monoxide oxidation on char particle combustion in actual and fictitious O_2/CO_2 environments. *Fuel, 217*, 59–65. https://doi.org/10.1016/j.fuel.2017.12.040
42. Winterbone, D. E. (1997). In D. E. B. T.-A. T. Winterbone & for E (Eds.), *Thermodynamics of combustion* (pp. 182–207). Butterworth-Heinemann.
43. Glassman, I., & Yetter, R. A. (2008). *Combustion*. Elsevier.
44. Ferguson, C. R. (1985). International combustion engines. In *Applied thermosciences*. Wiley.
45. Zeldovich, Y. B. (1946). The oxidation of nitrogen in combustion and explosions. *Acta Physicochim USSR, 4*, 577–628. https://doi.org/10.1515/9781400862979.364
46. Caton, J. A. (2015). *An introduction to thermodynamic cycle simulations for internal combustion engines*. Wiley.

The Emerging Role of Seaports as Maritime Hubs for Hydrogen

Michele Acciaro

1 Introduction

The global maritime industry is undergoing a radical transformation as it seeks to reduce its carbon footprint in response to the climate crisis, as well as growing customer and regulatory pressures [43]. With maritime transport accounting for nearly 3% of global greenhouse gas (GHG) emissions, international bodies, such as the International Maritime Organization (IMO), have established net-zero targets for decarbonisation [45]. Combined with the long economic life of vessels, this requires urgent action to develop alternatives that can operate with close to zero emissions [5]. The IMO's 2023 GHG Strategy has committed the industry to reach net zero by around 2050 compared to 2008 levels [19], and policy action at the EU and national level is likely to put even more pressure on shipping to decarbonise. Although technologies exist to reduce the carbon intensity of shipping, switching to alternative fuels seems to be the only viable alternative capable of delivering zero-carbon shipping. Still, no single alternative fuel has emerged as a clear winner [45].

Among the various fuel options being considered, hydrogen, alongside ammonia and methanol is gaining prominence due to its versatility and potential for zero emission when produced from renewable sources [33]. Each option presents advantages and challenges in relation to costs, technical maturity, availability, and safety, just to mention a few aspects [49]. Hydrogen is particularly attractive for its versatility and environmental benefits. When used in fuel cells, it emits no carbon dioxide or harmful particulates. However, its low energy density and the technical challenges associated with storing and handling hydrogen onboard ships complicate its large-scale adoption in the near term. Vessels using hydrogen require either compressed storage at high pressures or cryogenic storage at extremely low

M. Acciaro (✉)
Department of Strategy and Innovation, Copenhagen Business School, Frederiksberg, Denmark
e-mail: mac.si@cbs.dk

C. O. Colpan et al. (eds.), *Fuel Cell and Hydrogen Technologies in Maritime Transportation*, https://doi.org/10.1007/978-3-031-86110-9_9

temperatures, both of which involve significant capital investment, as difficult to manage, and have been tested only on pilots with relative success [2].

The adoption of hydrogen technology in shipping is currently at an early stage, with most developments focused on small-scale demonstrations and pilot projects. The high cost of producing green hydrogen, which relies on electrolysis powered by renewable energy, is a key barrier to widespread deployment. In the interim, dual-fuel engines capable of running on both liquefied natural gas (LNG) and hydrogen are being explored as a transitional solution, allowing for flexibility while infrastructure and technologies mature [27]. Proponents of hydrogen argue that existing LNG-powered vessels can be retrofitted to use hydrogen or hydrogen-derived fuels, such as ammonia or methanol, which helps mitigate some of the costs associated with the transition, and blends are also being investigated [33].

But hydrogen is not the only option available to decarbonise shipping. Ammonia, produced from hydrogen, offers certain advantages over pure hydrogen in terms of storage and transportation, particularly for long-distance voyages. It can be stored at lower pressures than hydrogen and does not require cryogenic storage. However, ammonia is toxic, and its combustion can lead to harmful nitrogen oxides, necessitating additional technology to manage these emissions [49]. Methanol, which can also be produced from hydrogen, is easier to store and handle due to its liquid state at ambient temperature but does not achieve the same level of GHG reduction as hydrogen or ammonia [37]. As a result, the choice between these fuels will depend on the specific requirements of different shipping routes and segments [46].

As the shipping industry deals with these technical and economic challenges, it has become clear that one of the main issues associated with the use at the scale of any alternative fuel is the development of adequate infrastructure for production, bunkering, and storage. Seaports are, at the moment, the primary locations where vessels refuel; hence, they play a central role in the storage, handling, and occasionally the production of bunker fuels [3]. They are critical nodes in the global fuel supply chain and energy trades. If any alternative fuel is to be used at scale in the shipping industry, seaport infrastructure will need to be developed or repurposed to serve the needs of the maritime industry [14].

The development of alternative fuels offers an opportunity for ports not only to maintain their role as energy hubs but also to expand their logistics and production capabilities as demand for alternative fuels grows. Ports that invest in hydrogen infrastructure now will not only support the decarbonisation of shipping but also position themselves as key players in the emerging hydrogen economy [1]. Several leading ports, particularly in Europe, are already taking steps to establish themselves as hydrogen hubs, integrating renewable energy sources with hydrogen production to create a supply chain that can serve both the maritime industry and other sectors [53].

The development of hydrogen infrastructure at seaports includes the installation of large-scale electrolysers for hydrogen production, storage facilities, and specialised bunkering stations capable of refuelling hydrogen-powered vessels [40]. Large ports such as Rotterdam and Hamburg have developed hydrogen production, transportation, storage, distribution, and bunkering plans, which will become operational in 2030. However, smaller ports have also made considerable investments [18]. Studies have looked at hydrogen transitions, for example, in Chile [32], Japan, Australia [7, 8], and South Korea [44], although the focus has been extensively on European ports [33]. These efforts are essential to scaling up the use of hydrogen in maritime transport, as they address the lack of bunkering infrastructure as one of the key bottlenecks in the hydrogen supply chain, but constitute also an opportunity for ports to actively contribute to the emerging hydrogen value chains [46].

In addition to infrastructure development, ports are also exploring the integration of hydrogen into their operations. By using hydrogen-powered equipment and vehicles, ports can reduce their emissions and serve as testbeds for hydrogen technologies [40]. This not only contributes to the decarbonisation of port activities but also demonstrates the viability of hydrogen as a fuel across various sectors. Furthermore, seaports that develop hydrogen infrastructure are well positioned to serve not just the shipping industry but also other industries, such as heavy road transport and aviation, which face similar decarbonisation challenges [26].

The economic viability of hydrogen in maritime transport, however, depends heavily on regulatory frameworks and government incentives [41]. The high upfront costs associated with hydrogen production, storage, and vessel retrofitting remain a significant hurdle. National and regional policies that promote the use of low-carbon fuels, such as carbon pricing or subsidies for green hydrogen production, will be essential in accelerating the adoption of hydrogen as a maritime fuel. Regulatory support from bodies such as the IMO and the EU will also be critical in setting emission standards that encourage investment in hydrogen technologies and infrastructure [33].

In what follows, we will review the role of hydrogen in ports, focusing on bunkering, hydrogen storage and handling, hydrogen value chains, hydrogen production, primarily through renewables, government intervention, and a few short case studies: Antwerp-Bruges, Rotterdam, Hamburg, Esbjerg, the North Tyrrhenian Port Network Authority, Singapore, the California Hydrogen hub, Kobe, Copenhagen/Malmø Ocean Valley, Constanta, North Sea Port, Valencia, Sines, Vancouver.

2 Hydrogen Bunkering

Hydrogen is a versatile energy carrier used in various industrial sectors, including the production of energy carriers such as ammonia and methanol, steelmaking, and as a feedstock in refining processes. Shipping is increasingly seen as a potential end-use sector for hydrogen, especially as the industry faces decarbonisation pressures. However, maritime transport competes with these established industries for hydrogen, which could limit the availability of hydrogen for shipping in the short term. Industrial sectors like ammonia production dominate demand [35].

Some of these energy carriers (i.e. primarily ammonia and methanol) are also used as alternative fuels, hence, hydrogen as a fuel competes with some of the products downstream in the production chain. Ammonia is often seen as an attractive option for maritime transport due to its higher energy density compared to hydrogen and less complex storage requirements [52]. Ammonia can be stored at lower pressures than hydrogen, and its global production already reaches approximately 270 million tons annually, creating a potential overlap between the energy and agricultural sectors.

As far as methanol's use in shipping is concerned, its appeal is linked to easier handling and storage requirements [37]. Hydrogen's low energy density and the need for cryogenic storage or high-pressure systems pose logistical challenges for maritime use. In contrast, methanol, which can be produced from hydrogen in combination with captured CO_2, is a liquid at ambient temperatures, making it much simpler to store and transport using existing infrastructure [4]. Methanol also requires less extensive modifications to vessels compared to hydrogen, further reducing upfront costs for shipping companies aiming to transition to cleaner fuel [49]. This dual-use of hydrogen complicates its availability for shipping.

Hydrogen's attractiveness as a fuel for bunkering is largely due to its potential to drastically reduce emissions from maritime transport. Hydrogen, when used directly in fuel cells, offers zero-emission propulsion, making it attractive for meeting future environmental regulations such as those mandated by the IMO [49]. However, its attractiveness for shipping decarbonisation depends on the production chain (similar to methanol and ammonia) and its price and availability.

Hydrogen bunkering infrastructure at ports is expected to play a key role in enabling the decarbonisation of shipping. Ports with established hydrogen production capabilities, or access to renewable energy sources, are particularly well positioned to capitalise on this opportunity [33]. Furthermore, hydrogen bunkering at ports aligns with broader efforts to integrate hydrogen across multiple sectors, including heavy industry and transportation [18].

Hydrogen bunkering requires specialised infrastructure, particularly because hydrogen must be stored either in compressed form at high pressures or in liquid form at cryogenic temperatures. The handling of hydrogen at ports involves safety challenges due to its flammability and low ignition energy. Therefore, hydrogen bunkering terminals must be equipped with advanced safety and monitoring systems to prevent accidents and ensure efficient operation [48].

Hydrogen demand in the maritime sector is projected to grow significantly from 2030 onwards. Demand in ports is expected to reach up to 50% of the total hydrogen demand by 2050, largely driven by the decarbonisation of shipping and heavy-duty transport. In total, hydrogen demand in European ports is projected to expand to around 1764 TWh by 2050 under ambitious scenarios. This growth highlights the need for rapid development of hydrogen bunkering facilities to meet future demand [10]. Green shipping corridors have also been identified as a tool to accelerate the uptake of hydrogen as a low-carbon fuel for shipping [22].

The hydrogen supply for bunkering will depend on local production and imports. Europe is projected to meet around 75% of its hydrogen demand through domestic production by 2050, primarily using green hydrogen from renewable sources like solar and wind. The remaining 25% is expected to come from imports, with North

Africa and the Middle East acting as major suppliers. However, the current rate of infrastructure development is insufficient to meet this demand without significant policy interventions [10].

Without strong policy support, the hydrogen supply chain is unlikely to develop at the pace required to meet future bunkering demand. Current projections indicate that Europe's renewable energy deployment needs to accelerate, and additional investments are required in hydrogen production, storage, and distribution infrastructure [10]. Policy measures such as subsidies, carbon pricing, and regulatory support are essential to stimulate the necessary investments [21, 41]. Without these, hydrogen bunkering supply is expected to lag behind demand.

There are significant synergies between hydrogen use in maritime shipping and other modes of transport, particularly in heavy-duty road transport and aviation [31]. Ports are well positioned to serve as hydrogen hubs for multiple sectors, facilitating hydrogen supply to not only ships but also trucks and, in some cases, even aircrafts. Developing hydrogen infrastructure that serves multiple sectors can enhance economies of scale, reduce costs, and create more robust business models for hydrogen in ports.

3 Hydrogen Storage and Handling

In addition to providing facilities for bunkering, many ports can capitalise from the increasing use of hydrogen (not necessarily only for maritime operations) acting as storage and trading points. Chen et al. [7] analysed the potential for increasing hydrogen-related energy trade and identified 20 ports that operate as first movers. They conclude that ports are still in the early stages of being prepared for hydrogen international trade, and significant efforts are needed in areas such as infrastructure development or renovation, risk management, regulatory frameworks, standards, as well as education and training.

Hydrogen requires specialised storage facilities at ports due to its unique properties, particularly its low density and flammability. Ports handling hydrogen must invest in infrastructure capable of storing and distributing hydrogen safely, including specialised pipelines, cryogenic tanks for liquid hydrogen, and high-pressure systems for gaseous hydrogen. For example, cryogenic storage tanks are necessary for liquid hydrogen, which must be stored at temperatures as low as −253 °C. The design of these facilities must also consider the risk of hydrogen leaks, which require robust safety measures to ensure secure storage and handling operations [10].

The characteristics of hydrogen handling at ports are shaped by the need for strict safety protocols and specialised equipment. Hydrogen is highly flammable and has a low ignition energy, meaning even small leaks can pose significant safety risks. Ports must ensure that hydrogen handling equipment, such as transfer lines and bunkering stations, is equipped with sensors to detect leaks and prevent ignition [28]. Moreover, hydrogen embrittlement—where metals exposed to hydrogen become brittle and crack—requires only specific materials, such as stainless steel, be used in storage and handling systems to avoid infrastructure damage [55].

Hydrogen storage at ports is complex due to the physical properties of hydrogen and the preferred storage form. Hydrogen, in fact, can be stored either as a compressed gas or in liquid form [28]. Compressed hydrogen is typically stored at pressures ranging from 350 to 700 bar, while liquid hydrogen, as mentioned before, requires cryogenic storage systems. Liquid hydrogen storage is more space-efficient but more expensive due to the need for ultra-low temperatures. Ports invest in both types of storage depending on their operational requirements, with a shift towards cryogenic systems expected as hydrogen demand increases [10, 40]. Additionally, ports will need to allocate considerable space to ensure safe separation between hydrogen storage areas and other port activities.

Developing hydrogen storage and handling infrastructure at ports can be expensive, with costs varying based on the scale and technology used. For example, building a large-scale hydrogen storage facility with cryogenic tanks can cost upwards of €30 million, depending on the capacity and safety requirements. In comparison, smaller scale compressed hydrogen systems are less expensive but offer lower storage efficiency. Moreover, the cost of retrofitting existing port infrastructure to accommodate hydrogen handling, including modifications to pipelines and safety systems, can further increase overall investment requirements, making cost-effective planning essential for port authorities [10].

Ports are critical nodes in the hydrogen value chain, serving as hubs for hydrogen production, storage, and distribution [33]. Ports not only enable the bunkering of hydrogen-powered ships but also serve as central points for hydrogen import and export [7]. This makes them essential in connecting local hydrogen production—often from offshore wind farms or other renewable sources—to broader domestic and international markets. Several ports are already positioning themselves as key hydrogen hubs, establishing large-scale electrolysis plants and storage facilities, which will enable them to supply hydrogen to multiple sectors, including shipping, road transport, and industry.

Ports can adopt various business models to capitalise on the handling and storage of hydrogen, focusing on their roles as logistical hubs and service providers in the hydrogen supply chain, depending on their location, for example, proximity to demand or production [33]. One model involves ports acting as infrastructure landlords, where they lease storage and handling facilities to private hydrogen companies. In this model, ports provide the necessary land and specialised infrastructure, such as cryogenic tanks and high-pressure storage systems, while private operators manage the day-to-day handling and distribution of hydrogen [10].

Another approach is the terminal operator model, where ports directly invest in and operate hydrogen storage and bunkering facilities, charging fees for storage, transfer, and bunkering services. This model allows ports to generate revenue from handling operations while maintaining full control over safety and logistical efficiency [10].

Collaborative partnerships with industrial stakeholders represent another strategy, where ports form joint ventures with hydrogen producers and logistics companies, sharing the capital expenditure and operational risks associated with developing hydrogen infrastructure. This approach ensures alignment between supply chain actors and enhances the port's position in the hydrogen value chain [10]. Further considerations on the attractiveness of hydrogen handling and storage relate to the acceptability from local communities, technological capabilities in the proximity of the port, access to finance, and of course potential for hydrogen uses, for example, connectivity to shipping and land transport [33]. In order to better understand the business opportunities offered by hydrogen in ports, it is valuable to analyse the position of ports in the hydrogen value chain.

4 Hydrogen Value Chain

The value chain for an energy carrier like hydrogen refers to a sequence of stages through which hydrogen passes from production to consumption, with each stage adding distinct value. These stages include production, transformation into transportable forms, transportation and storage, and ultimately consumption across sectors like industry and transport [13]. Value chain theories emphasize how value is incrementally added at each stage [54]. In the hydrogen context, value addition occurs through the infrastructure, technical processes, and logistical coordination required to efficiently produce, convert, store, and deliver hydrogen to end users.

The hydrogen value chain is structured into four major segments: production, conversion, transportation and storage, and consumption [13, 16]. In production, hydrogen is generated through processes like electrolysis, using renewable energy, or through steam methane reforming with carbon capture [13]. Conversion refers to the transformation of hydrogen into more easily transportable carriers like ammonia or methanol. Transportation and storage involve moving hydrogen in pipelines or cryogenic tankers and holding it in compressed or liquefied forms [50]. Finally, consumption refers to the use of hydrogen in end sectors like shipping, heavy industry, and transport, as illustrated in the sections above.

At ports, hydrogen production typically relies on green hydrogen generation through electrolysis, powered by renewable energy sources like offshore wind. Several ports, for example, North Sea Port (Belgium/Netherlands) or Port La Nouvelle (France), are investing in electrolyser capacities, with some like Rotterdam, targeting up to 1 GW by 2030, to supply local industries and shipping with green hydrogen. Producing hydrogen at ports significantly reduces transportation needs, placing hydrogen production close to key demand sectors. This localised production strategy lowers costs and energy losses, particularly for industries and maritime activities directly tied to port infrastructure. Ports add significant value by streamlining the production and distribution process, ensuring that renewable hydrogen can be efficiently integrated into the energy ecosystem [10].

Conversion at ports involves transforming hydrogen into derivatives like ammonia or methanol to enhance its storage and transport properties [23]. These carriers are easier to store and ship, especially over long distances, and can also serve as fuels for sectors like shipping. The Port of Antwerp, for instance, has focused on converting hydrogen into ammonia for export, leveraging ammonia's higher energy density and ease of handling compared to hydrogen. This conversion step allows ports to serve international markets more effectively, adding value by overcoming the inherent difficulties of transporting gaseous hydrogen, thus facilitating the global trade of hydrogen-based products [8].

Hydrogen transport involves moving the fuel from production sites to consumers, and ports are critical nodes in this network. Ports with developed hydrogen transport infrastructure, such as pipelines or shipping terminals, enhance the hydrogen value chain by ensuring efficient movement between production centres and areas of demand [10]. For example, the Port of Hamburg is investing in hydrogen pipelines that will connect it to inland industrial hubs, ensuring reliable supply chains for Germany's heavy industries [24]. By acting as transport hubs, ports support the broader distribution of hydrogen, linking local production to both domestic and international markets and adding value by reducing logistical bottlenecks.

Hydrogen storage at ports plays a crucial role in stabilising the supply chain, addressing the intermittency of renewable energy used in hydrogen production. Storage facilities, particularly those for compressed or liquefied hydrogen, allow ports to maintain a buffer supply, ensuring hydrogen availability even when production fluctuates [40]. For instance, the Port of Rotterdam's storage infrastructure includes cryogenic tanks that can hold liquefied hydrogen, enabling it to meet fluctuating industrial and maritime demand. By providing large-scale storage, ports increase the reliability of the hydrogen supply chain, adding value by ensuring the continuity of hydrogen availability to various consumers.

In terms of consumption, ports are increasingly serving hydrogen-powered ships and transport vehicles, driving demand in the maritime and logistics sectors, as illustrated before. The development of hydrogen bunkering facilities at ports like Rotterdam, Singapore, Antwerp-Bruges, and many others (see case studies) allows for the refuelling of hydrogen-powered vessels (deep-sea and short-sea) or for port operations (e.g. tugboats). Terminal operations can also benefit from hydrogen fuels, although electrification seems a more convenient approach [50]. Direct access to hydrogen fuel in ports enables the transition to cleaner transport, thus adding significant value by supporting the decarbonisation of global shipping, local port operations, and heavy-duty road transport [33].

Beyond shipping, hydrogen consumption in industrial applications near ports is a key value driver [40]. Industries like steel production and chemicals, often located near ports, are transitioning to hydrogen to reduce their carbon emissions. For example, hydrogen can replace coal in Direct Reduced Iron processes, significantly lowering emissions in steelmaking. Additionally, hydrogen supports decarbonisation in urban areas around ports, with potential uses in public transportation and municipal services. Ports that supply hydrogen to both industrial and urban areas can increase their value by creating synergies across sectors, supporting wide scale decarbonisation while leveraging their existing infrastructure (see, for example, California Hydrogen Hub project in Sect. 8).

5 Hydrogen Production and Conversion

Hydrogen can be produced directly in ports due to ports' strategic location near renewable energy sources and industrial hubs. It is important here to highlight the distinction between port facilities and the industrial cluster that is often located in the proximity of or within port areas. It can be argued that industrial activities are not port activity in the strict sense, even though effectively ports are suitable locations for many of them, from refineries, powerplants, steel mills, and chemical industrial facilities. These facilities not only represent important demand sources for energy but are also important in determining the type of materials and products that are imported or exported through the port.

With the transition towards renewable power generation, energy management within the port is becoming an even more important issue [1]. Given the close relationship between power generation and energy storage on one side, and conversion, trade, and use of energy carriers on the other side, it is valuable to consider the synergies between hydrogen production, conversion, storage, handling, and bunkering.

Ports, especially those in coastal areas, have the potential to produce hydrogen using electrolysis powered by offshore wind or other renewable energy sources. This local production reduces the need for long-distance hydrogen transport, cutting costs and emissions. For instance, many European ports located in the North Sea and in the proximity of large power generation facilities are developing electrolysis projects aimed at supplying green hydrogen to various industries within the port's vicinity, including transport and industrial applications.

Ports are evolving into energy hubs [1], making them central to the future hydrogen economy. This shift includes the integration of renewable energy generation such as wind, solar, wave and tidal energy, and hydrogen production (e.g. [25, 30, 32]). Ports (see cases later) are investing in infrastructure to not only support traditional logistical functions but also act as centres for energy production. These developments position ports as critical nodes in both power generation and distribution, enabling them to supply local industries and contribute to the broader decarbonisation of transport and heavy industry.

This transformation means that ports are increasingly entering the power generation business. Ports are becoming self-sufficient energy hubs, and in some cases, exporting excess electricity back into the grid. This shift is not only a result of the increased demand for low-carbon fuels for maritime and land transport but also of the growing number of industrial activities that are shifting to cleaner energy sources. Port-related activities, including the energy-intensive processes involved in hydrogen production through electrolysis and conversion into other energy carriers, are also driving the need for ports to generate their renewable electricity [36].

Increased demand for electricity, especially from renewable sources, creates a challenge for energy management in ports. As ports become transit points for an increasing amount of renewable energy, they are confronted with the need to manage power peaks [20]. Hydrogen can act as an energy buffer by smoothing out fluctuations in renewable power generation. When renewable energy production peaks, hydrogen can be produced and stored, ensuring a consistent energy supply during periods of low renewable output. This flexibility is expedient for balancing the intermittent nature of renewables that depend on weather, like wind and solar power, providing ports with a reliable energy solution that integrates with their expanding role as energy hubs [6].

Hydrogen production at ports requires significant investment, especially in building the necessary infrastructure for electrolysis, storage, and distribution. Estimates from recent projects indicate that developing a hydrogen production facility can require several hundred million euros in capital expenditures [3]. These investments are focused primarily on establishing large-scale electrolysers, storage tanks, and pipelines, all of which are essential to supporting hydrogen as a key energy carrier [16].

For ports, these investments should add value by creating new revenue streams and enhancing their role in the energy transition. Hydrogen production can generate income through the sale of hydrogen to various industries and transport sectors, while also positioning ports as key players in the emerging hydrogen economy. Ports that develop the necessary infrastructure will be able to attract industries looking for low-carbon energy solutions, thereby boosting their competitiveness and long-term economic sustainability [38, 40].

Production investments are closely linked to specific port functions, such as their role as landlords, community builders, and enablers of the local economy. Ports can provide land and infrastructure to companies involved in hydrogen production, acting as facilitators in the broader hydrogen ecosystem. By investing in hydrogen production facilities, ports fulfil their roles as community builders, contributing to local job creation and supporting regional industrial growth. These activities also align with the ports' responsibilities as enablers of sustainable development and environmental goals [10].

The level of investment in hydrogen production at ports will depend on various factors, including the size, archetype, and strategic goals of each port [10]. Larger ports with high levels of industrial activity, such as Rotterdam or Singapore, will likely invest more heavily in hydrogen production infrastructure, given their capacity to handle large-scale projects and the ability to consolidate demand and take advantage of economies of scale in construction and production. Smaller ports or those with different operational focuses may adopt more targeted strategies, investing in niche hydrogen applications that align with their specific needs and regional opportunities. Understanding how ports can develop successful business cases for participating in the hydrogen value chains still requires study. It is clear, however, that this will not happen without government intervention.

6 Government Actions

While hydrogen has the potential to play an important role in decarbonising hard-to-abate sectors like transport and heavy industry, its uptake faces challenges that cannot be resolved solely by market forces [21, 41]. The high initial costs of hydrogen infrastructure, lack of widespread production, and competition with established fossil fuels, among other things, create barriers to its development [3]. Without targeted government interventions, such as subsidies or regulatory mandates, hydrogen's cost competitiveness and market penetration are likely to remain limited. Government support is essential to create the conditions necessary for hydrogen to thrive [7].

Firstly, clear and ambitious decarbonisation goals set by governments are critical to creating a predictable policy environment for hydrogen investments [41]. For example, the EU's REPowerEU plan outlines the production of ten million tonnes of renewable hydrogen by 2030, alongside importing an additional ten million tonnes [10]. By specifying such targets, governments provide the private sector with confidence that hydrogen will be part of the long-term energy mix, encouraging investments in infrastructure and technology development. This is necessary for aligning hydrogen deployment with broader climate objectives.

For countries that depend on energy exports, such as natural gas, hydrogen can represent a transitional fuel, as most hydrogen today is produced through fossil fuels. The decarbonisation strategies of energy-export countries such as Saudi Arabia, Australia, and Qatar, include a strong focus on hydrogen as a potential intermediate fuel to help diversify their exports and leverage on the energy transition. Hydrogen will play a strategic role in the countries' economy [17] and it would be critical to establish partnerships with importing countries to facilitate hydrogen trade [8].

It is important to note that not all hydrogen is produced sustainably, and without strict government regulations, hydrogen production could increase emissions rather than reduce them [29]. 'Grey' hydrogen, produced from natural gas without carbon capture, results in significant carbon emissions, while 'blue' hydrogen, which incorporates carbon capture, still relies on fossil fuels and carbon capture technologies that have not yet reached maturity. Only 'green' hydrogen, produced using renewable energy, contributes to true decarbonisation. But the market for green hydrogen is in its infancy. Governments must implement policies that prioritise green hydrogen production to avoid inadvertently promoting hydrogen derived from fossil fuels.

Secondly, hydrogen, and in particular 'green' hydrogen, remains more expensive than conventional fuels and other renewable alternatives such as wind and solar energy [21]. Government subsidies, such as those outlined in the European Hydrogen Bank's commitment to pay a fixed price per kilogram of hydrogen produced for 10 years, are necessary to bridge the cost gap between hydrogen and fossil fuels [10]. Without such financial mechanisms, hydrogen will most likely struggle to achieve cost competitiveness, hindering its widespread adoption [25].

Third, hydrogen infrastructure, including electrolysers, storage facilities, and pipelines, requires significant upfront investments that are generally not feasible without government support. Public–private partnerships and direct investments in infrastructure projects are critical. For example, the EU's projected investment of €6–11 billion in hydrogen storage infrastructure by 2030 highlights the scale of commitment required to build the necessary facilities [10]. Governments need to coordinate and support these efforts to ensure the rapid build-out of hydrogen infrastructure.

There are various instruments that governments can adopt to incentivise the use of hydrogen across industries through mechanisms such as carbon pricing, fuel subsidies, or mandatory blending quotas [21]. The European Commission's FuelEU Maritime initiative, which aims to ensure that 2% of the maritime fuel mix is made up of hydrogen-based fuels by 2030, demonstrates how policy can drive demand for hydrogen [10]. Other countries, such as Japan and the USA, have also developed policies aimed at advancing the uptake of hydrogen [21]. In China, a policy mix has been adopted by provincial governments to accelerate its deployment countrywide [56]. Such incentives encourage early adoption and help create market demand, which, in turn, encourages further infrastructure development and technological advancements [3].

Moreover, the absence of common technical standards for hydrogen production, storage, and transport presents a major barrier to its adoption. Governments and international bodies like the International Organization for Standardization (ISO) need to develop harmonised regulations to ensure safety and interoperability across borders. Standardisation efforts, such as those initiated by the EU in coordination with CEN and CENELEC, help reduce technical uncertainties and enable the large-scale deployment of hydrogen technologies across sectors and regions [36].

It should be further noted that ports are large energy consumers and can use hydrogen to decarbonise their operations. Government-run port authorities can lead by example, integrating hydrogen into bunkering, cargo-handling equipment, and port vessels. For example, hydrogen-powered terminal vehicles and handling equipment could drastically reduce emissions from terminals [18]. Hydrogen tugboats can result in a reduction of emissions of 85% in comparison to diesel-propelled ones [9]. By incentivising the adoption of hydrogen technologies, authorities can demonstrate the viability of hydrogen, encourage broader industry adoption, and increase demand.

In addition, ports are obvious gateways for hydrogen and hydrogen-based fuels, and government policies can support their role as transit platforms. Strategic government planning for hydrogen import and export terminals, as well as regulatory alignment between different transport modes (pipeline, ship, truck), ensures efficient hydrogen flows through ports [7]. European ports (see cases later) are set to play critical roles in importing green hydrogen to Europe, with government backing needed to streamline the logistics and regulatory frameworks governing these flows [10].

Governments will also play a role in the building and financing of hydrogen infrastructure needed not only for import and export but also for downstream use at ports, such as bunkering stations and multimodal refuelling stations, in addition to the related storage and transportation facilities. The EU has already recognised the importance of ports in its hydrogen strategy, providing funding opportunities for ports to accelerate investments in hydrogen bunkering and refuelling infrastructure though financial instruments such as the Connecting Europe Facility (CEF), and similar initiatives, for example, the Hydrogen Station Construction Support Program in Japan, aimed at accelerating the use of hydrogen in transport. Government coordination is crucial for ensuring this infrastructure meets future demand while maintaining compatibility with other sectors and international standards [10].

As mentioned, hydrogen, particularly in its liquefied form, poses significant safety risks, including flammability and cryogenic hazards for bunkering [48] and transportation and storage [28]. The role of governments then is critical in establishing adequate safety regulations, adopting international standards, and enforcing their implementation at ports. This involves updating existing safety frameworks like the Seveso Directive to accommodate hydrogen storage and handling, and mandating training programs for workers handling hydrogen, including during bunkering [10].

Finally, the shift to hydrogen must include a focus on social equity, ensuring that workers displaced by decarbonisation efforts are retrained for jobs in the new hydrogen economy. Governments must also engage with local communities near ports to address concerns related to safety, land use, and job losses [50]. Just transition funds, which are designed to support regions and workers impacted by the energy transition, can play an important role in ensuring that hydrogen adoption benefits all stakeholders without leaving vulnerable parts of society behind.

7 Case Studies

This section provides an overview of hydrogen projects and policies in ports. These short case studies are based on publicly available information and the author's own research. The objective is to highlight the great diversity of action in ports and the large amount of investment, policy, and support actions that can be observed in European and global ports.

7.1 Port of Hamburg, Germany[1]

The Port of Hamburg has a strategy to develop as a hydrogen hub, driven by its strategic location, existing industrial ecosystem, and ambitious hydrogen

[1] Based on https://www.hghh.eu/en and https://www.hafen-hamburg.de/en/port-of-hamburg-magazine/gateway-hamburg/leinen-los-fuer-gruenen-wasserstoff/

development projects in alignment with the hydrogen strategy of the State of Hamburg and the German federal government. One of the flagship initiatives is the Hamburg Green Hydrogen Hub (HGHH), where large-scale green hydrogen production is set to begin by 2025. This hub, located at the former Moorburg power plant site, aims to produce hydrogen using renewable energy sources to supply local industries, the port, and transportation sectors in the region. The port is expected to play a critical role in decarbonising not only Hamburg's industrial base but also its logistics and transport systems.

Hamburg's hydrogen infrastructure development is closely tied to its industrial demand for clean energy. The Port of Hamburg serves as a major industrial hub with potentially high hydrogen demand, particularly for sectors such as steelmaking, chemical production, and heavy industry. This demand creates significant opportunities for green hydrogen production and storage within the port. By fostering a hydrogen ecosystem, the port is aligning its strategy with the broader German and European decarbonisation goals, which aim for substantial emission reductions by 2030 and full carbon neutrality by 2050. Hydrogen is a central pillar in these decarbonisation strategies, and Hamburg's industrial infrastructure positions it well for large-scale hydrogen deployment.

The port's logistics and transport network will also benefit significantly from hydrogen developments. As one of Europe's largest ports, Hamburg handles extensive cargo and container traffic, with ambitions to integrate hydrogen into its transport operations. Hydrogen-fuelled heavy-duty vehicles and port equipment, such as cranes and forklifts, are being explored as part of ports' decarbonisation efforts. These initiatives not only reduce emissions from port operations but also contribute to the broader adoption of hydrogen in logistics, making Hamburg one of the first movers in hydrogen-powered transportation.

In terms of storage, the Port of Hamburg is developing extensive hydrogen storage capabilities to complement its production and distribution efforts. The construction of hydrogen storage facilities is essential to ensuring a consistent supply of hydrogen, especially as the port seeks to manage the fluctuating demand of its industrial and transport customers. The port is also part of a larger regional hydrogen pipeline network that will connect local hydrogen producers to industrial consumers in and around Hamburg. These storage and pipeline developments are designed to ensure that hydrogen is readily available for both immediate and long-term use.

Hamburg is positioning itself as a key player in the global hydrogen trade. As part of its hydrogen strategy, the port is planning for large-scale hydrogen imports, particularly from regions with abundant renewable energy resources, such as North Africa and the Middle East. The port is developing infrastructure to handle both the import of hydrogen and hydrogen carriers, such as ammonia, ensuring that it can serve as a central hub for hydrogen distribution across Europe. With the support of local and national government policies, the Port of Hamburg aims to become a leading hydrogen import and export hub, facilitating the global transition to clean energy.

7.2 California Hydrogen Hub[2]

California's hydrogen strategy includes a $1.2 billion allocation as part of the US federal government's hydrogen hub initiative. The ports of Los Angeles, Long Beach, and Oakland are integral to this effort, particularly given their status as major cargo and logistics centres. These ports aim to replace diesel-powered cargo-handling equipment with hydrogen fuel cell equivalents. The Los Angeles Department of Water and Power will transition power plants to 100% renewable hydrogen, creating a scalable, zero-emissions model for port operations.

The ports are also expected to serve as key logistical nodes for hydrogen fuel distribution. Southern California Gas Company is planning a dedicated clean hydrogen pipeline, the Angeles Link, to distribute hydrogen to critical sectors, including transport. The plan will see over 60 hydrogen fuelling stations constructed, enabling the operation of more than 5000 hydrogen-powered trucks and over 1000 hydrogen buses, substantially reducing emissions from the region's heavy-duty transport sector.

At the Port of Oakland, hydrogen will be integral in supporting logistics operations by powering heavy cargo handling equipment. This will reduce emissions and improve air quality in surrounding communities, which have been disproportionately affected by pollution from port activities. The transition to hydrogen-powered equipment is part of California's broader goal to decarbonise its ports, where ARCHES—the state's public-private hydrogen partnership—will coordinate investments and hydrogen infrastructure development.

Storage and transportation of hydrogen across California's ports will require substantial infrastructure upgrades. The ports will need to build specialised facilities to store both gaseous and liquid hydrogen, ensuring that sufficient supply is available for fuelling operations. Trucks and pipelines will play critical roles in the distribution network. Environmental and safety considerations are high priorities, as hydrogen's small molecular size makes it prone to leakage, posing challenges in storage and transport.

California's hydrogen strategy also aims to address economic and health disparities. ARCHES estimates that by 2031, the state's hydrogen projects will save $2.95 billion in health costs annually, primarily by reducing pollution from diesel engines in disadvantaged communities near ports. Furthermore, $380 million will be invested in workforce development in these communities, ensuring that hydrogen development provides both economic benefits and improves public health outcomes.

[2] Based on https://www.gov.ca.gov/2024/07/17/california-launches-world-leading-hydrogen-hub/

7.3 Port of Antwerp-Bruges, Belgium[3]

The Port of Antwerp-Bruges is becoming a major hydrogen hub, actively investing in both import and production capabilities. As part of the Hydrogen Import Coalition, the port is positioning itself to be a key entry point for green hydrogen imports from regions rich in renewable energy, including North Africa and the Middle East. To accommodate this, Antwerp-Bruges is expanding terminal capacity specifically for hydrogen carriers such as ammonia and methanol. These efforts are designed to meet Europe's growing demand for clean energy, with the port planning significant infrastructure upgrades to ensure efficient storage and distribution. The port's most ambitious hydrogen production initiative is the HyOffWind project, which aims to generate large-scale green hydrogen using offshore wind energy. Antwerp-Bruges is also playing a central role in the Power-2-Methanol initiative, which involves the local production of sustainable methanol using hydrogen and captured CO_2. These projects not only align with Belgium's national decarbonisation goals but also contribute to the port's broader strategy of becoming a leading centre for green hydrogen production and storage.

Antwerp is advancing in hydrogen logistics, as well. It has become a key node in the HyTrucks consortium, aiming to deploy 300 hydrogen-powered trucks in Belgium by 2025. Additionally, the port is integrating hydrogen into its maritime operations, with the construction of the world's first hydrogen-powered tugboat by Compagnie Maritime Belge (CMB). These efforts show Antwerp-Bruges' commitment to decarbonising logistics, both on land and at sea, leveraging hydrogen to reduce emissions in heavy transport sectors.

Antwerp-Bruges is developing the NextGen District, an 88-hectare area dedicated to sustainability initiatives, including hydrogen-related projects. This business park will host companies working on hydrogen technologies, enabling the port to attract investment and innovation. Moreover, the port is actively expanding its hydrogen storage capabilities to ensure a consistent supply for future demands in shipping and industry.

7.4 Port of Valencia, Spain[4]

The Port of Valencia is taking significant steps to integrate hydrogen into its operations as part of its goal to become carbon-neutral by 2030. The port has pioneered the H2PORTS project, which is testing and validating hydrogen technologies for port machinery to achieve zero local emissions. This initiative, coordinated by the Valenciaport Foundation and funded by the Fuel Cells and Hydrogen Joint

[3] Based on https://parkwind.eu/r-d-projects/hyoffwind and https://www.portofantwerpbruges.com/en/our-port/climate-and-energy-transition/hydrogen

[4] Based on https://h2ports.eu/

Undertaking (FCH JU), includes testing hydrogen-powered equipment such as a reach stacker and a 4x4 tractor unit. These technologies aim to decarbonise cargo handling operations, one of the major emission sources at the port.

A key feature of the H2PORTS project is the installation of a mobile hydrogen refuelling station, which is essential for supporting the deployment of hydrogen-powered equipment. The mobile station ensures the necessary fuel is available to the tested machinery, allowing the port to experiment with a flexible hydrogen refuelling infrastructure. This is part of Valencia's broader ambition to decarbonise its terminal operations, which aligns with the port's 2030 Net Zero Emissions strategy. This infrastructure development is essential for scaling hydrogen applications across other areas of the port, creating a basis for widespread adoption.

The Port of Valencia is not only a hydrogen user but also plays the role of a community builder. It has established an advisory group, which includes over 65 members and representatives from 30 ports worldwide. The goal of this group is to facilitate the exchange of knowledge and best practices for integrating hydrogen technologies within port operations. By fostering collaboration on hydrogen initiatives, the port is helping to accelerate the hydrogen value chain across the maritime industry. This community-oriented approach is also helping to attract new investments and partnerships, further establishing Valencia as a leader in port decarbonisation.

In addition to its efforts in hydrogen-powered machinery and collaboration, the Port of Valencia is exploring the broader decarbonisation of logistics and transport. Hydrogen is seen as a potential solution for reducing emissions from both domestic shipping and other vehicles operating within the port. The port's commitment to integrating hydrogen technology across various transport modes is an example of a comprehensive strategy to address emissions from multiple sectors.

7.5 *The North Tyrrhenian Port Network Authority (Port of Livorno, Piombino and Portoferraio), Italy*[5]

The North Tyrrhenian Port Network Authority, managing the ports of Livorno, Piombino, and Portoferraio, is actively investing in hydrogen as part of its strategy to decarbonise port operations and contribute to regional energy transition efforts. A central part of its hydrogen initiative involves creating infrastructure for hydrogen production using renewable energy sources like offshore wind and solar, which are available in the region. This includes plans to develop electrolysis plants in Livorno that will produce green hydrogen locally, leveraging the port's existing industrial ecosystem and its proximity to key renewable energy sites.

In terms of logistics, the Port of Livorno is positioning itself as a key player in hydrogen distribution. The port is working on creating a logistics hub that will

[5] Based on Chapter 10 in SRM Centro Studi e Ricerche [47].

facilitate the storage and movement of hydrogen and its derivatives, such as ammonia, to both domestic and international markets. This infrastructure is designed to support the integration of hydrogen into various transport modes, including maritime shipping. Additionally, plans are underway to connect the port to a broader European hydrogen pipeline network, ensuring that Livorno plays a role in the regional and European hydrogen supply chain.

The Port of Piombino is focused on hydrogen's potential to decarbonise its industrial sector, especially the steel industry, which is one of the region's largest emitters. Hydrogen will be used in the Direct Reduced Iron (DRI) process, replacing coal in steelmaking to significantly reduce carbon emissions. Piombino is also integrating hydrogen into its energy and logistics operations by investing in hydrogen-powered equipment and vehicles to replace traditional diesel-powered units. These initiatives are part of the port's larger strategy to align with Italy's national hydrogen roadmap and the EU's decarbonisation targets for 2050.

Hydrogen storage infrastructure is also part of the North Tyrrhenian ports' strategy, with significant investments in both liquefied and compressed hydrogen storage facilities. The ability to store large volumes of hydrogen will enable the ports to ensure a stable supply for industrial use and shipping, even during periods of fluctuating demand. By creating advanced storage solutions, the port aims to secure its role in the European hydrogen supply chain, supporting both local consumption and exports to other regions of Europe. The development of these storage facilities is expected to attract further investment in hydrogen technologies and infrastructure across the region.

7.6 Port of Rotterdam, the Netherlands[6]

The Port of Rotterdam has been positioning itself as a central hub for hydrogen developments, production, logistics, and storage. Rotterdam is home to some of the most ambitious hydrogen projects in Europe. The port's hydrogen strategy revolves around becoming a central hub for the large-scale import, production, and distribution of hydrogen. Rotterdam aims to facilitate up to 20 million tonnes of hydrogen by 2050, with approximately 90% imported via ship and 10% produced locally. This plan includes partnerships with global players to develop infrastructure for hydrogen transport and storage, as well as a robust local demand network for hydrogen use in various industries.

The Port of Rotterdam's role as an energy hub is evident through its investment in electrolysis projects. A key example is the Maasvlakte 2 area, where a 2-GW

[6] Based on: https://www.uniper.energy/news/cooperation-uniper-and-port-of-rotterdam-authority-in-production-green-hydrogen and https://www.portofrotterdam.com/en/port-future/energy-transition/ongoing-projects/hydrogen-rotterdam

hydrogen production park is planned in collaboration with major industry players like Shell. This facility is expected to produce green hydrogen, utilising renewable energy from wind and solar sources. Additionally, the port has partnered with Uniper to develop a green hydrogen plant on the Uniper Maasvlakte site, which is expected to come online in 2025. These efforts align with Rotterdam's long-term vision to lead Europe's hydrogen economy and transition to sustainable energy sources.

In terms of logistics and storage, Rotterdam is developing specialised terminals to handle both hydrogen and hydrogen carriers like ammonia. A noteworthy example is the Transhydrogen Alliance, a consortium consisting of Protonne Ventures, Trammo, and VARO, which is working with the port to create facilities that import and store green hydrogen and ammonia. These terminals are important for Rotterdam to achieve its goal of becoming Europe's largest hydrogen import hub. Plans also include the development of a 40-km hydrogen pipeline connecting the port to refineries in Pernis, a joint effort between the port and GasUnie.

To complement its production and logistics strategies, Rotterdam is also advancing hydrogen storage capabilities. The port is investing in large-scale hydrogen storage facilities, including underground caverns and tanks for liquefied hydrogen. These facilities are critical for ensuring the supply of hydrogen during periods of high demand, especially for industrial applications. Furthermore, the port is exploring synergies between hydrogen and other renewable energy forms, such as wind and solar, to create a resilient energy system that can buffer energy supply and demand fluctuations. The efforts of Rotterdam can be seen with its focus on the energy transition, and the development of several pilot projects (see, for example, the MAGPIE project and its demos).

The integration of hydrogen into the maritime sector is also a significant focus in Rotterdam. As the largest bunkering port in Europe, Rotterdam, is carefully considering hydrogen as a marine fuel. The port's goal is to establish itself as a leader in hydrogen bunkering, positioning itself as a key enabler of zero-emission shipping. This effort is supported by Rotterdam's ongoing partnerships with maritime companies to develop the necessary infrastructure for hydrogen bunkering, as well as the participation of the port in the green shipping corridors [22].

7.7 Port of Esbjerg, Denmark[7]

The Port of Esbjerg, on the North Sea coast of Denmark, has become an important part of the country's hydrogen and energy transition strategy, with several key projects focused on green hydrogen production, infrastructure development, and integration with renewable energy sources. One of the major initiatives is the HØST PtX Esbjerg project, a large-scale Power-to-X facility that aims to produce green hydrogen and ammonia using electrolysis powered entirely by renewable energy.

[7] See also: https://hoestptxesbjerg.dk/

This facility is expected to reduce CO_2 emissions by approximately 1.5 million tonnes annually. The facility will also provide emission-free district heating to around 15,000 households in the Esbjerg region, further enhancing its sustainability impact. Excess heat from hydrogen production is fed into the local district heating network, demonstrating an efficient sector coupling approach by integrating power, heating, and maritime transport. The project is anticipated to create 100–150 permanent jobs, contributing to local economic development [46].

Esbjerg's green hydrogen strategy is heavily tied to its offshore wind capacity. European Energy began supplying green hydrogen to the port in early 2023, utilising power generated from nearby wind farms. This green hydrogen is used in fuel cells to provide clean electricity to vessels docked at the port. The port is also set to become a logistics hub for hydrogen exports and imports. It is strategically located near future hydrogen pipelines from the North Sea, positioning it as a critical node in Denmark's and Europe's hydrogen supply chain. Plans include connecting the port to the European Hydrogen Backbone, which will facilitate the transportation of hydrogen across borders. This infrastructure will allow for the seamless distribution of hydrogen via pipelines and ship-to-ship transport.

The port's contribution to the maritime and heavy industrial sectors is further supported by the installation of mobile hydrogen fuel cells, which can be deployed across various parts of the port. These mobile units supplement the port's onshore power supply and can be relocated based on demand.

7.8 North Sea Ports, Belgium/the Netherlands[8]

The North Sea Ports, which includes the ports of Ghent (Belgium), Vlissingen, and Terneuzen (Netherlands), have received significant investments in hydrogen production and logistics. The port has set ambitious goals, with plans to reach 500 MW of green hydrogen production capacity by 2025 through projects like VoltHydrogen Terneuzen and Vlissingen, which will include the construction of 25 MW electrolysis units at both locations. This capacity is expected to grow further as part of the port's long-term strategy to meet Europe's growing demand for renewable hydrogen, particularly for industrial applications in its immediate hinterland.

The port is also working on the SeaHydrogenLand project, which aims to establish a 1 GW green hydrogen plant by 2030. This project, led by Ørsted, is designed to cover up to 20% of the current hydrogen demand within the North Sea Port area, targeting both local industrial users and broader export markets. These efforts are part of the port's dual strategy to produce green hydrogen locally and to serve as a gateway for hydrogen imports, with an import potential of 2–million tonnes per year by 2050.

[8] Based primarily on: https://orsted.com/en/media/news/2021/03/451073134270788

In terms of infrastructure, North Sea Port is partnering with GasUnie and Fluxys to develop the Hydrogen Delta Network, a regional hydrogen network that will connect local producers to end-users within the industrial cluster. This network will also be linked to the broader European Hydrogen Backbone, facilitating cross-border hydrogen trade. Furthermore, the ELYgator project in Terneuzen, developed by Air Liquide, will construct a large-scale hydrogen production plant to meet the needs of both domestic and international markets, reinforcing the port's role in Europe's future hydrogen logistics. Another significant initiative is the reorientation of the North-C-Methanol project, which initially focused on methanol production. Now, the project will pivot towards green hydrogen production with two demo plants planned, including a 65-MW electrolyser.

7.9 Port of Vancouver, Canada[9]

The Port of Vancouver is central for the hydrogen development in Canada, focusing on decarbonising its operations and reducing emissions in the transport sector. One of the major initiatives is the BC Hydrogen Ports Project (BCH2 Ports Project), which received $four million in funding from the government of British Columbia. This project, led by HTEC, a clean hydrogen production company, is one of the first large-scale efforts in the country aimed at using hydrogen and fuel cells to power port operations and heavy-duty vehicles at the port.

The hydrogen initiative in Vancouver includes the installation of a high-capacity hydrogen fuelling station at Tsawwassen First Nation land, which will initially support heavy-duty hydrogen-electric vehicles. The station will fuel hydrogen-powered trucks and yard equipment, including four fuel-cell electric yard trucks and two Class 8 drayage trucks that will be operated by major port operators such as Harbour Link Container Services and BC Ferries Commercial Services. This project is expected to reduce emissions by 110 tonnes of CO_2 per year.

In addition to these operational upgrades, the hydrogen strategy at Vancouver aligns with broader provincial and federal decarbonisation targets, which aim to reduce transport emissions significantly. The hydrogen projects are positioned to demonstrate the viability of hydrogen for large-scale port and logistics operations, with future plans to expand the use of hydrogen across more areas of port activity.

With hydrogen-powered trucks and fuel-cell technology being tested in the port, Vancouver is focussing on decarbonising heavy-duty transport, one of the sectors most difficult to electrify. These advancements are part of British Columbia's efforts to become a leader in the hydrogen economy, with over 50% of Canada's hydrogen companies based in the province. These efforts are being actively supported through both provincial programs like CleanBC and collaborations with companies such as HTEC.

[9] Based on: https://www.htec.ca/hydrogen-will-fuel-clean-transportation-at-b-c-ports/

7.10 Port of Constanza, Romania

The Port of Constanza has significantly increased its focus on developing a hydrogen economy by leveraging its strategic location in the Black Sea region to establish itself as a hub for hydrogen production, logistics, and storage. The port's position as a gateway to Eastern Europe, combined with its proximity to substantial renewable energy sources, such as offshore wind in the Black Sea, makes it well suited to integrating hydrogen into its operations. Additionally, the European Union's REPowerEU strategy, which aims to import up to four million tonnes of hydrogen annually by 2030, underlines the potential of Constanza to play an important role in facilitating green hydrogen imports, particularly from renewable sources like wind and solar energy. The Romanian hydrogen strategy is similarly ambitious, aiming to produce over 150,000 tonnes of renewable hydrogen per year by 2030, with significant investments to support this effort [12].

One of Constanza's key initiatives involves building the infrastructure necessary to support the storage and distribution of hydrogen. The port is developing specialised facilities for both compressed gaseous hydrogen and liquid hydrogen, ensuring the capacity to handle hydrogen safely and efficiently. The port is also evaluating the potential to store and manage hydrogen derivatives such as ammonia and methanol, which offer alternative pathways for hydrogen use in maritime transport. These infrastructure advancements are necessary for Constanza to serve as a logistics hub for Eastern Europe, where the demand for clean energy solutions in industry and transportation is growing rapidly [39].

Infrastructure development at the Port of Constanza includes the integration of hydrogen transport infrastructure, such as pipelines and shipping routes, into the port's existing logistical framework. These efforts will allow the port to connect local and regional hydrogen producers to major industrial hubs throughout Europe. This integration is key to reducing reliance on fossil fuels, as hydrogen pipelines will enable efficient and large-scale distribution to energy-intensive industries, aligning with the EU's overall decarbonisation objectives [39].

The scale of investment required to realise these hydrogen projects is substantial, with Romania aiming to invest nearly €4.8 billion by 2030 to develop its hydrogen economy. This includes infrastructure for production, storage, and transportation. The Port of Constanza's role as a primary hub for hydrogen imports and distribution will be crucial for achieving these targets, facilitating Romania's broader transition to a low-carbon economy and contributing to Eastern Europe's energy independence and sustainability efforts [12].

7.11 Copenhagen Malmö Port, Denmark and Sweden[10]

One of the projects that the Copenhagen Malmö Port (CMP) involved in is the Ocean Valley innovation hub, primarily with the focus on integrating advanced technologies to decarbonise maritime operations and improve logistics efficiency. Through projects like Carbon Hub and E-port, CMP is exploring carbon capture and storage (CCUS) and the electrification of port infrastructure, aiming to transition away from fossil fuel dependency. These initiatives are backed by practical, research-driven goals, such as mapping future electricity demand and developing interim storage for captured CO_2 to reduce emissions within port activities and the wider maritime industry.

CMP's focus on sustainable innovation is aligned with the decarbonisation targets of Denmark and Sweden. The Carbon Hub project aims to identify economically viable use cases for CO_2 storage and reuse, a critical step in reducing emissions at port facilities. The E-port project, meanwhile, is assessing the electrification potential of the port's heavy infrastructure to reduce operational emissions. This research-backed approach ensures CMP's initiatives are not only experimental but also designed to be scalable and replicable across other ports in Europe.

The CMP's role within Ocean Valley demonstrates how ports can act as testing grounds for decarbonisation technologies in real-world settings with a focus on long-term sustainability and scalability. By collaborating with academic institutions, private sector stakeholders, and regional governments, CMP ensures that the technological advancements made through Ocean Valley have practical applications that can be scaled beyond the Øresund region. These projects not only focus on the port's own sustainability but also contribute to broader climate goals, creating a blueprint for ports globally to reduce emissions and support greener maritime logistics.

7.12 Port of Sines, Portugal[11]

The Port of Sines in Portugal aims at strengthening its position in the hydrogen economy, capitalising on its strategic location and infrastructure to become a hub for green hydrogen production and logistics. The port is part of the Sines Green Hydrogen Valley project, which aims to produce large volumes of green hydrogen using renewable energy. This project, supported by both national and European funding, is expected to significantly contribute to Portugal's goal of producing 5 GW of electrolysis capacity by 2030. The hydrogen produced at Sines is anticipated to

[10] Ocean Valley [34]. *Ocean Valley Innovation Hub*. Retrieved August 17, 2024, from https://www.oceanvalley.com/; Copenhagen Malmö Port [11]. *Copenhagen Malmö Port – A modern port with global reach*. Retrieved August 17, 2024, from https://www.cmport.com/

[11] Mostly based on https://portugalenergia.pt/ and https://www.apsinesalgarve.pt/

serve not only domestic markets but also to be exported to other European countries via pipeline or by sea.

The strategy of the Port of Sines in the production of hydrogen was initially strengthened by its partnership with the Port of Rotterdam through the H2Sines-Rdam project. This joint initiative aimed to establish a green hydrogen corridor between Portugal and the Netherlands, facilitating the export of hydrogen produced at Sines to Rotterdam, one of Europe's largest energy hubs. The project was designed to leverage Sines' potential for large-scale hydrogen production from renewable sources, particularly offshore wind, and Rotterdam's advanced infrastructure for receiving, storing, and distributing hydrogen across Europe. The collaboration was seen as a crucial step in positioning Sines as a major player in the European hydrogen market and advancing decarbonisation efforts across the continent.

However, despite the strategic importance of the H2Sines-Rdam project, ENGIE: the French multinational recently confirmed its termination. The cancellation reflects the complexity and challenges associated with building international hydrogen supply chains, particularly in the early stages of the hydrogen economy. The decision highlights the uncertainties surrounding large-scale hydrogen infrastructure projects [42].

In addition to production, the port is also working on integrating hydrogen into its own operations, including hydrogen-powered machinery and equipment, aligning with broader decarbonisation efforts in the maritime sector. This strategic focus on hydrogen allows Sines to leverage its existing LNG infrastructure and port facilities to transition towards cleaner energy sources, thus potentially playing a central role in the hydrogen supply chain in Southern Europe.

7.13 Port of Singapore

The Port of Singapore is integrating hydrogen into its strategic framework to support decarbonisation across its extensive maritime operations. Singapore's Maritime Singapore Decarbonisation Blueprint outlines hydrogen as a key element in reducing the carbon footprint of the port, which handles over 50 million tonnes of marine bunker fuel annually [15]. One of the major focuses is developing hydrogen bunkering infrastructure to support the transition of international shipping to cleaner fuels like ammonia and hydrogen. This is crucial for achieving global decarbonisation targets, especially with projections that hydrogen-based fuels could meet up to 45% of global maritime energy demand by 2050 [15].

Singapore's hydrogen strategy is underpinned by the development of hydrogen production and storage facilities within the port. The government is investing in the establishment of hydrogen import and storage capabilities, leveraging its position as a major energy hub to import hydrogen from regions with abundant renewable energy sources, such as Australia and the Middle East. The port is integrating hydrogen infrastructure into its existing LNG facilities, facilitating the import, storage, and distribution of hydrogen and its carriers, such as ammonia and LOHCs [53]. These developments aim to ensure that Singapore remains competitive in the global energy transition while supporting the growing hydrogen demand in Asia.

Additionally, Singapore is collaborating with global partners to develop hydrogen refuelling stations within the port, allowing vessels to refuel with hydrogen as they transit through one of the busiest maritime crossroads in the world. Hydrogen-powered equipment is also being deployed in the port's operations, with pilot projects focusing on hydrogen fuel cells to power cranes and other cargo-handling machinery, further reducing operational emissions [15].

The port's long-term vision involves expanding the role of hydrogen beyond maritime bunkering to broader industrial applications within Singapore's energy system. By 2050, hydrogen is expected to contribute significantly to Singapore's energy mix, providing up to 50% of electricity demand depending on advancements in hydrogen technologies [15]. This comprehensive strategy aligns with Singapore's overall net-zero ambitions and enhances its role as a leader in the global hydrogen economy.

7.14 *Port of Kobe, Japan*[12]

The Port of Kobe has developed a comprehensive hydrogen strategy as part of Japan's national push towards decarbonisation, with the goal of establishing itself as a key hydrogen hub by 2050. As part of this initiative, the port is focusing on building the infrastructure necessary to facilitate hydrogen imports, storage, and supply for industrial and maritime use. A significant milestone was reached in 2021 when the first shipment of liquefied hydrogen from Australia arrived at Kobe's liquefied hydrogen terminal, demonstrating the feasibility of hydrogen imports at a commercial scale [8, 51]. This infrastructure is essential for supporting Japan's broader hydrogen strategy, which aims to promote hydrogen as a central pillar of the energy transition.

The Port of Kobe is not only working on the import and storage of hydrogen but is also actively participating in efforts to develop hydrogen supply chains. These supply chains involve converting and distributing hydrogen for use in various sectors, including shipping and local industries. By facilitating hydrogen distribution through pipelines and other logistical infrastructure, the port supports Japan's efforts to reduce greenhouse gas emissions across multiple sectors, including heavy industry and transport [8]. Additionally, Kobe's strategic location and existing maritime infrastructure make it a natural hub for future hydrogen imports, particularly as Japan looks to scale up its hydrogen consumption to meet its 2050 carbon neutrality targets.

[12] Based on https://www.hydrogenenergysupplychain.com/ and https://www.hystra.or.jp/en/project/

In terms of future developments, the Port of Kobe is integrating plans for the expanded use of hydrogen, including the potential development of hydrogen bunkering services for maritime vessels. As the maritime sector looks to transition away from conventional fossil fuels, Kobe is positioning itself as a leader in hydrogen refuelling, with a target of supporting hydrogen-powered vessels by 2030. The port's ongoing investments in hydrogen infrastructure will be critical to achieving these objectives, ensuring that it remains a central player in Japan's hydrogen economy [8, 51].

8 Discussion

The 14 cases presented in Sect. 7 provide an overview of various activities and approaches that ports are taking to become hydrogen hubs and are examples of how ports can meet the challenges of the hydrogen economy by transforming their infrastructure and expanding their role beyond traditional logistics. Case studies from large ports like Rotterdam, Singapore, Hamburg, and Antwerp-Bruges demonstrate the large-scale investments being made in hydrogen production, storage, and distribution. These ports are becoming central nodes for hydrogen, and in particular 'green' hydrogen, using wind and solar energy to fuel electrolysis plants, while also developing advanced storage systems (cryogenic tanks and high-pressure storage) to handle hydrogen effectively. The investments made in specialised hydrogen terminals are designed to ensure that hydrogen can be stored and distributed at scale, addressing the lack of infrastructure that has historically slowed hydrogen adoption. These efforts provide practical examples of how ports are solving the technical and logistical challenges posed by hydrogen and reflect the increasing importance of ports in facilitating the broader energy transition. Also, smaller ports can leverage on their strategic location for this purpose (e.g. Esbjerg, Sines, Constanza).

Ports located in energy-exporting countries are in a good position to redevelop and expand their infrastructure for supplying also hydrogen. Some ports (e.g. Sines and North Sea Ports) are taking advantage of the increasing demand for hydrogen (especially 'green') to redevelop their industrial clusters. Other ports (e.g. North Tyrrhenian Port Network Authority, Port of Kobe) are also expecting an increase in imports.

All ports are addressing key bottlenecks in the hydrogen supply chain, particularly related to the infrastructure needed for bunkering and land transport refuelling. Especially ports that for economic and geographical reasons already play an important role in bunkering, are developing strategies to manage the decarbonisation of the shipping sector resulting from regulation and changes in customer demands.

Government support and regulatory frameworks are essential in enabling these developments. Case studies from Vancouver and California, and in general, in Europe, show how public–private partnerships, along with targeted funding, are helping to overcome the financial barriers to hydrogen infrastructure development.

In Vancouver, for example, federal and provincial funding has been critical to establishing hydrogen refuelling and distribution networks, while similar initiatives in Europe have been supported by EU-wide and national policies promoting hydrogen adoption. The role of ports is also important for testing new technologies (e.g. Copenhagen-Malmö). These are good examples of the role of governments in facilitating the transition by providing the necessary financial and regulatory backing, or opportunities for piloting and testing new technologies.

In addition to their evolving role in hydrogen handling, trading, bunkering, and logistics, ports are increasingly working to supply hydrogen locally produced to port activities (e.g. terminals) and other industries to support their decarbonisation efforts. Hamburg and Antwerp-Bruges, for instance, are linking hydrogen production to heavy industry, such as steelmaking and chemical manufacturing, creating synergies that allow hydrogen to play a broader role in reducing emissions beyond maritime transport. This collaboration across sectors not only strengthens the position of ports as hubs for hydrogen but also highlights their role in driving decarbonisation across multiple industries, leveraging their strategic locations and infrastructure.

Smaller ports, such as Esbjerg, Valencia, and the North Tyrrhenian Network Port Authority, also demonstrate how regional hubs can contribute to the hydrogen economy. These ports are making significant investments in hydrogen production and storage, despite not having the scale of larger ports like Rotterdam. By focusing on niche applications and regional distribution, these ports are showing that the hydrogen economy is not limited to large ports. Instead, it is becoming a more decentralised and flexible system, with a wide range of ports contributing to the development and expansion of hydrogen infrastructure on both regional and global levels.

9 Conclusions

Seaports are becoming indispensable in the hydrogen value chain, functioning as key hubs for production, storage, and distribution. Large ports like Singapore, Rotterdam, Hamburg, and Antwerp-Bruges are leading the way by integrating renewable energy sources with large-scale electrolysis facilities and developing large storage facilities for hydrogen. Their strategic locations and existing infrastructure enable them to facilitate the decarbonisation of shipping while serving as critical transit points for hydrogen and its derivatives, such as ammonia. These developments reflect the role of ports not only in enabling cleaner fuel alternatives for the maritime sector but also in connecting hydrogen supply to industrial and transport applications, enhancing their function in the global energy transition.

However, realising the full potential of seaports as hydrogen hubs requires clear government policy, targeted investments, and cross-sector collaboration. Government support through subsidies, carbon pricing, and infrastructure investment plays an important role in offsetting the high costs of hydrogen production and storage.

Governments need also to provide foresight and a strategic vision for decarbonisation and the role that hydrogen plays within this vision.

Ports must collaborate with industries and energy providers to build efficient hydrogen ecosystems, integrating renewable energy sources and aligning with decarbonisation goals. Moreover, investment in green hydrogen offers opportunities for economic growth, particularly in the creation of high-value jobs within the hydrogen sector. By strategically investing in hydrogen infrastructure, ports can drive a just transition, balancing environmental goals with socio-economic benefits, and ensuring sustainable growth in regions dependent on maritime and industrial activities.

This transition remains uncertain, and further research is needed to better understand how ports can maximise their benefits and contributions to the hydrogen economy. This involves exploring how to develop more resilient and responsible business models, identifying effective strategies for accelerating value creation and capture, and addressing the external effects associated with the hydrogen economy. Additional areas warranting further investigation include public acceptance, labour transition, and the preservation of the natural environment beyond decarbonisation, such as biodiversity regeneration.

References

1. Acciaro, M., Ghiara, H., & Cusano, M. I. (2014). Energy management in seaports: A new role for port authorities. *Energy Policy, 71*, 4–12.
2. Agarwala, N. (2024). Is Hydrogen a decarbonising fuel for maritime shipping? *Maritime Technology and Research, 6*(4), 271244. https://doi.org/10.33175/mtr.2024.271244
3. Alamoush, A. S., Dalaklis, D., Ballini, F., & Ölcer, A. I. (2023). Consolidating port decarbonisation implementation: Concept, pathways, barriers, solutions, and opportunities. *Sustainability, 15*(19), 14185. https://doi.org/10.3390/su151914185
4. Ammar, N. R. (2019). An environmental and economic analysis of methanol fuel for a cellular container ship. *Transportation Research Part D, 69*, 66–76. https://doi.org/10.1016/j.trd.2019.02.001
5. Ampah, J. D., Yusuf, A. A., Afrane, S., Jin, C., & Liu, H. (2021). Reviewing two decades of cleaner alternative marine fuels: Towards IMO's decarbonization of the maritime transport sector. *Journal of Cleaner Production, 320*, 128871–128871. https://doi.org/10.1016/j.jclepro.2021.128871
6. Buonomano, A., Del Papa, G., Giuzio, G. F., Palombo, A., & Russo, G. (2023). Future pathways for decarbonization and energy efficiency of ports: Modelling and optimization as sustainable energy hubs. *Journal of Cleaner Production, 420*, 138389.
7. Chen, P. S. L., Fan, H., Enshaei, H., Zhang, W., Shi, W., Abdussamie, N., et al. (2023). A review on ports' readiness to facilitate international hydrogen trade. *International Journal of Hydrogen Energy, 48*(46), 17351–17369.
8. Chen, P. S.-L., Fan, H., Enshaei, H., Zhang, W., Shi, W., Abdussamie, N., et al. (2024). Opportunities and challenges of hydrogen ports: An empirical study in Australia and Japan. *Hydrogen, 5*(3), 436–458. https://doi.org/10.3390/hydrogen5030025
9. Chen, Z. S., & Lam, J. S. L. (2022). Life cycle assessment of diesel and hydrogen power systems in tugboats. *Transportation Research Part D: Transport and Environment, 103*, 103192.

10. Clean Hydrogen Joint Undertaking. (2023). *Study on hydrogen in ports and industrial coastal areas*. European Commission. https://www.clean-hydrogen.europa.eu/media/publications/study-hydrogen-ports-and-industrial-coastal-areas-reports_en
11. Copenhagen Malmö Port. (n.d.). *Copenhagen Malmö Port – A modern port with global reach*. Retrieved August 17, 2024, from https://www.cmport.com/
12. Energy Policy Group. (2024). *The future of hydrogen in Romania: Dispelling myths from reality*. Retrieved from enpg.ro.
13. Garcia-Navarro, J., Isaacs, M. A., Favaro, M., Ren, D., Ong, W. J., Grätzel, M., & Jiménez-Calvo, P. (2024). Updates on hydrogen value chain: A strategic roadmap. *Global Challenges, 8* (6), 2300073.
14. Gordon, J. A., Balta-Ozkan, N., & Nabavi, S. A. (2023). Socio-technical barriers to domestic hydrogen futures: Repurposing pipelines, policies, and public perceptions. *Applied Energy, 336*, 120850–120850. https://doi.org/10.1016/j.apenergy.2023.120850
15. Government of Singapore. (2023). *Singapore's national hydrogen strategy*. Ministry of Trade and Industry.
16. Ma, H., Sun, Z., Xue, Z., Zhang, C., & Chen, Z. (2023). A systemic review of hydrogen supply chain in energy transition. *Frontiers in Energy, 17*(1), 102–122. https://doi.org/10.1007/s11708-023-0861-0
17. Hjeij, D., Biçer, Y., & Koç, M. (2022). Hydrogen strategy as an energy transition and economic transformation avenue for natural gas exporting countries: Qatar as a case study. *International Journal of Hydrogen Energy, 47*(8), 4977–5009. https://doi.org/10.1016/j.ijhydene.2021.11.151
18. Holder, D., Percy, S. D., & Yavari, A. (2024). A review of port decarbonisation options: Identified opportunities for deploying hydrogen technologies. *Sustainability, 16*(8), 3299.
19. International Maritime Organization (IMO). (2023). *IMO strategy on reduction of greenhouse gas emissions from ships*. Resolution MEPC.377(80), Annex 15.
20. Iris, Ç., & Lam, J. S. L. (2019). A review of energy efficiency in ports: Operational strategies, technologies and energy management systems. *Renewable and Sustainable Energy Reviews, 112*, 170–182.
21. Islam, A., Islam, T., Mahmud, H., Raihan, O., Islam, M. S., Marwani, H. M., et al. (2024). Accelerating the green hydrogen revolution: A comprehensive analysis of technological advancements and policy interventions. *International Journal of Hydrogen Energy, 67*, 458–486.
22. Jesus, B., Ferreira, I. A., Carreira, A., Erikstad, S. O., & Godina, R. (2024). Economic framework for green shipping corridors: Evaluating cost-effective transition from fossil fuels towards hydrogen. *International Journal of Hydrogen Energy, 83*, 1429–1447.
23. Klopott, M., Popek, M., & Urbanyi-Popiołek, I. (2023). Seaports' role in ensuring the availability of alternative marine fuels—A multi-faceted analysis. *Energies, 16*(7), 3055. https://doi.org/10.3390/en16073055
24. Kolb, S., Müller, J., Luna-Jaspe, N., & Karl, J. (2022). Renewable hydrogen imports for the German energy transition – A comparative life cycle assessment. *Journal of Cleaner Production, 373*, 133289.
25. Kumar, S., Baalisampang, T., Arzaghi, E., Garaniya, V., Abbassi, R., & Salehi, F. (2023). Synergy of green hydrogen sector with offshore industries: Opportunities and challenges for a safe and sustainable hydrogen economy. *Journal of Cleaner Production, 384*, 135545.
26. Le, T. T., Sharma, P., Bora, B. J., Tran, V. D., Truong, T. H., Le, H. C., & Nguyen, P. Q. P. (2024). Fueling the future: A comprehensive review of hydrogen energy systems and their challenges. *International Journal of Hydrogen Energy, 54*, 791–816.
27. Lee, K., Murakami, S., Ölçer, A. I., Dong, T., Estebanez, G., & Schönborn, A. (2024). Hydrogen enriched LNG fuel for maritime applications – A life cycle study. *International Journal of Hydrogen Energy, 78*, 333–343.
28. Li, H., Cao, X., Liu, Y., Shao, Y., Nan, Z., Teng, L., et al. (2022). Safety of hydrogen storage and transportation: An overview on mechanisms, techniques, and challenges. *Energy Reports, 8*, 6258–6269.

29. Lindstad, E., Lagemann, B., Rialland, A., Gamlem, G. M., & Valland, A. (2021). Reduction of maritime GHG emissions and the potential role of E-fuels. *Transportation Research Part D: Transport and Environment, 101*, 103075.
30. Macingwane, Z., & Schönborn, A. (2024). Techno-economic analysis of green hydrogen production as maritime fuel from wave energy. *Energies, 17*(18), 4683.
31. Martin, J., Neumann, A., & Ødegård, A. (2023). Renewable hydrogen and synthetic fuels versus fossil fuels for trucking, shipping and aviation: A holistic cost model. *Renewable and Sustainable Energy Reviews, 186*, 113637.
32. Masip Macía, Y., Rodríguez Machuca, P., Rodríguez Soto, A. A., & Carmona Campos, R. (2021). Green hydrogen value chain in the sustainability for port operations: Case study in the region of Valparaiso, Chile. *Sustainability, 13*(24), 13681.
33. Notteboom, T., & Haralambides, H. (2023). Seaports as green hydrogen hubs: Advances, opportunities and challenges in Europe. *Maritime Economics & Logistics, 25*(1), 1–27. https://doi.org/10.1057/s41278-023-00253-1
34. Ocean Valley. (n.d.). *Ocean Valley Innovation Hub*. Retrieved August 17, 2024, from https://www.oceanvalley.com/
35. Okolie, J. A., Patra, B. R., Mukherjee, A., Nanda, S., Dalai, A. K., & Kozinski, J. A. (2021). Futuristic applications of hydrogen in energy, biorefining, aerospace, pharmaceuticals and metallurgy. *International Journal of Hydrogen Energy, 46*(13), 8885–8905.
36. Parhamfar, M., Sadeghkhani, I., & Adeli, A. M. (2023). Towards the application of renewable energy technologies in green ports: Technical and economic perspectives. *IET Renewable Power Generation, 17*(12), 3120–3132.
37. Parris, D., Spinthiropoulos, K., Ragazou, K., Giovou, A., & Tsanaktsidis, C. (2024). Methanol, a plugin marine fuel for green house gas reduction—A review. *Energies, 17*(3), 605.
38. Perna, A., Jannelli, E., Micco, S. D., Romano, F., & Minutillo, M. (2023). Designing, sizing and economic feasibility of a green hydrogen supply chain for maritime transportation. *Energy Conversion and Management, 278*, 116702–116702. https://doi.org/10.1016/j.enconman.2023.116702
39. Pioneers Consortium. (2024). *State of the art of the European green ports master plans*. https://pioneers-ports.eu/
40. Pivetta, D., Dall'Armi, C., Sandrin, P., Bogar, M., & Taccani, R. (2024). The role of hydrogen as enabler of industrial port area decarbonization. *Renewable and Sustainable Energy Reviews, 189*, 113912.
41. Pomaska, L., & Acciaro, M. (2022). Bridging the maritime-hydrogen cost-gap: Real options analysis of policy alternatives. *Transportation Research Part D: Transport and Environment, 107*, 103283.
42. PV Magazine. (2024, April 12). *The Hydrogen Stream: ENGIE confirms termination of H2Sines-Rdam project*. Retrieved from https://www.pv-magazine.com/2024/04/12/the-hydrogen-stream-engie-confirms-termination-of-h2sines-rdam-project/
43. Romano, A., & Yang, Z. (2021). Decarbonisation of shipping: A state of the art survey for 2000–2020. *Ocean & Coastal Management, 214*, 105936.
44. Shin, J. E. (2022). Hydrogen technology development and policy status by value chain in South Korea. *Energies, 15*(23), 8983.
45. Song, D. P. (2024). A literature review of seaport Decarbonisation: Solution measures and roadmap to net zero. *Sustainability, 16*(4), 1620.
46. Sornn-Friese, H., Sofev, P., & Kondratenko, K. (2023). The port authority as system builder in cross-border regionalization: An exploratory study of port Esbjerg in the development of North Sea wind. *Maritime Transport Research, 4*, 100084. https://doi.org/10.1016/j.martra.2023.100084
47. SRM Centro Studi e Ricerche. (2024). *Italian Maritime Economy Report 2024: Le nuove sfide dei porti dell'area euro-mediterranea*. SRM.
48. Tofalos, C., Jeong, B., & Jang, H. (2020). Safety comparison analysis between LNG/LH2 for bunkering operation. *Journal of International Maritime Safety, Environmental Affairs, and Shipping, 4*(4), 135–150.

49. Tomos, B. A. D., Stamford, L., Welfle, A., & Larkin, A. (2024). Decarbonising international shipping – A life cycle perspective on alternative fuel options. *Energy Conversion and Management, 299*, 117848.
50. Vichos, E., Sifakis, N., & Tsoutsos, T. (2022). Challenges of integrating hydrogen energy storage systems into nearly zero-energy ports. *Energy, 241*, 122878–122878. https://doi.org/10.1016/j.energy.2021.122878
51. WSP. (2023). *Hydrogen in maritime: Opportunities and challenges*. WSP.
52. Yang, M., & Lam, J. S. L. (2023). Operational and economic evaluation of ammonia bunkering – Bunkering supply chain perspective. *Transportation Research Part D: Transport and Environment, 117*, 103666.
53. Yap, W. Y., & Notteboom, T. (2023). Renewable energy options for seaport cargo terminals with application to mega port Singapore. *Maritime Business Review, 12*(3), 1–17. https://doi.org/10.1108/MABR-12-2023-0083
54. Zamora, E. A. (2016). Value chain analysis: A brief review. *Asian Journal of Innovation and Policy, 5*(2), 116–128. https://doi.org/10.7545/ajip.2016.5.2.116
55. Zhang, L., Jia, C., Bai, F., Wang, W., An, S., Zhao, K., Li, Z., Li, J., & Sun, H. (2024). A comprehensive review of the promising clean energy carrier: Hydrogen production, transportation, storage, and utilization (HPTSU) technologies. *Fuel, 355*, 129455–129455. https://doi.org/10.1016/j.fuel.2023.129455
56. Zhang, Q., Chen, W., & Ling, W. (2022). Policy optimization of hydrogen energy industry considering government policy preference in China. *Sustainable Production and Consumption, 33*, 890–902.

Sustainable Waterborne Transportation with Hydrogen: An Ecosystem Approach to Design

Arianna Bionda, Laura Pirrone, Nikos Sakellaridis, and Erik-Jan Boonen

1 Introduction

The transition towards sustainable transportation and its decarbonization represents one of the major challenges worldwide. Waterborne transportation presents a valuable opportunity to reduce road traffic and decrease air and noise pollution. Ships can transport larger volumes of people and goods, offering flexible routing options, and adapting to varying logistical needs, while requiring less infrastructure, such as roads or bridges [1].

In recent years the sector has experienced an increase in demand and the number of vessels, and the majority of port calls in the EU were attributed to Ro-pax ships (41%) and passenger vessels (18%), including ferries, cruise ships, and smaller craft. Combined, these two categories accounted for over half of all port activities (59%) [2].

Despite the sector's potential, waterborne transportation still faces challenges in reducing resource exploitation, emissions, and waste. These include reliance on hydrocarbon combustion, discharge of pollutants, fouling of waterways, and inadequate docking facilities, all contributing to climate change, air pollution, and water pollution [3, 4]. To tackle this issue, the European Commission has introduced a series of legislative measures. The international goals set by the European Green Deal, the Fit for 55 Package, the EU's Emissions Trading System, and the International Maritime Organization's (IMO) strategy aim at achieving carbon-neutral shipping and significant reductions in greenhouse gas (GHG) emissions by 2050

A. Bionda (✉) · L. Pirrone
Design Department, Politecnico di Milano, Milan, Italy
e-mail: Arianna.Bionda@polimi.it; Laura.Pirrone@polimi.it

N. Sakellaridis · E.-J. Boonen
Damen Research, Development & Innovation B.V., Gorinchem, The Netherlands
e-mail: Nikos.Sakellaridis@damen.com; Erik.Jan.Boonen@damen.com

C. O. Colpan et al. (eds.), *Fuel Cell and Hydrogen Technologies in Maritime Transportation*, https://doi.org/10.1007/978-3-031-86110-9_10

[5]. The IMO introduced a strategic plan to reduce GHG emissions by over 50% by 2050 compared to the levels in 2008 [6].

Alternative fuels and energy sources such as biofuels, synthetic fuels, hydrogen (H_2), ammonia, and batteries are emerging as viable options to replace conventional fossil fuels. These alternatives are not only suitable for passenger ships but also for tanker vessels with high-power demands. Research has highlighted the potential benefits of hydrogen fuel cells in the maritime sector. Despite cost challenges, hydrogen stands out as the most promising clean fuel option for the global shipping industry [7]. Both blue and green hydrogen offer pathways for significant reductions in GHG emissions. Moreover, hydrogen fuel cells operate quietly with few moving parts and high efficiency, produce minimal noise pollution, and release only water vapour and oxygen as by-products, thereby virtually eliminating air and water pollutants during fuel consumption.

The introduction of zero-emission fuels and energy converters like hydrogen and fuel cell stacks in the waterborne transportation sector offers a promising solution to addressing sustainability challenges. However, the current approach to hydrogen vessel design is hindered by the absence of a clear design methodology and a dedicated normative framework. While the technology readiness level supports the adoption of hydrogen and fuel cell stacks on board, stakeholders and designers still lack a systematic approach that spans from the concept design phase to the engineering design stage [8].

By introducing an ecosystemic approach, this chapter aims to propose a methodology for designing hydrogen passenger vessels. This methodology seeks to pave the way for developing a systematic normative framework and guidelines for adopting hydrogen as a zero-emission alternative fuel in waterborne passenger transportation.

2 Need for an Ecosystemic Approach to Design

The adoption of hydrogen in waterborne transportation represents a complex transformation that impacts not only the vessels but also the entire system in which hydrogen is produced and operated.

Vessels need to be viewed as part of a complex transportation system rather than as a single ship. This issue requires an ecosystemic approach that considers the network of stakeholders mutually interacting to realize a common value proposition [9], that is, waterborne transportation means operating by hydrogen and fuel cell technology.

This ecosystem of stakeholders encompasses various companies, industries, organizations, and institutions involved through the entire design process, from the vessel design and risk assessment, starting from the early stages to the supply chain of the hydrogen, including its provision and bunkering operations. These stakeholders come from the maritime and transportation sectors, spanning from basic research environments to mature industrial sectors. They include shipyards, transport operators, hydrogen providers, fuel cell technology providers, research centres, port authorities, and classification and standardization bodies.

Therefore, designing a hydrogen-fuelled vessel requires considering the entire systemic context and its related stakeholders.

Designers and naval architects need to look beyond the vessel's boundaries, following a methodology that incorporates the extensive knowledge and insights from all these ecosystem's actors and their interactions. Although the nautical and maritime sectors have traditionally been approached with a vertical perspective, treating the vessel as an isolated element, this study introduces a holistic perspective, largely unexplored within the sector, and essential for addressing the complexity of the issue.

3 The e-SHyIPS Experience

The e-SHyIPS project aims to define the new guidelines for an effective introduction of hydrogen in the maritime passenger transport sector and to boost its adoption within the global and EU strategies for a clean and sustainable environment, towards the accomplishment of a zero-emission maritime transport scenario. The goal of e-SHyIPS is to move from the idea to the application, filling the existing gaps in normative and technical knowledge concerning all the related aspects of hydrogen in the maritime transport sector. By means of an ecosystemic approach, e-SHyIPS proposes theoretical pre-normative research activities on standards, simulation and laboratory experiments, design of an appropriate certification process, and spot future standardization activities to enhance the EU normative and regulatory landscape.

The ecosystemic framework developed within the project, called e-SHyIPS Learning Ecosystem, aims to create formalized and shareable knowledge within its specific research context. A learning ecosystem consists of stakeholders who incorporate learning processes and utilities within specific environmental boundaries [10]. These stakeholders interact with each other and external sources to populate a dynamic system, generating, managing, and exploiting comprehensive knowledge.

The e-SHyIPS project is structured based on five pillars each one representing a line of research and experimental activities.

- **Pillar 1—State of the art and theoretical studies:** e-SHyIPS begins by conducting an in-depth analysis of the current state of the art and practice in the use of hydrogen-based fuels in the maritime, transport, and energy sectors. This involves performing a normative and technical knowledge gap analysis. In addition to the expertise within the consortium, a wide ecosystem of stakeholders is engaged. The resulting state-of-the-art assessment served as the baseline for defining new theoretical knowledge to close existing gaps and identify new ones. Furthermore, the gathered knowledge enabled the consortium to draft general scenarios for simulations and laboratory experiments.

- **Pillar 2—Ship design experiments:** This pillar investigates new arrangement designs to achieve the best performance configurations for hydrogen-fuelled passenger ships and optimal propulsion system configurations for various operational profiles and energy requirements. This analysis facilitated the definition of pre-standards for integrating fuel cell power generation systems into the ship's network. Additionally, a digital twin model of the ship's machinery system was developed to evaluate different configuration alternatives and simulate various operating condition scenarios, allowing for the prediction of overall system performance, including fuel storage needs.
- **Pillar 3—Safety system experiments:** This pillar focuses on studying safety concerns related to hydrogen leakage and vapourization aboard hydrogen-fuelled vessels. It aims to establish emergency discharge procedures and determine optimal piping arrangements and exit points. Additionally, the behaviour of hydrogen explosions was thoroughly investigated, linking safety measures with ship design arrangements explored in Pillar 2. This approach enabled a detailed examination of passenger compartment layouts in areas least affected by potential explosions. Furthermore, an extensive safety plan for onboard safety systems is developed.
- **Pillar 4—Material and component experiments:** This Pillar aims to establish guidelines for the selection of materials used in hydrogen containment, handling, and distribution onboard vessels. It involves conducting experimental testing of chosen materials and components for a scaled-down ship hydrogen powertrain. These tests are based on realistic onboard operating profiles and conducted under various ambient conditions in a climate chamber. The quality of hydrogen is analysed to detect any potential impurities that could leach from materials and components, potentially affecting the longevity of the fuel cell system.
- **Pillar 5—Port-bunkering–ship interface experiments:** This pillar focuses on conducting specific experiments related to the bunkering process to assess safety distance requirements and hazardous zones at ports. It involves conducting risk assessments and identifying the most suitable interfaces and equipment both onboard and at port sites, considering hydrogen in both gaseous and liquid forms. A digital twin model of the refuelling station was developed to analyse the refuelling and bunkering processes, aiding in the risk assessment by quantifying threats and identifying mitigation measures. Furthermore, drawing from experience with liquefied natural gas (LNG) as a fuel, a comparison is made regarding safety issues to understand the similarities and compatibility of processes and equipment. The outcomes of this research are contributing to accelerating infrastructure development in this area.

In the project early phase, three scenarios of vessels have been defined. These scenarios differ from each other mainly according to their operation profile and were shaped by data collected during contextual inquiry activities and subsequently validated through structured and semi-structured interviews with a focus group. This group included industry experts engaged in vessel design, development, maintenance, operation, and the creation of hydrogen fuel solutions. Despite the project and its developed methodology being vessel-independent, with differentiation based on the type of hydrogen (liquid or gaseous) rather than the type of vessel, the scenarios play a crucial role in contextualizing the experimental activities being undertaken.

- **Scenario Small: Fast Water Buses.** These small, fast ferries primarily operate in inland waters, such as rivers and fluvial areas, providing regular daily water-based public transport services. They are common in lagoon or fjord areas where daily transport occurs mainly on internal waters. The reference vessel for this scenario is a 24.4-meter fast water bus by Damen shipyard, currently operating in Rotterdam and designed for urban mobility along inland waterways. Thanks to their flexible design, these vessels can be configured with different internal and external layouts to meet market demands.
- **Scenario Medium: Roll-On/Roll-Off Ferries.** Medium-sized ferries mainly operate as roll-on/roll-off passenger ships, capable of transporting commercial vehicles and private cars while accommodating large numbers of passengers for shorter voyages. These ships must meet stringent safety and technical standards due to their mixed passenger and cargo features. The reference vessel is a 118.8-meter ferry operating daily routes to the islands of Zakynthos, Kefalonia, and Ithaca by Levante Ferry. It offers amenities such as bars, restaurants, children's facilities, and a shopping arcade, alongside decks for cars and trucks and passenger accommodations.
- **Scenario Large: Small Luxury Cruise Ship**. This scenario focuses on small- to mid-sized luxury cruise ships known for providing high level of services to a select group of users. The reference project is the Celebrity Flora, launched in 2019 as the first ship specifically built for the Galápagos Islands. Its design ensures the destination remains the focal point of the sailing experience and combining features of both pleasure yachts and passenger ships. It operates weekly cruises among islands, primarily in coastal waters.

4 A Design Approach to H_2 Passenger Vessels

Compared to fossil fuels, which the shipping industry has optimized over decades in terms of design, the introduction of hydrogen-based fuels and related technologies presents new design challenges directly linked to the specific safety risks that must be mitigated or managed. The current risk management framework is tailored to the needs of traditional propulsion systems and employs three primary strategies, derived from IMO guidelines.

- The International Code of Safety for Ships Using Gases or Other Low-Flashpoint Fuels (IGF Code) sets forth obligatory standards for the arrangement and installation of machinery, equipment, and systems on vessels utilizing gas or low-flashpoint liquids as fuel, aiming to mitigate risks to the vessel, its crew, and the environment [11].

- The IMO Alternative Design process mandates that the safety, reliability, and dependability of these systems must demonstrate equivalence to those of traditional fuel and power generation systems. As outlined in the Safety of Life at Sea (SOLAS) regulations, this process applies to the entire vessel concept or can be targeted at specific systems, subsystems, or individual components.
- Formal safety assessment (FSA) is a systematic methodology that enhances maritime safety by protecting life, health, and the marine environment through structured risk analysis and cost–benefit assessment [12].

However, the properties and safety challenges associated with hydrogen-fuelled vessels differ significantly from those of conventional fuels [13]. Due to a lack of prescriptive rules on H2-fuelled ferries, the alternative design approach is prioritized. As already demonstrated in the currently ongoing H_2 vessels design processes [14], the alternative design approach is expected to create a comprehensive and rather expensive design and approval process. For this reason, it asks designers and naval architects to deal with a high degree of uncertainty, especially during the early design phase, when requirements are translated into the layout and configuration of compartments. The following methodology flow (Fig. 1) was designed, adopted, and refined in e-SHyIPS project with the aim of giving a methodological approach to maritime professionals in the initial design stage dealing with general arrangement design, feasibility analysis, and preliminary risk assessment processes.

The e-SHyIPS developed methodology must be seen as a self-assessment process that is more flexible than prescriptive. It applies to the first three steps of the implementation of a hydrogen system onboard a vessel, as described by the *DNV Handbook for Hydrogen-fuelled Vessels* [15]: feasibility studies, specification outline, and design and procurement. It guides professionals from the operational requirement definition to the general arrangement design for preliminary design approval (alternative design approach).

The Design methodology flow includes three main steps:

- Phase 1: From operational requirements to technical requirements definition.
- Phase 2: From technical requirements to technical specifications definition.
- Phase 3: Technical specifications to general arrangement (GA) design for the approval process.

Phase 1 implies defining the operational requirements, which qualitatively describe the system functions or tasks to be performed in operations, essentially outlining the basic vessel behaviours to meet user and market demands. They include

- Environmental requirements: e.g. type of voyage (inshore, offshore, coastal, etc.), wind speed, significant waves height and periods, depth of waters.
- Performance and logistic requirements: For example, human factors, ergonomics, availability, maintainability, and reliability, cruise speed, range, number of round trips to be performed, fuel autonomy, energy demand.
- Design constraints and market requirements: e.g. number. of passengers/crew members, vehicles, freight.

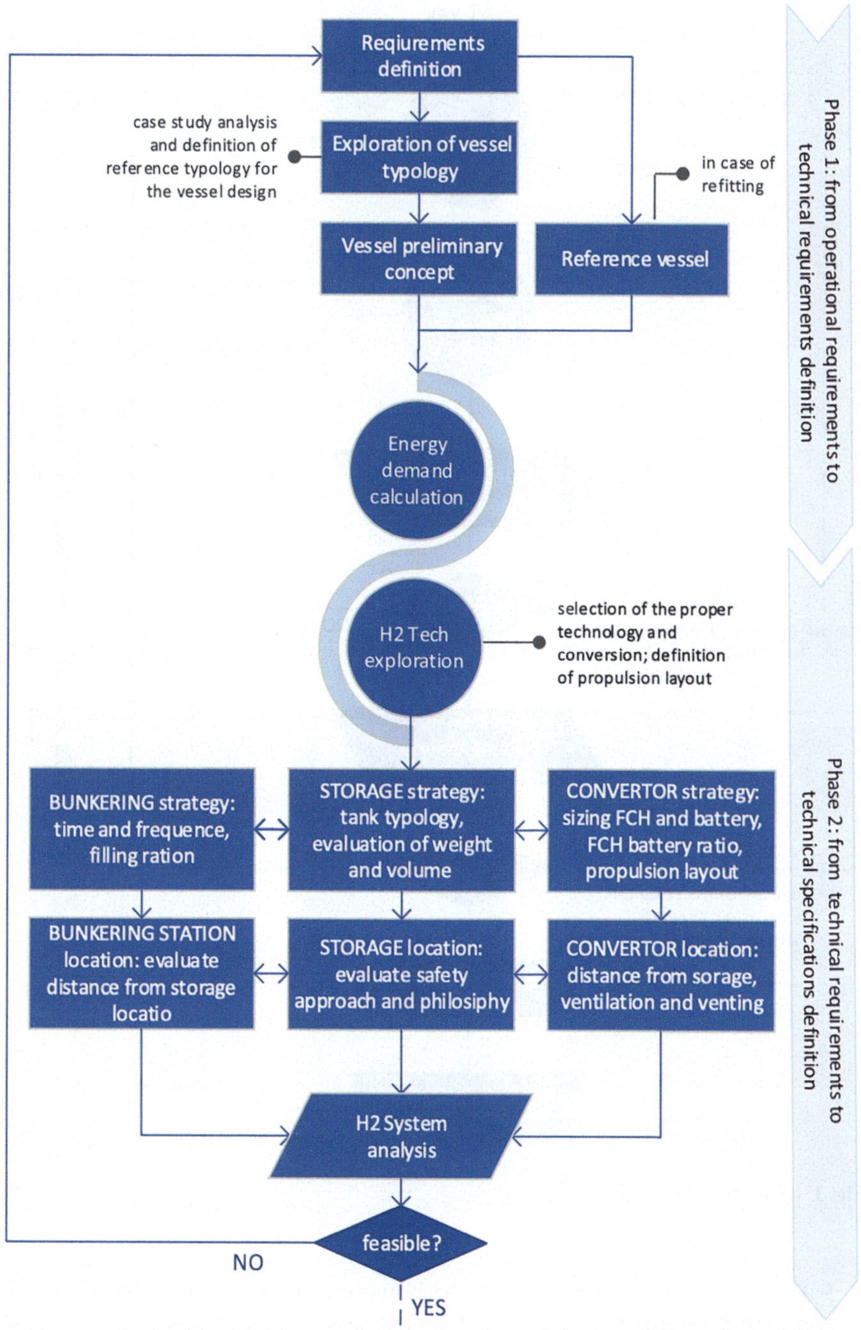

Fig. 1 e-SHyIPS Design methodology flow

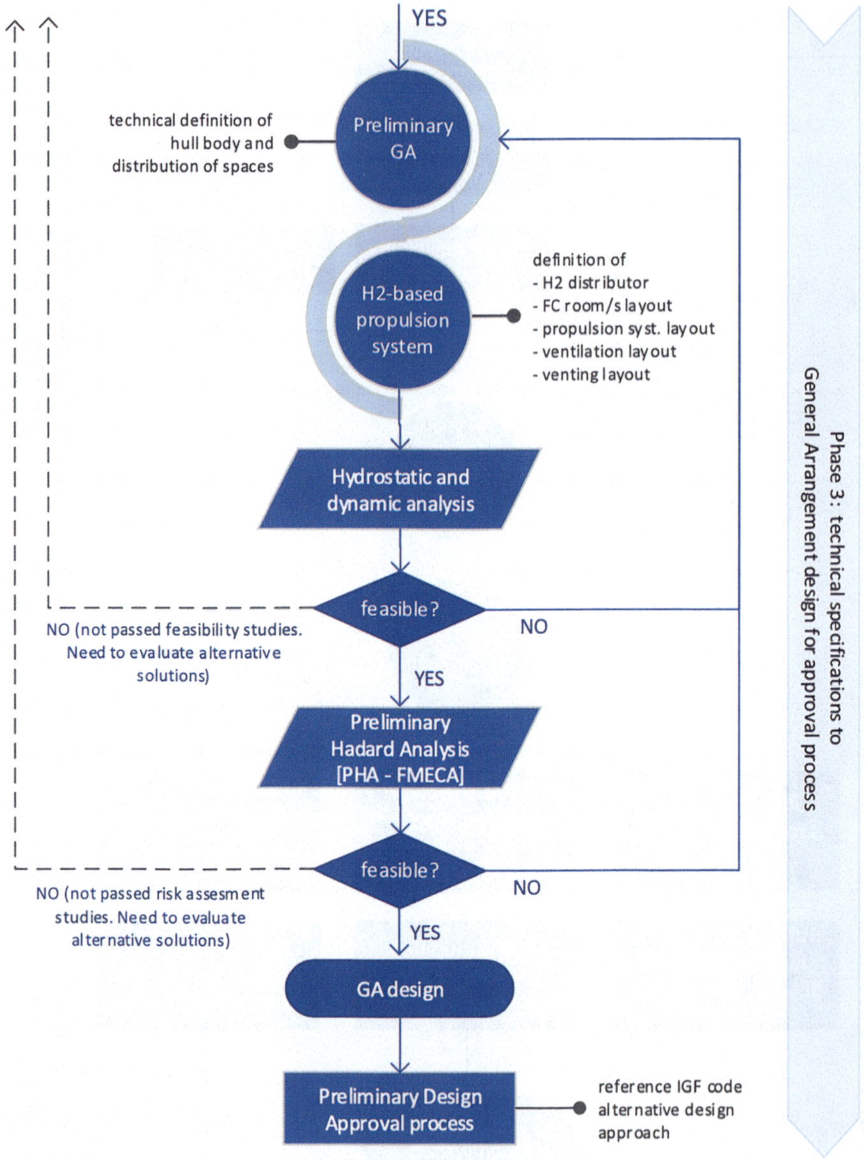

Fig. 1 (continued)

- Safety requirements: According to the IGF code, the vessel shall be as safe as conventionally powered vessels and the reliability of the complete system must be at least as good as a conventional vessel.

The process of designing an H_2-fuelled passenger vessel generally follows normal practice for maritime applications where the stakeholder expectations are translated into a definition of the problem and then into a complete set of validated technical requirements through case studies analysis and exploration of different vessel typology options. Reference vessels could be adopted in case of refitting or re-engineering of previous ferry models. In phase 1, the primary stakeholders involved are ship owners and ferry operators, shipbuilders, ship designers, and port authorities.

Phase 1 is in connection with Phase 2 through a loop of investigation for the calculation of power/energy demand and selection of proper technology for H_2 storage and energy conversion.

Phase 2 aims to perform a feasibility study before making the final decision regarding the possible design of a hydrogen-fuelled vessel. It starts from the definition of technical requirements: traceable and describes factors that must be met for the vessel to respond to users/market needs (operational requirements). They describe how the system's needs should be fulfilled. Technical requirements specify the parameters that should be present in the system to resolve technical challenges. During this stage, it is essential to consider the expected operational modes and profiles, relevant energy load variations, the targeted system lifespan, and possible bunkering strategy.

Technical requirements serve as the foundational guidelines that drive the following design stage. For this reason, it is essential to perform a feasibility study on the selected H_2 technology before moving to the general arrangement preliminary design. The methodological flow suggests performing evaluation on bunkering, storage strategy, and convertor strategy in parallel, defining at first the size (volume and weight needs) and then the possible locations onboard. In this step, applicable standards and regulations must be considered, including the applicability of the alternative design. Furthermore, the safety and reliability of the hydrogen systems need to be prioritized. In phase 2, the main stakeholders involved are ferry operators, shipbuilders, ship designers, technology providers, and H_2 providers. The following aspects must be considered:

- The degree of hybridization and the strategies for redundancy: for both purely hydrogen-powered vessels and hybrid-powered vessels (e.g. those with batteries as an alternative power source) two completely independent fuel-containment systems and energy converters may need to be considered.
- The positions of hydrogen components and rooms—e.g., fuel cell (FC) spaces or tank hold spaces—must be inaccessible to the public but accessible for maintenance and replacement. Additionally, these hydrogen system spaces must be protected against external hazards such as adverse weather conditions, human interference, fires, and mechanical impact.
- Distance between H_2 components (e.g. between bunkering station and tanks or between tanks and FC rooms) must be controlled to reduce piping distribution.
- All H_2 system spaces must be provided with a safety system (e.g. H_2 detection, ventilation, fire prevention).
- Loss of power shall not affect critical vessel function. Furthermore, the loss of power must not impact the vessel's critical hydrogen systems, ensuring that

essential hydrogen-related operations and safety systems remain fully operational even in the event of a power failure (e.g. emergency discharge). Single failure of critical modules shall not compromise the integrity of the vessel.

The dimensioning of the hydrogen systems is strongly influenced by the vessel power demand, the degree of hybridization, and the characteristics of available bunkering infrastructure. For this reason, the use of digital twin platforms could contribute to running punctual simulations for different sailing conditions. In case of negative results of feasibility studies, a new evaluation loop of power demand and H_2 technology choices must be considered, eventually followed by a diverse operational profile definition.

The result of Phase 2 is the definition of a preliminary general arrangement and H_2 propulsion system through a loop of design refinements. While technical requirements are the input to Phase 2, technical specifications are the output.

Phase 3 is the direct consequence of the technical specification with the evaluations of vessel stability, performance, and risk assessment that is prior to the alternative design approval process. However, it is recommended to initiate the alternative design approval process and the related contact with the administration during this stage, to discuss the initial approach to the vessel design. In phase 3, the main stakeholders involved are ferry operators, shipbuilders, ship designers, technology providers, and certification and flag bodies.

In the e-SHyIPS project, virtual towing tank simulation tools were developed to perform hydrostatic and hydrodynamic evaluation under calm and wave conditions. H_2 systems shall demonstrate robustness for exposure in a marine environment, including trim, roll, and acceleration due to wave effect. Stability simulations have been used to verify that the preliminary arrangement and main compartment definition were acceptable in terms of ship stability. This study aimed to verify that the reliability of the entire vessel is equivalent to that achieved with conventional fuel. Calm waters simulation was used mainly to compute the power resistance curve in order to verify the energy demand of engines. Wave hull interaction simulations for both zero hull velocity (bunkering) and non-zero hull velocity (seakeeping) are essential to verify whether stresses related to environmental conditions propagate through the supports to the tanks and affect tank and FC conditions.

Risk assessment studies play an important role in this phase. As passenger ship safety and reliability are affected the most by decisions made during an early-stage design phase—when technical specifications are translated into shape and location of compartments—a risk assessment that can be performed with less mature data has critical importance. The proposed methodological flow suggests preliminary hazard analysis (PHA) as a tool to identify hazards, their associated causal factors, effects, level of risk, and mitigating design measures when detailed information is not available. The PHA must be performed, including all the relevant stakeholders for ship design, construction, operation, and certification. Furthermore, FMECA, according to IEC 60812, could be introduced as a systematic analysis of the H_2 systems to identify potential failure modes, their causes, and their effects on performance. Risk assessment is essential to gain critical knowledge for the development of passenger ship concepts—especially in the translation of functional and safety requirement specifications into a general arrangement design. Furthermore, it serves as a guide to rank and prioritize the focus of quantitative simulations to be performed at the vessel and system level in the alternative design approval process.

4.1 Scenario Small: Damen Waterbus 2907

The Damen Waterbus 2907 is a high-speed catamaran with a composite hull designed for inland waterway passenger transport. For the Small scenario vessel, the philosophy involves a retrofit process, starting with the existing diesel-hybrid version of the 2907 Waterbus (a new development from the baseline Waterbus 2407). This retrofit process replaces the existing diesel system with the corresponding hydrogen systems [16]. Specifically, the existing diesel genset and fuel systems are removed and replaced with a proton exchange membrane fuel cell hydrogen (PEM FC) system within the existing genset compartment. No changes are made to the ship's dimensions or hull.

A 350-bar compressed H_2 storage system is chosen due to its high technology readiness level (TRL), commonly used in heavy-duty road transport applications, and its favourable weight/volume footprint. The H_2 storage capacity is determined based on energy and power calculations for a demanding inland route, aiming for a maximum of one bunkering/refilling operation per day. The designated number of tanks is placed on the roof as part of the vessel's safety philosophy, allowing the H_2 storage and distribution system to benefit from natural ventilation.

In the fuel cell room, the necessary FC capacity is installed to ensure that even at the end-of-life condition of the FC modules (defined as a 20% reduction in power output under full load conditions), the net power output matches the current diesel genset rating. This ensures that the fuel cell system meets the power requirements of the propulsion layout defined for the diesel-hybrid version. The total installed FC capacity is divided into two separate systems to meet redundancy requirements. A concept for the distribution system from the storage to the modules is defined as input for risk assessment studies within WP2. The modules are connected with DC/DC converters to the existing DC grid of the vessel.

After replacing the diesel genset with fuel cells and the diesel system with compressed hydrogen, a weight check is performed to ensure the system weight remains within restrictions, as excessive weight increases vessel wash and power consumption. It was found that the hydrogen-converted waterbus has a weight comparable to the diesel system while also respecting volume limitations, demonstrating the technical feasibility of the design.

The battery capacity remains the same as in the baseline hybrid version. The battery fulfills load levelling, boosting, and peak shaving functions, as well as providing an energy reserve in case of complete power conversion system failure. As system weight is unaltered, the propulsion power and energy calculations of the

Fig. 2 Scenario small concept: impression of H_2 powered waterbus

baseline hybrid version remain valid, and therefore, the thrusters, DC/AC drives, and general electrical/propulsion systems do not change.

A final illustration of the concept is provided in Fig. 2. Since technical feasibility is established using available technology, safety becomes the focal point of the follow-up, with superstructure geometry and the fuel cell room used in computational fluid dynamics (CFD) dispersion/explosion calculations.

4.2 Scenario Medium: Fior di Levante ROPAX

The medium scenario focuses on medium-sized ferries, specifically ROPAX (roll-on/roll-off passenger) vessels. The reference vessel is the Fior di Levante, operated by Levante Ferry, which transports commercial vehicles, private cars, and numerous passengers on shorter routes. The ship features various amenities, including bars, restaurants, children's facilities, and a small shopping arcade, and operates daily routes to Zakynthos, Kefalonia, and Ithaca.

In this scenario, designers integrate hydrogen tanks and fuel cells into the existing spaces of the vessel without altering the number of decks or the hull shape. The design process involved multiple iterations to ensure safety, stability, and commercial viability. Initially, compressed hydrogen storage at 700 bar was considered, but issues with stability, bunkering time, and volume led to the adoption of liquid hydrogen storage. This approach, informed by the FreeCO2ast project by Havyard Group ASA, allows machinery rooms to remain on the lower decks and maintains the current bunkering frequency. Furthermore, modifications to the passenger decks and garage area are minimized (Fig. 3).

The proposed system features a 7.9-MW setup with 36 PEM FC modules, divided into two rooms of 18 modules each. FC modules are arranged in two lines of 18 each,

Fig. 3 Scenario medium concept: impression of H_2 Fior di Levante ROPAX profile

in separate rooms designed for ventilation, with direct connections to the outer vessel, excluding down-flooding areas. The PEM FC produces unregulated DC voltage, fed into a DC/DC converter and a DC/AC inverter. The power output supports the propulsion system (electric engine, bow thrusters, electric network, and switchboards) and auxiliary equipment (fuel handling system, batteries, safety systems, and utility appliances). The energy management system regulates available electric power for the engines and utilities, while a frequency converter enables a full range of speeds for the two electric engines. Daily liquid hydrogen consumption is estimated at 2.3 tons, requiring 32.5 m^3 for daily bunkering and 97.5 m^3 for 3-day bunkering. Three MAN Cryo LH2 FGSS 95 m^3 tanks are selected to store 2.5 tons of liquid hydrogen each, adhering to IGF code deterministic rules for tank location [11]. The LH_2 tank connects to a tank connection system (TCS), including a pressure build-up vapourizer, valves, a pressure building unit, a nitrogen generator for inert gas, and a control and safety system. Safety barriers in the GH_2 line towards FC rooms include automatic valves with pressure transducers, pressure relief devices, pressure regulators, and a vent system.

4.3 Scenario Large: Celebrity Flora Cruise

Scenario large refers to a small/mid-size length luxury cruise ship primarily a high level of services offered to a select group of users. The reference project for this scenario is the Celebrity Flora, launched in 2019. For the large scenario vessel, the methodology primarily focuses on new design development. This approach is based on the general arrangement of the reference vessel and the operational profiles provided by the Advisory Board companies. To evaluate the feasibility of replacing the diesel-based powertrain with a PEM FC-based system, both at-berth and sailing consumption were considered, along with the mass and space requirements of components (Figs. 4 and 5).

According to Directive 2014/94/EU and the COM/2021/562, the H2 conversion strategy choice for scenario large design was 100% on-shore power energy supplied at berth. Thus, the H_2-based propulsion system was designed to meet the needs of sailing, cruising, and maneuvering only.

Similar to scenario medium, the H_2-based system design leveraged the expertise of the e-SHyIPS partner on the FreeCO2ast project and MAN Cryo's LH2 storage

Fig. 4 Hull triangulated geometry, used for scenario large simulations

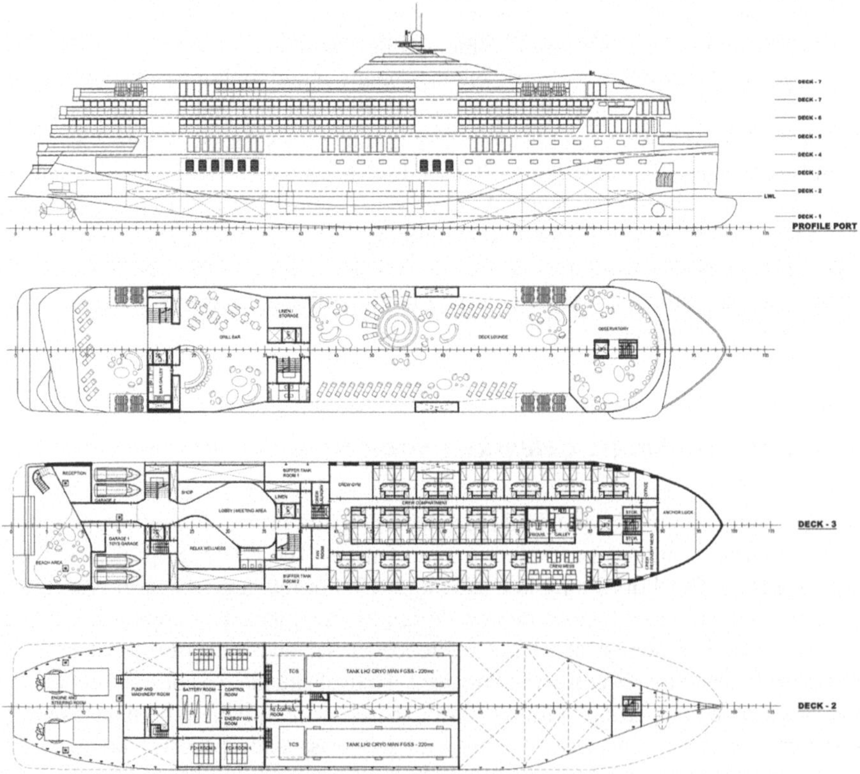

Fig. 5 Extract of GA used for scenario large simulations

and fuel gas supply system (FGSS). The proposed system includes a 7.1 MW PEM FC, divided into four FC rooms of eight modules each, occupying a total volume of 290 mq (16% of the current engine room space). Fuel cell rooms are located on deck

2 and are designed to ensure proper ventilation with a direct connection to the outer vessel above water at deck 3, excluding down-flooding areas. Two rooms are on the starboard side and two on the port side, with each pair connected to two MAN Cryo LH2 FGSS tanks with 2.3 tons of H_2 capacity. The two H_2-based system lines ensure redundancy. Both lines are connected to an energy management module that can switch power lines and recharge the energy storage system. The power output supports the propulsion system (electric engine, bow thrusters, electric network, and switchboards), auxiliary equipment (fuel handling system, batteries, safety systems, utility appliances), and all hospitality systems, including galleys, restaurants/bars, crew and guest services, and pool management.

5 Design Achievements and Conclusions

The three case studies demonstrate the potentiality of the presented methodology in driving the design activities of different categories of passenger vessels, from operational requirement definition to the preliminary general arrangement and H_2 system configuration useful in the Alternative design process. The proposed feasibility studies were instrumental in fulfilling the intended methodology objectives.

The case study results and the achievements on the design of the vessel's general arrangement and H_2-based fuel propulsion system basic design are depicted. The primary area of interest results in:

- Hull design/GA for performance and safety reasons, weight in position (centre of gravity) and amount, hydrostatic and hydrodynamic behaviours.
- Design and location of H_2 storage according to GA and safety in principle requirements.
- Design and location of fuel cell modules for optimal configuration of propulsion systems; evaluation of different operational profiles and energy requirements, including the possibility of battery hybridization.
- Vessel seakeeping, acceleration profiles, and acceleration forces on the systems due to wave effect in sailing and bunkering.

Operational and technical requirements were carefully considered in the design activities for scenarios small, medium, and large. Experiments on the stability, hydrostatic, and hydrodynamic behaviours of the vessel were conducted to ensure that the floating parameters of the hydrogen-fuelled vessel are at least equivalent to those of conventionally-fuelled vessels.

The design of PEM FC and H_2 storage systems for passenger vessels involved strategic considerations to ensure operational efficiency, safety, and regulatory compliance. All three scenarios were designed with primary safety strategies in mind, as indicated in the IGF code [11] and DNV's *Handbook for Hydrogen-fuelled Vessels* [15]. The results for all the case studies highlight that both the volume and weight of H_2 storage tanks (CGH_2 and LH_2) are considerably high with respect to the reference conventionally fuelled vessel. In designing the new storage tank solution, proper feasibility studies for bunkering frequency and strategy need to be considered prior to general arrangement and stability calculations. The feasibility analysis for replacing the diesel-based power-train system with a PEMFC-based one is assessed considering the size of components, weight, and space requirements. An evaluation of battery hybridization for redundancy of the propulsion system was performed. The design results show that from a vessel requirements perspective, all FC configurations result in a slight payload reduction, remaining competitive and comparable with the reference vessels. Risk assessments for FC spaces and tank spaces reveal potential criticalities due to the proximity to passenger compartments. This is a characteristic issue in passenger ferries, where machinery spaces are typically limited and close to crew or passenger areas, confirming the necessity of anticipating safety-related concepts at the first stage of the proposed methodology.

The analysis of seakeeping wave-hull interaction simulations aimed to determine whether stresses related to wave conditions propagating through component supports could affect fuel containment tanks and FC systems. In all tested case studies, the maximum vertical acceleration of the centre of gravity and selected points was below 3.67 m/s^2, in line with IMO resolution MSC.501. Furthermore, the LH_2 selected storage systems feature a 5-bar pressure vessel capable of handling low-temperature liquids and preventing sloshing in rough conditions. No safety issues related to acceleration profiles emerged.

The transition to sustainable waterborne transportation through the adoption of hydrogen as a fuel is a multifaceted challenge that necessitates a comprehensive and systematic approach. This chapter has presented an ecosystemic methodology for the design and implementation of hydrogen-powered vessels, drawing insights from the e-SHyIPS project.

Acknowledgement This study is part of the e-SHyIPS project. The project has received funding from the Fuel Cells and Hydrogen 2 Joint Undertaking (JU) under grant agreement no. 101007226. This JU receives support from the European Union's Horizon 2020 research and innovation programme, Hydrogen Europe and Hydrogen Europe Research. The project consortium is composed by Politecnico di Milano, UNI Ente Italiano Di Normazione, Teknologian Tutkimuskeskus VTT OY, CINECA Consorzio Interuniversitario, ATENA scarl—Distretto Alta Tecnologia Energia Ambiente, Proton Motor Fuel Cell, Levante Ferries Naftiki Etaireia, Ghenova Ingenieria sl Danaos Shipping Company Limited, OY Woikoski AB, DF—Ingegneria del Fuoco srl, Dimos Andravidas-kyllinis DNV Hellas sa, Scheepswerf Damen Gorinchem. The authors thank all e-SHyIPS partners for the valuable contribution in both methodology discussion and vessel design.

References

1. Johansson, M. (2020). *Waterborne mobility service in sustainable regional planning—a holistic study approach for new mobility services in North Western Skåne*. Department of Service Management, Lund University, Campus Helsingborg.
2. Ansaloni, G. M. M., Bionda, A., & Rossi, M. (2022). EcoDesign strategies for zero-emission hydrogen fuel vessels scenarios', in 2022 7th International Conference on Smart and

Sustainable Technologies (SpliTech). 2022. In *7th international conference on smart and sustainable technologies (SpliTech)* (pp. 1–6). IEEE.
3. Campillo, J., Dominguez-Jimenez, J. A., & Cabrera, J. (2019). Sustainable Boat Transportation Throughout Electrification of Propulsion Systems: Challenges and Opportunities. In *2019 2nd Latin American Conference on Intelligent Transportation Systems (ITS LATAM). 2019 2nd Latin American conference on intelligent transportation systems (ITS LATAM)* (pp. 1–6). IEEE.
4. Anon. (2023). *Inland Water Passenger Transport Global Market Report 2023* (p. 175). The Business Research Company.
5. Guven, D., & Ozgur Kayalica, M. (2023). Life-cycle assessment and life-cycle cost assessment of lithium-ion batteries for passenger ferry. *Transportation Research Part D: Transport and Environment, 115*, 103–586.
6. Waterborne Technology Platform. (2023). *Strategic research and innovation agenda for the partnership on zero-emission waterborne transport*. Waterborne Technology Platform. https://www.waterborne.eu/images/230505_SRIA_Zero_Emission_Waterborne_Transport_2.0_clean_public_consultation_final.pdf
7. Stark, C., Xu, Y., Zhang, M., Yuan, Z., Tao, L., & Shi, W. (2022). Study on applicability of energy-saving devices to hydrogen fuel cell-powered ships. *Journal of Marine Science Engineering, 10*(38), 34.
8. van Biert, L., Godjevac, M., Visser, K., & Aravind, P. V. (2016). A review of fuel cell systems for maritime applications. *Journal of Power Sources, 327*, 345–364.
9. Adner, R. (2017). Ecosystem as structure: An actionable construct for strategy. *Journal of Management, 43*(1), 39–58.
10. Chang, V., & Guetl, C. (2007). E-learning ecosystem (ELES) - a holistic approach for the development of more effective learning environment for small-and-medium sized enterprises (SMEs). In *2007 inaugural IEEE-IES digital EcoSystems and technologies conference* (pp. 420–425). Cairns.
11. International Maritime Organization. (2016) International code of safety for ships using gases or other low-flashpoint fuels (IGF code). .
12. Kontovas, C. A., & Psaraftis, H. N. (2009). Formal safety assessment: A critical review. *Marine Technology and SNAME News, 46*(01), 45–59.
13. Atilhan, S., Park, S., El-Halwagi, M. M., Atilhan, M., Moore, M., & Nielsen, R. B. (2021). Green hydrogen as an alternative fuel for the shipping industry. *Current Opinion in Chemical Engineering, 31*, 100668.
14. Ustolin, F., Campari, A., & Taccani, R. (2022). An extensive review of liquid hydrogen in transportation with focus on the maritime sector. *Journal of Marine Science and Engineering, 10*(9), 1222.
15. Dnv, G. (2017). *Study on the use of fuel cells in shipping*. European Maritime Safety Agency (EMSA).
16. Minutillo, M., Cigolotti, V., Di Ilio, G., Bionda, A., Boonen, E. J., & Wannemacher, T. (2022). Hydrogen-based technologies in maritime sector: Technical analysis and prospective. In *E3S web of conferences* (Vol. 334, p. 06011). EDP Sciences.

Life Cycle Assessment of Hydrogen-Powered Vessels

Levent Bilgili and Mauricio Latapí

1 Introduction

The International Maritime Organization (IMO)'s struggle for maritime decarbonisation began in 1997 with a brief but instructive text, which was an invitation for the IMO and the Marine Environment Protection Committee (MEPC) to take action against climate change and develop strategies to be implemented for carbon dioxide (CO_2) reduction [1]. Since then, the IMO has adopted four greenhouse gas (GHG) reports (in 2000, 2008, 2014, and 2020) to measure the impacts of international shipping on global climate change. The fourth and latest report indicated that international shipping activities were responsible for 2.89% of total anthropogenic CO_2 emissions during 2018 [2].

In 2003, the IMO took a more concrete step in the fight against global climate change by authorising the MEPC to develop mechanisms for the reduction of GHG emissions and requested the work to start as soon as possible [3]. The scientific literature on ship emissions has developed, and studies on this subject have become widespread during the 2000s.

The IMO adopted some important methodologies in 2013: Ship Energy Efficiency Management Plan (SEEMP), Energy Efficiency Design Index (EEDI), and Energy Efficiency Operational Indicator (EEOI). While SEEMP consists of numerous processes to increase efficiency during operation and, thus, to reduce GHG

L. Bilgili (✉)
Department of Naval Architecture and Marine Engineering, Maritime Faculty, Bandirma Onyedi Eylul University, Balikesir, Turkey
e-mail: lbilgili@bandirma.edu.tr

M. Latapí
Department of General Management and Human Resources, Faculty of Business, Communication Studies, and Aviation, Mount Royal University, Calgary, AB, Canada
e-mail: malatapi@mtroyal.ca

C. O. Colpan et al. (eds.), *Fuel Cell and Hydrogen Technologies in Maritime Transportation*, https://doi.org/10.1007/978-3-031-86110-9_11

emissions, EEDI and EEOI are used to classify ships according to their energy efficiency levels. EEDI calculations were mandatory for all newly built ships, and these ships must have a score below pre-determined values.

Then, after the Sustainable Development Goals and the Paris Agreement were adopted respectively in 2015 and 2016, the IMO drafted an initial GHG strategy, which consists of some tangible targets for the reduction of GHG emissions from shipping. The main targets were to reduce CO_2 emissions per transport work by at least 40% by 2030 and pursue efforts towards 70% by 2050. It was also aimed that GHG emissions would be reduced by at least 50% by 2050 and phased out by the end of this century [4]. One of the most significant outcomes of the Initial Strategy was the decision to draft life cycle assessment (LCA) Guidelines for alternative fuels, which are considered one of the main solutions to fight against climate change. Finally, in 2023, the IMO adopted the 2023 GHG Strategy, which has set indicative checkpoints that are committed statements for the reduction of GHG emissions. The checkpoints indicated that the total annual GHG emissions from international shipping will be reduced by at least 20% and 70%, respectively, by 2030 and 2040. The ultimate target of the strategy is to reach net-zero emissions by or around 2050 [5].

Both the Initial Strategy and the 2023 Strategy recognise alternative fuels as a lasting and promising solution for the reduction of GHG emissions. Hydrogen, among numerous alternative fuels, is considered one of the most viable options to reach the IMO's targets [6]. Besides, ammonia and hydrogen are expected to correspond to 70% of total ship fuel demand by 2035 [7]. It is also expected that a 5% increase in hydrogen use would result in a 9% reduction in total global GHG emissions [8]. On the other hand, while using hydrogen can achieve zero emissions in the tank-to-wake phase, producing hydrogen with energy from fossil fuels may cause a large quantity of GHG emissions during the production phase. For instance, if hydrogen is produced through the energy from renewables and transported to the end-user without emissions, it can be considered that the fuel is green and clean. However, if coal (or any other fossil fuel) is used as the main energy source, then hydrogen would not be as clean as expected. In this case, it would not be possible to reach IMO's ultimate targets. Therefore, it is recommended to utilise a life cycle approach to identify and mitigate GHG emissions generated throughout the complete life cycle of hydrogen. This chapter provides information on the place of hydrogen in the process of decarbonisation of shipping and includes an examination of hydrogen applications from a life cycle perspective.

2 Hydrogen as a Promising Solution

Hydrogen is the most abundant element in the universe. However, it is not often found in pure form, H_2. Hydrogen has a low flash point, high flame speed, and a wide flammability range, and it is non-toxic. On the other hand, difficulties in handling, storage, and transport (hydrogen is stored in liquid form, and it liquefies

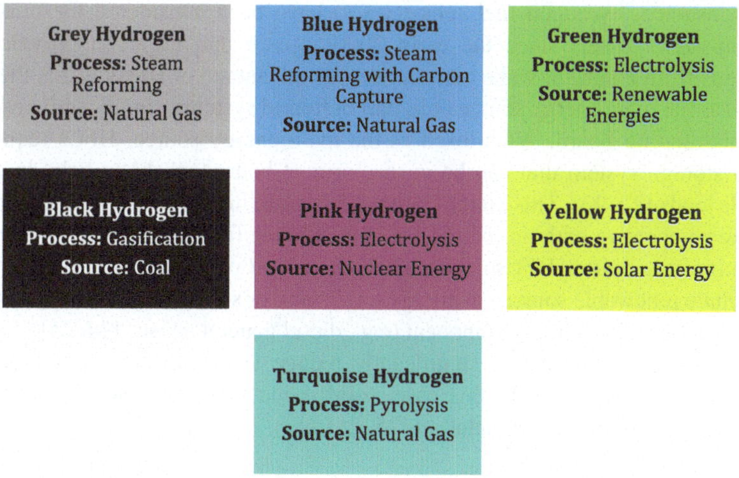

Fig. 1 Hydrogen production pathways and colour codes (adapted from [13])

at pressures higher than 700 bar or temperature below −253 °C) may cause significant problems for the operation and utilisation of hydrogen. Besides, the high energy consumption for the liquefaction of hydrogen is also considered a troubling issue [9]. Hydrogen produces zero emissions (not only GHG emissions but also carbon monoxide (CO), sulphur oxides (SO_x), nitrogen oxides (NO_x), and particulate matter (PM)) [10]). In the case of using hydrogen in fuel cells, the only emission to the atmosphere is water vapour [11].

The main environmental concern on hydrogen shows itself in the life cycle perspective. Hydrogen is a zero-emission fuel in case of producing from renewable sources such as wind and solar energy [12]. Since most of the hydrogen is produced via electrolysis, the method of generation of the energy used also determines the life cycle emissions of hydrogen. Figure 1 presents the production pathways of hydrogen and the corresponding colour codes for each method.

In some recent studies, hydrogen is considered the best option in terms of various aspects such as environment, safety, global availability, supply chain, applicability, compliance with current and future regulations, and economy [14–16]. Nevertheless, hydrogen carriers such as ammonia and dimethyl ether are considered to provide an optimal solution to the concerns arising from hydrogen itself [17, 18]. Although there are some serious concerns about high costs and difficulties in using long-range international voyages due to the cost of tanks, low density, and flammability [19, 20], hydrogen might be a lasting solution to reach the IMO's 2050 targets. Numerous hydrogen-powered ships are already completed and some are under construction/development. The majority of these ships are equipped with fuel cells, while some use wind energy, solar energy, and hybrid engines. The ships are of various types and sizes, such as sailboats, yachts, ferries, cruise ships, and transport ships, and are used in different countries, especially in the Nordic countries [21].

For instance, hydrogen fuel cells (HFCs) have been considered for small and medium-sized vessels since the early 2000s, given that they can provide zero emissions from a tank-to-wake perspective. An advantage of HFCs is that they can be used as the main energy source or as part of hybrid systems, significantly reducing the emissions on board. When used as the main energy source, HFCs require an energy storage system that can be in the form of battery stacks to help deal with variable loads [22, 23]. This kind of system can result in zero emissions from a tank-to-wake perspective with proton exchange membrane fuel cells (PEMFC) being the most promising option. HFCs can also be used as part of a larger hybrid system that can include renewable sources of energy (e.g., wind or solar), heat recovery systems, and sometimes a fossil fuel component (e.g., diesel generator) (see [24, 25]). Using a hybrid system is expected to significantly reduce vibration, noise pollution, and GHG emissions [26, 27]. In this case, solid oxide fuel cells (SOFC) have been considered for future implementation [25, 28].

3 Life Cycle Assessment of Hydrogen Implementation

Life cycle assessment (LCA) is an innovative and holistic approach to measure the cumulative environmental, economic, and social impacts of any kind of product system. LCA is considered a very useful tool to observe the impacts of different life cycle stages of a product and thus, it enables a comparison between sub-processes and provides data on which process should be focused for a comprehensive improvement. Figure 2 presents the general framework of LCA. Besides, Fig. 3 presents a sample of the hydrogen value chain.

LCA has been frequently used in maritime decarbonisation so far, and it is projected that it will be an important method for the maritime sector in the coming years with the guidance of the LCA Guidelines drafted and updated by the IMO in 2023 and 2024, respectively.

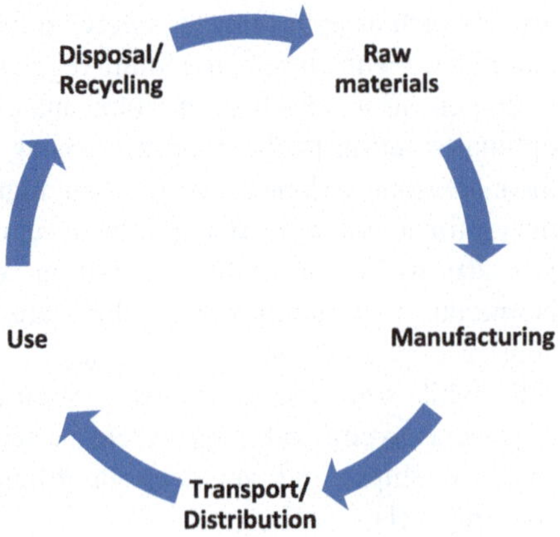

Fig. 2 LCA phases of any product system

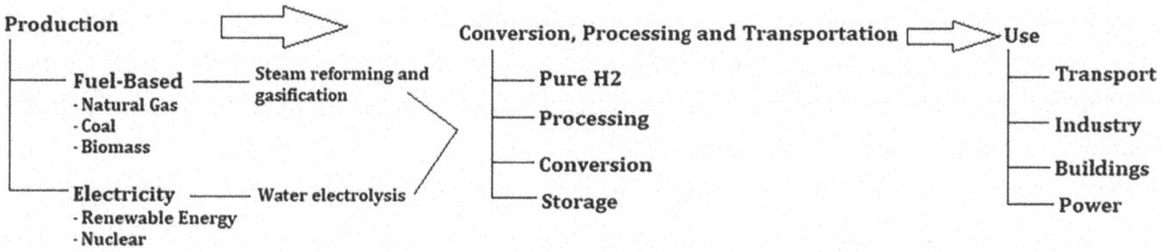

Fig. 3 Hydrogen value chain (adapted from [29])

The guidelines consist of more than 120 fuel and energy system pathways and present emission factors for 14 of them [30]. In this way, the IMO aims to compare and assess the environmental impacts of different fuel and energy systems from a life cycle perspective. In this context, studies on life cycle emissions of fuels are of great importance, and the number of these studies is increasing.

Since fuel cells are one of the main tools that use hydrogen, there are remarkable studies in the literature on the assessment of hydrogen fuel cells from the life cycle perspective. The LCA of a hydrogen-powered vessel consists of the manufacturing of fuel cells, hydrogen production and liquefaction, transportation and bunkering of hydrogen, and operation [31]. A recent study, which focuses on solid oxide fuel cells (SOFC) using different fuels as primary energy sources, concluded that hydrogen reaches less global warming potential (GWP) in the life cycle [32]. However, a life cycle cost (LCC) analysis is also required to provide a better understanding of the potential use of hydrogen. Thus, another recent study considered LCA and LCC together and investigated SOFC and proton exchange membrane fuel cell (PEMFC) using hydrogen or ammonia. The study found that although green hydrogen performs best in terms of environmental concerns, it is more expensive than others [33]. On the other hand, Wang et al. [34] concluded that using hydrogen may also reduce LCC. This is mainly because of the carbon credits, which are specialised credits that allow the owner to buy or sell a certain amount of CO_2 or other GHGs. Cumulative environmental impacts are undoubtedly much less in the case of using hydrogen [34]. A recent study also concluded that using green or blue hydrogen may reduce GWP by 93% and 48%, respectively [35]. Besides, further developments in fuel cell technologies and using of fuel cells combined with wind systems may significantly reduce costs [36].

Hydrogen fuel cells not only reduce GWP (or GHG emissions) but also have significant reduction impacts on other LCA midpoint categories. For instance, blue hydrogen, PEMFC, and SOFC systems significantly reduce aerosol formation and acidification potential [33]. A more recent study also concluded that using hydrogen can reduce acidification potential, eutrophication potential, and photochemical ozone creation potential in comparison with using marine diesel oil [34]. However, a different study, which compares diesel, PEMFC, and hydrogen, showed that PEMFC caused more environmental impacts in GWP, human toxicity potential, terrestrial ecotoxicity potential, abiotic depletion potential, ozone layer depletion potential, and eutrophication potential. On the other hand, PEMFC presents better environmental performance in photochemical ozone creation potential, acidification potential, freshwater aquatic ecotoxicity potential, and marine ecotoxicity potential [37]. A similar study also concluded that LCA of hydrogen represents more environmental impacts in acidification potential, eutrophication potential, and photochemical ozone creation potential [38].

Since storage and bunkering of hydrogen represent a significant challenge, it is generally preferred for short sea shipping and inland waterways. A study conducted on a 170 GT nearshore ferry operating in South Korean waters indicated that hydrogen emits more GHGs than marine gas oil (MGO) and liquefied natural gas (LNG) due to higher emissions in the well-to-tank phase [38, 39]. Similarly, although hydrogen is considered a promising solution, due to the high emissions during production, it lags behind methanol [31]. Besides, it is also argued that hydrogen does not meet environmental expectations when used alone [40].

Although some studies indicated that bio-methanol shows a greater environmental performance than green hydrogen in terms of GHG emissions [41], there is a general consensus that green hydrogen is much more environmentally friendly than conventional fuels [42]. For instance, if hydrogen is produced from wind energy, the impacts are considered negligible [43]. Furthermore, focusing more on manufacturing materials during hydrogen production is strongly recommended [44].

4 Conclusions

Hydrogen is considered one of the most promising alternatives in maritime decarbonisation. It has a wide range of applications as it can be used directly on-board as well as through fuel cells. As pure hydrogen contains no other elements, no emissions are produced. A ship propelled solely by hydrogen would achieve the zero-emission target. Hydrogen is, therefore, in a crucial position to achieve IMO‘s 2050 net-zero target. Since the net-zero target is determined by considering the life cycle emissions, it covers not only the emissions produced during the operation phase but also the emissions generated in the processes such as production, distribution, and storage of the fuel. The net-zero target should be well understood. Although the first target is zero emissions in the operational phase, the production process of fossil fuels is well known and very efficient. Thus, they may have lower life cycle emissions. In this case, for example, hydrogen produced using fossil fuels can result in a lifecycle emission production equivalent to fossil fuels. Thus, due to the possible emissions in the upstream process of hydrogen, these targets should be examined comprehensively and carefully. The fact that hydrogen, which has an undeniable reducing effect on GWP, causes an increase in other LCA indicators should also be questioned. This is where the life cycle perspective comes into play, enabling an assessment of hydrogen’s cumulative and holistic impacts in maritime applications.

Green hydrogen is mostly produced by the electrolysis of water, and the energy source used during this process is the main determinant of life cycle emissions. If fossil fuels are used as the primary energy source, a greater amount of emissions may occur. This could make things worse and undermine the decarbonisation challenge.

The reviewed studies reveal that the life cycle zero emission target is successfully achieved if hydrogen is produced using renewable energy sources (green hydrogen). However, it is not possible to achieve the targeted results if fossil fuels are used as the primary energy source. In addition, extensive utilisation is not expected in the short term due to the high life cycle cost of hydrogen. Yet, hydrogen is predicted to have widespread use in the medium and long term (2040 and beyond).

References

1. IMO. (1997). *MP/CONF.3/35 Resolution 8*. IMO.
2. Faber, J., Hanayama, S., Zhang, S., Pereda, P., Comer, B., Hauerhof, E., van der Loeff, W., Smith, T., Zhang, Y., Kosaka, H., Adachi, M., Bonello, J., Galbraith, C., Gong, Z., Hirata, K., Hummels, D., Kleijn, A., Lee, D., Liu, Y., Lucchesi, A., Mao, X., Muraoka, E., Osipova, L., Qian, H., Rutherford, D., la Fuente, S., Yuan, H., Perico, C., Wu, L., Sun, D., Yoo, D., & Xing, H. (2020). *Fourth IMO GHG study*. IMO.
3. IMO. (2003). *Resolution A.963(23)-IMO policies and practices related to the reduction of greenhouse gas emissions from ships*. IMO.
4. IMO. (2018). *Adoption of the initial IMO strategy on reduction of ghg emissions from ships and existing IMO activity related to reducing GHG emissions in the shipping sector*. IMO.
5. IMO. (2023a). *Resolution MEPC.377(80), 2023 IMO strategy on reduction of GHG emission from ships*. IMO.
6. Kim, K., Roh, G., Kim, W., & Chun, K. (2020). A preliminary study on an alternative ship propulsion system fueled by ammonia: Environmental and economic assessments. *Journal of Marine Science and Engineering, 8*, 1–21. https://doi.org/10.3390/jmse8030183
7. Halim, R. A., Kirstein, L., Merk, O., & Martinez, L. M. (2018). Decarbonization pathways for international maritime transport: A model-based policy impact assessment. *Sustainability, 10*, 2243. https://doi.org/10.3390/su10072243
8. Atilhan, S., Park, S., El-Halwagi, M. M., Atilhan, M., Moore, M., & Nielsen, R. B. (2021). Green hydrogen as an alternative fuel for the shipping industry. *Current Opinion in Chemical Engineering, 31*, 100668. https://doi.org/10.1016/j.coche.2020.100668
9. Bilgili, L. (2023). A systematic review on the acceptance of alternative marine fuels. *Renewable and Sustainable Energy Reviews, 182*, 113367. https://doi.org/10.1016/j.rser.2023.113367
10. Wang, W., Herreros, J. M., Tsolakis, A., & York, A. P. E. (2013). Ammonia as hydrogen carrier for transportation; investigation of the ammonia exhaust gas fuel reforming. *International Journal of Hydrogen Energy, 38*, 9907–9917. https://doi.org/10.1016/j.ijhydene.2013.05.144
11. Brynolf, S. (2014). *Environmental assessment of present and future marine fuels*. Chalmers University of Technology.
12. Ryste, J. A. (2019). *Comparison of alternative marine fuels*. DNV GL.
13. Acciona. (2024). Retrieved July 1, 2024, from https://www.acciona.com.au/updates/stories/what-are-the-colours-of-hydrogen-and-what-do-they-mean/?_adin=02021864894
14. Ahn, J., You, H., Ryu, J., & Chang, D. (2017). Strategy for selecting an optimal propulsion system of a liquefied hydrogen tanker. *International Journal of Hydrogen Energy, 42*, 5366–5380. https://doi.org/10.1016/j.ijhydene.2017.01.037
15. Deniz, C., & Zincir, B. (2016). Environmental and economical assessment of alternative marine fuels. *Journal of Cleaner Production, 113*, 438–449. https://doi.org/10.1016/j.jclepro.2015.11.089

16. Ren, J., & Liang, H. (2017). Measuring the sustainability of marine fuels: A fuzzy group multi-criteria decision making approach. *Transportation Research Part D: Transport and Environment, 54*, 12–29. https://doi.org/10.1016/j.trd.2017.05.004
17. Dimitriou, P., & Javaid, R. (2020). A review of ammonia as a compression ignition engine fuel. *International Journal of Hydrogen Energy, 45*, 7098–7118. https://doi.org/10.1016/j.ijhydene.2019.12.209
18. Takeishi, K. (2016). Dimethyl ether (DME): A clean fuel/energy for the 21st century and the low carbon society. *International Journal of Energy Environment, 10*, 238–242.
19. Cheliotis, M., Boulougouris, E., Trivyza, N. L., Theotokatos, G., Livanos, G., Mantalos, G., et al. (2021). Review on the safe use of ammonia fuel cells in the maritime industry. *Energies, 14*, 1–20. https://doi.org/10.3390/en14113023
20. Durbin, D. J., & Malardier-Jugroot, C. (2013). Review of hydrogen storage techniques for on board vehicle applications. *International Journal of Hydrogen Energy, 38*, 14595–14617. https://doi.org/10.1016/j.ijhydene.2013.07.058
21. Wang, Z., Li, M., Zhao, F., Ji, Y., & Han, F. (2024). Status and prospects in technical standards of hydrogen-powered ships for advancing maritime zero-carbon transformation. *International Journal of Hydrogen Energy, 62*, 925–946. https://doi.org/10.1016/j.ijhydene.2024.03.083
22. Cavo, M., Rivarolo, M., Gini, L., & Magistri, L. (2023). An advanced control method for fuel cells-metal hydrides thermal management on the first Italian hydrogen propulsion ship. *International Journal of Hydrogen Energy, 48*(54), 20923–20934. https://doi.org/10.1016/j.ijhydene.2022.07.223
23. Xie, P., Asgharian, H., Guerrero, J. M., Vasquez, J. C., Araya, S. S., & Liso, V. (2024). A two-layer energy management system for a hybrid electrical passenger ship with multi-PEM fuel cell stack. *International Journal of Hydrogen Energy, 50*, 1005–1019. https://doi.org/10.1016/j.ijhydene.2023.09.297
24. Dolatabadi, S. H., Ölçer, A. I., & Vakili, S. (2023). The application of hybrid energy system (hydrogen fuel cell, wind, and solar) in shipping. *Renewable Energy Focus, 46*, 197–206. https://doi.org/10.1016/j.ref.2023.05.008
25. Oh, D., Cho, D.-S., & Kim, T.-W. (2023). Design and evaluation of hybrid propulsion ship powered by fuel cell and bottoming cycle. *International Journal of Hydrogen Energy, 48*(22), 8273–8285. https://doi.org/10.1016/j.ijhydene.2022.11.157
26. Ghenai, C., Bettayeb, M., Brdjanin, B., & Hamid, A. K. (2019). Hybrid solar PV/PEM fuel cell/diesel generator power systems for cruise ship: A case study n Stockholm, Sweden. *Case Studies in Thermal Engineering, 14*, 100497. https://doi.org/10.1016/j.csite.2019.100497
27. Gianni, M., Pietra, A., & Taccani, R. (2021). Outlook of future implementation of PEMFC and SOFC onboard cruise ships. In *E3S web of conferences*.
28. Lee, J. I., Yoon, B. Y., & Cha, S. W. (2023). Analysis of solid oxide fuel cell hybrid power system in marine application for CO_2 reduction. *Energy Reports, 9*, 3072–3081. https://doi.org/10.1016/j.egyr.2023.01.123
29. UNECE. (2022). *Technology brief-hydrogen*. United Nations Economic Commission for Europe.
30. IMO. (2023b). *Resolution MEPC.376(80), guidelines on life cycle GHG intensity of marine fuels*. IMO.
31. Percic, M., Vladimir, N., & Fan, A. (2021). Techno-economic assessment of alternative marine fuels for inland shipping in Croatia. *Renewable and Sustainable Energy Reviews, 148*, 111363. https://doi.org/10.1016/j.rser.2021.111363
32. Liao, C., Tang, Y., Liu, Y., Sun, Z., Li, W., & Ma, X. (2023). Life cycle assessment of the solid oxide fuel cell vehicles using ammonia fuel. *Journal of Environmental Chemical Engineering, 11*(5), 110872. https://doi.org/10.1016/j.jece.2023.110872
33. Percic, M., Vladimir, N., Jovanovic, I., & Korican, M. (2022). Application of fuel cells with zero-carbon fuels n short-sea shipping. *Applied Energy, 309*, 118463. https://doi.org/10.1016/j.apaenergy.2021.118463

34. Wang, H., Aung, M. Z., Xu, X., & Boulougouris, E. (2023). Life cycle analysis of hydrogen powered marine vessels-case ship comparison study with conventional power systems. *Sustainability, 15*, 12946. https://doi.org/10.3390/su151712946
35. Dong, D. T., Schönborn, A., Christodoulou, A., Ölçer, A. I., & Gonzalez-Celis, J. (2023). Life cycle assessment of ammonia/hydrogen-driven marine propulsion. *Proceedings of the Institution of Mechanical Engineers, Part M: Journal of Engineering for the Maritime Environment, 238*, 531. https://doi.org/10.1177/14750902231207159
36. Köhler, J. (2020). Zero carbon propulsion in shipping-scenarios for the development of hydrogen and wind technologies with the MATISSE-SHIP model. *International Shipbuilding Progress, 67*, 79–95. https://doi.org/10.3233/ISP-190269
37. Fernandez-Rios, A., Santos, G., Pinedo, J., Santos, E., Ruiz-Salmon, I., Laso, J., Lyne, A., Ortiz, A., Ortiz, I., Irabien, A., Aldaco, R., & Margallo, M. (2022). Environmental sustainability of alternative marine propulsion technologies powered hydrogen-a life cycle assessment approach. *Science of the Total Environment, 820*, 153189. https://doi.org/10.1016/j.scitotenv.2022.153189
38. Lee, G. N., Kim, J. M., Jung, K. H., Park, H., Jang, H. S., Lee, C. S., & Lee, J. W. (2022). Environmental life-cycle assessment of eco-friendly alternative ship fuels (MGO, LNG, and hydrogen) for 170 GT nearshore ferry. *Journal of Marine Science and Engineering, 10*, 755. https://doi.org/10.3390/jmse10060755
39. Hwang, S. S., Gil, S. J., Lee, G. N., Lee, J. W., Park, H., Jung, K. H., & Suh, S. B. (2020). Life cycle assessment of alternative ship fuels for coastal ferry operating in Republic of Korea. *Journal of Marine Science and Engineering, 8*, 660. https://doi.org/10.3390/jmse8090660
40. Percic, M., Vladimir, N., & Fan, A. (2020). Life-cycle cost assessment of alternative marine fuels to reduce the carbon footprint in short-sea shipping: A case study of Croatia. *Applied Energy, 279*, 115848. https://doi.org/10.1016/j.apenergy.2020.115848
41. Tomos, B. A. D., Stamford, L., Welfle, A., & Larkin, A. (2024). Decarbonising international shipping-a life cycle perspective on alternative fuel options. *Energy Conversion and Management, 299*, 117848. https://doi.org/10.1016/j.enconman.2023.117848
42. Elrhoul, D., Gomez, M. R., & Naveiro, M. (2023). Review of green hydrogen technologies application in maritime transport. *International Journal of Green Energy, 20*(15), 1800–1825. https://doi.org/10.1080/15435075.2023.2194384
43. Evers, V. H. M., Kirkels, A. F., & Godjevac, M. (2023). Carbon footprint of hydrogen-powered inland shipping: Impacts and hotspots. *Renewable and Sustainable Energy Reviews, 185*, 113629. https://doi.org/10.1016/j.rser.2023.113629
44. Aydin, M. I., & Dincer, I. (2022). An assessment study on various clean hydrogen production methods. *Energy, 245*, 123090. https://doi.org/10.1016/j.energy.2021.123090

Adoption of Hydrogen Use in Marine Applications

Omer Berkehan Inal

1 Introduction

The maritime industry is a cornerstone of global trade, responsible for approximately 90% of the total share by volume. However, maritime transportation is under increasing scrutiny because of its essential greenhouse gas emissions (GHG) and air pollution. The maritime industry contributes significantly to global greenhouse gas emissions, primarily by the combustion of heavy fuel oils [1]. International Maritime Organization (IMO) regulations, such as the IMO 2020 Sulphur Cap, have been implemented to reduce sulphur oxide (SOx) emissions. However, further measures are necessary to meet the ambitious targets set by the Paris Agreement. The substantial environmental impact of maritime transportation is mainly sourced from the reliance on heavy fuel oil (HFO) and marine diesel oil (MDO) [2]. The usage of fossil fuels results in high emissions of carbon dioxide (CO_2), sulphur oxides (SOx), nitrogen oxides (NOx), and particulate matter (PM) [3]. As global efforts to combat climate change intensify, the industry faces increasing pressure to transition to sustainable, cleaner energy sources. As countries strive to meet international climate goals, the shift towards sustainable and renewable energy sources by switching conventional fuels to alternative ones such as hydrogen, ammonia, or methanol has become a promising solution. Hydrogen, known for its zero-emission potential, presents a substantial alternative to conventional maritime fuels [4].

The maritime transport sector accounts for roughly 3% of global anthropogenic greenhouse gas (GHG) emissions, with projections indicating a 50% increase in CO_2 emissions in the maritime industry from 2008 to 2050 [5]. The revised 2023 strategy significantly sharpens these objectives, setting expectations for a 30% reduction in

O. B. Inal (✉)
Department of Marine Engineering, Faculty of Maritime, Istanbul Technical University, Tuzla, Istanbul, Turkey
e-mail: inalo@itu.edu.tr

C. O. Colpan et al. (eds.), *Fuel Cell and Hydrogen Technologies in Maritime Transportation*, https://doi.org/10.1007/978-3-031-86110-9_12

total annual GHG emissions by 2030, an 80% reduction by 2040, and ultimately achieving net-zero emissions by 2050 [5]. Between 2012 and 2018, emissions from maritime transportation—including CO_2, methane (CH_4), particulate matter (PM), sulphur oxides (SO_x), and nitrous oxides (N_2O)—increased from 977 to 1076 million tons, with CO_2 emissions alone rising from 962 to 1056 million tons. CO_2 emissions constitute the majority of GHG emissions from maritime transport [6].

The International Maritime Organization (IMO), the industry's regulatory authority, continuously implements regulations and standards for environmental protection in global shipping. The IMO's primary motivation is to minimize ship-sourced GHG emissions by enforcing emission limitation regulations. The primary strategy for controlling ship-sourced emissions globally is the International Convention for the Prevention of Pollution from Ships (MARPOL) and Annex VI Regulations for the Prevention of Air Pollution from Ships. Additionally, the IMO sets regulations and introduces vital measures to achieve emission reduction and zero-carbon goals in the global maritime industry. These measures include the Energy Efficiency Design Index (EEDI), Ship Energy Efficiency Management Plan (SEEMP), Energy Efficiency Existing Ship Index (EEXI), and Carbon Intensity Index (CII), all aimed at limiting emissions from ships [7]. The EEDI, introduced in 2011, sets a CO_2 limit per ship capacity mile for vessels over 400 gross tons. The SEEMP, introduced subsequently, serves as an operational tool for monitoring and enhancing the energy efficiency of ships and fleets, although it is not a direct measure itself. However, these actions needed to be revised to meet the IMO's ultimate targets, prompting the introduction of two new measures, EEXI and CII, which came into force in January 2023. EEXI applies to all merchant ships over 400 gross tons, similar to EEDI in reducing CO_2 emissions per nautical mile, though ship-specific and varying by ship type and power [8]. CII, effective January 2023, applies to ships over 5000 gross tons, assessing annual CO_2 emissions and categorizing them from A to E [9].

Despite these regulations, achieving significant emission reductions requires transitioning to alternative fuels like hydrogen. Hydrogen fuel, with its potential for zero emissions due to its carbon-free structure, is emerging as a promising alternative fuel for maritime applications.

As international regulations, such as those set by the International Maritime Organization (IMO), continue to push for stringent emissions reductions, hydrogen has emerged as a viable solution for meeting decarbonization targets [1]. However, its benefits extend far beyond its zero-emission potential. One key advantage of hydrogen lies in its ability to enhance energy security. As global energy markets face growing instability, the maritime industry's reliance on fossil fuels makes it vulnerable to fluctuations in oil prices and supply chains. Hydrogen, mainly when produced domestically from renewable resources (green hydrogen), offers a pathway toward diversifying the energy mix and reducing reliance on imported fuels. This can contribute to a more resilient and secure energy supply for shipping, especially for nations with abundant renewable resources. Additionally, the hydrogen economy promises economic opportunities through job creation and industrial growth. Developing hydrogen fuel technologies, infrastructure, and supply chains will require investment in new industries, ranging from hydrogen production to ship retrofitting and new vessel construction. This shift could create numerous jobs in renewable energy, engineering, and transportation sectors, contributing to economic growth and supporting a transition to low-carbon shipping [4]. Moreover, hydrogen adoption fosters technological innovation in the maritime industry. The need for hydrogen-compatible vessels and fuel infrastructure will drive research and development in fuel cell technology, energy storage systems, and ship design. These advancements could improve the efficiency and performance of hydrogen-powered ships and lead to breakthroughs in related areas such as autonomous shipping and digitalization, positioning the maritime industry as a leader in the green technology revolution. In this context, hydrogen represents more than a solution for emissions reduction; it is a transformative fuel that can reshape the future of maritime transport by enhancing energy security, generating economic growth, and driving technological advancements. Understanding and promoting hydrogen's full range of benefits will be crucial for achieving a sustainable and competitive global maritime industry.

This chapter aims to provide an in-depth analysis of the current legislative landscape governing hydrogen use in the maritime industry, explore technological advancements, assess economic implications, and offer policy recommendations for improving hydrogen adoption.

2 Hydrogen as a Maritime Fuel

Hydrogen offers several advantages over traditional maritime fuels. It produces zero carbon emissions when used in fuel cells, reducing the environmental footprint of maritime operations [10, 11]. Moreover, renewable sources can produce hydrogen, aligning with global sustainability goals. Despite these benefits, hydrogen's widespread adoption faces several challenges, including high production costs, limited infrastructure, and technical difficulties related to storage and handling [12, 13]. A pros and cons table for hydrogen use in the maritime industry is formed in Table 1.

- Zero emissions: Hydrogen fuel cells produce only water vapour as a byproduct, eliminating CO_2, SOx, and NOx emissions.
- Energy density: While hydrogen has a high energy density by weight, its low volumetric energy density poses storage challenges.

Table 1 Pros and cons of hydrogen as a fuel in the maritime industry

Hydrogen as a Maritime Fuel	
Pros	Cons
Zero emission	Cost
High energy density	Storage and distribution
Renewable production	Safety concerns
Energy storage medium	Technological maturity
Various production method	Market weakness—Bunkering net

- Renewable production: Hydrogen can be produced from renewable sources, reducing dependency on fossil fuels.
- Energy storage medium: Hydrogen can act as an energy storage medium, allowing excess renewable energy to be stored.
- Various production methods: Hydrogen can be produced using electrolysis or natural gas reforming methods.

However, the adoption of hydrogen faces several challenges:

- Production costs: Current hydrogen production methods, mainly green hydrogen, are expensive compared to conventional fossil fuels.
- Storage and distribution: Hydrogen's low energy density requires advanced storage solutions, such as compressed or liquid hydrogen, and a robust distribution infrastructure.
- Safety concerns: Hydrogen's flammability requires stringent safety measures.
- Technological maturity: Lack of experience in application in the maritime sector.
- Market weakness: A highly marine competitive market with low-carbon fuels.

Technological advancements are addressing these challenges. For instance, proton exchange membrane (PEM) fuel cells offer high efficiency and durability for maritime applications. Innovations in storage, such as metal hydrides and cryogenic tanks, are improving hydrogen's viability as a marine fuel [14]. Technological advancements are crucial for successfully integrating hydrogen into the maritime industry. Key areas of development can be categorized as hydrogen production, storage, fuel cell technology, and infrastructure.

2.1 Hydrogen Production

Hydrogen can be produced via different production methods from various sources, and the production emissions, costs, and efficiencies mainly vary depending on the technique used. Therefore, according to the production method, hydrogen is categorized by different colours like grey, blue, turquoise, green, purple, and yellow, as shown in Table 2 [15, 16]. Electrolysis is the most common way to produce hydrogen, but the source of electric energy is becoming critical. Renewable and nuclear energy are the most promising methods for environmentally friendly

Table 2 Hydrogen colours according to different production methods

Colour	Source
Grey	Fossil fuels
Blue	Fossil fuels with carbon capture system
Turquoise	Pyrolysis of fossil fuels
Green	Electrolysis from renewable energy source
Purple	Electrolysis from nuclear power plants
Yellow	Electrolysis from grid electricity

hydrogen production [17]. Innovations in solid oxide electrolysis cells (SOEC) and proton exchange membrane (PEM) electrolyzers also enhance efficiency and reduce costs.

Production-dependent colour variations in hydrogen also result in varying impacts on the maritime industry. Green hydrogen, produced through electrolysis powered by renewable energy, is the most environmentally friendly option, emitting no CO_2 during production. However, its high cost due to expensive renewable energy sources and the complexity of electrolyzer technology makes it less economically competitive for widespread adoption in the maritime industry. Blue hydrogen, produced from natural gas with carbon capture and storage (CCS) technology, significantly reduces CO_2 emissions but does not eliminate them. While it is cheaper than green hydrogen, it still relies on fossil fuels, limiting its long-term sustainability benefits. Grey hydrogen, produced from natural gas without CCS, is the most affordable and widely available but results in significant CO_2 emissions, undermining its environmental benefits. Turquoise hydrogen, produced through methane pyrolysis, generates solid carbon instead of CO_2, offering an intermediate environmental impact, though its commercial viability and scalability are still in development. Brown and black hydrogen, made from coal, have the highest carbon footprint and are unsuitable for decarbonization efforts in the maritime sector. Purple hydrogen is like green hydrogen; it emits no CO_2 during production, but the debate around nuclear energy's environmental impact, waste management, and public perception makes it a complex topic. Yet, it offers a potentially stable and significant energy source that could aid the hydrogen transition. Yellow hydrogen is entirely eco-friendly because it leverages the sun's abundant energy without emitting CO_2 during production. However, its scalability depends heavily on geographic and weather conditions. Finally, market prices for hydrogen vary significantly, with green hydrogen being the most expensive and grey hydrogen the cheapest. As the maritime industry seeks to reduce emissions, the balance between cost and environmental impact will guide the adoption of different hydrogen colours. While fossil fuel-based hydrogen production is budget-friendly, blue and green hydrogen will likely play critical roles as production technologies mature and prices decline over time with their environmental-friendliness.

2.2 *Storage*

The hydrogen storage options can be categorized under four titles: compressed, liquid, solid state, and alternative carriers, which are ammonia, CO_2 based, and organic liquids. However, if the maritime industry and the nature of the ships are considered, liquid and compressed hydrogen storage techniques are worth elaborating on compared to other methods with current technology. Meanwhile, chemical carriers, specifically ammonia, are also gaining importance, especially with the burning research on marine diesel engines.

Compressed hydrogen: Stored at high pressures (up to 700 bar) to increase energy density [18, 19]. Compressed hydrogen technology is generally used in fuel cell electric cars, and this technology is well proven and experienced.

Liquid hydrogen (LH_2): Hydrogen is stored at cryogenic temperatures (−253 °C) to increase volumetric energy density, suitable for long-term storage and transport [20].

Chemical carriers: Hydrogen can be stored in chemical carriers like ammonia, methanol, and metal hydrides, which offer higher energy densities and safer storage options. Ammonia seems to be the most promising solution for the near future.

2.3 Fuel Cell Technology

Fuel cells are electrochemical devices that use hydrogen as fuel. They offer high efficiency, quick start-up, and adaptability for maritime applications [21]. They are viable and promising solutions for using hydrogen in maritime transportation and are involved in many research and development projects worldwide [22]. The major obstacles to the fuel cell for widespread shipping are the limited lifetime, difficulties in hydrogen storage, and production cost compared to marine diesel engines [23].

2.4 Infrastructure

Developing hydrogen refuelling stations at ports and bunkering facilities is essential for adopting hydrogen-powered vessels. Investments in hydrogen pipelines, storage facilities, and distribution networks are also critical. Furthermore, some hydrogen fuel cell applications can accelerate hydrogen spreading, contributing to the industry's decarbonization efforts. Some recent applications are summarized as follows: In Asia, Japan's HydroBingo project exemplifies the integration of hydrogen fuel cells in larger vessels. Initiated in 2023, HydroBingo is a research vessel equipped with hydrogen fuel cells developed by Japanese companies such as Toshiba and Toyota. This vessel aims to validate the performance and reliability of hydrogen propulsion systems in various maritime conditions, paving the way for broader commercial applications. HydroBingo's deployment underscores Japan's commitment to advancing hydrogen technology as part of its national strategy for a hydrogen-based society [24, 25]. Norled, a leading Norwegian ferry operator, has also launched the Ellen Ferry, one of the world's first hydrogen-powered ferries. Deployed in 2022 on the Hirtshals to Kristiansand route, Ellen uses PEM fuel cells to provide clean and efficient propulsion. The ferry's booming operation has demonstrated the practicality of hydrogen fuel cells in medium-sized passenger vessels, highlighting significant reductions in CO_2, NO_x, and SO_x emissions compared to traditional diesel-powered ferries [26].

3 Global and Regional Legislation and Policies

International maritime regulations play a crucial role in shaping the adoption of hydrogen as a marine fuel. The IMO has been at the forefront, setting guidelines and frameworks to support the transition to low- and zero-emission fuels. Key initiatives include the IMO Initial Strategy on reducing GHG emissions and establishing the Global Industry Alliance (GIA) to Support Low Carbon Shipping [27]. These efforts promote research, development, and deployment of alternative fuels, including hydrogen. On the other hand, for efficient use of hydrogen in the maritime industry, an interim guideline has been published by the IMO, which is the most detailed document on the installation of fuel cells in ships. Considering safety issues, the guideline determines the required criteria for arrangement and installation onboard the vessels [28]. The IMO primarily shapes the international regulatory framework for maritime hydrogen use. Key initiatives include the IMO Initial Strategy on GHG Emissions, adopted in 2018; this strategy aims to reduce GHG emissions from international shipping and phase them out as soon as possible. GIA is a public–private partnership to support low-carbon shipping by researching and developing alternative fuels, including hydrogen. IMO FuelEU Maritime, part of the European Green Deal, this initiative aims to increase the uptake of sustainable maritime fuels. European Green Deal aims to make Europe the first climate-neutral continent by 2050, with hydrogen playing a vital role in the energy transition. The European Green Deal includes the EU Hydrogen Strategy, which outlines plans to install at least 6 GW of renewable hydrogen electrolysers and produce up to one million tons of renewable hydrogen by 2024 [29]. FuelEU Maritime, a proposed regulation under the European Green Deal, aimed at increasing the uptake of sustainable maritime fuels to reduce GHG emissions.

International cooperation and harmonization of standards are crucial for the global adoption of hydrogen. The IMO's regulatory framework provides a foundation, but individual countries and regions also play a vital role in implementing supportive policies and incentives.

- EU zone: The EU has positioned itself as a leader in the energy transition, particularly with hydrogen, by aligning its policies with international climate goals, such as those outlined in the Paris Agreement and the European Green Deal. Germany's National Hydrogen Strategy, launched in 2020, includes plans to invest €9 billion in hydrogen technology and infrastructure, focusing on green hydrogen. Norway's Hydrogen Roadmap emphasizes the development of hydrogen infrastructure and technology, with a particular focus on the maritime sector. The Dutch government supports hydrogen adoption through its National Climate Agreement, which includes hydrogen production and utilization targets.
- North America: The United States Department of Energy (DOE) Hydrogen and Fuel Cell Technologies Office supports research, development, and demonstration projects for hydrogen production, storage, and utilization. The Maritime Administration (MARAD) Sustainable Ports Program focuses on reducing emissions from port operations, including using alternative fuels like hydrogen.

Canada's Hydrogen Strategy, released in 2020, aims to position Canada as a global leader in hydrogen production and utilization. The strategy outlines plans for hydrogen infrastructure development, including ports and maritime applications.
- Asia: Japan's Basic Hydrogen Strategy, established in 2017, aims to establish a hydrogen-based society by 2050. The strategy includes developing hydrogen-powered ships and port infrastructure, supported by substantial government and industry investment. China's 14th Five-Year Plan recognizes hydrogen as a strategic industry, with significant investments in hydrogen production and technology. The plan includes targets for hydrogen infrastructure development and adopting hydrogen-powered vessels.
- Rest of world: Australia's National Hydrogen Strategy, launched in 2019, focuses on leveraging the country's abundant renewable energy resources to become a significant hydrogen exporter. The strategy includes hydrogen production, storage, and utilization plans in the maritime sector. From the Middle East, countries like Saudi Arabia and the United Arab Emirates (UAE) are investing in green hydrogen projects to diversify their economies and reduce reliance on oil. The Middle East's potential for large-scale renewable energy production positions it as a critical player in the global hydrogen market.

The EU's strategy is ambitious and aligned with international goals, but a coordinated approach between member states is necessary to overcome infrastructure challenges and reduce the costs of green hydrogen. The maritime sector, while included in broader energy strategies, could benefit from more targeted funding and infrastructure development to promote hydrogen adoption on a large scale. While Asia's hydrogen strategies are ambitious, particularly regarding infrastructure and investment, they are still too reliant on fossil fuels since their main focus is grey hydrogen. Green hydrogen production must be emphasized to align with global emission reduction goals. Japan's maritime hydrogen projects are promising, but the region must overcome significant hydrogen production and supply logistics challenges. North America's policies regarding investment in hydrogen research and infrastructure development are ambitious, but there is a noticeable gap in addressing the maritime sector's specific needs. Future strategies must focus on scaling up green hydrogen production and developing port infrastructure to facilitate its adoption in shipping.

Lastly, significant progress has been made in hydrogen production, storage, and distribution technologies. Advances in electrolysis, especially using renewable energy sources, have improved the efficiency and sustainability of hydrogen production. Innovations in storage solutions, such as compressed and liquid hydrogen, address some of the challenges related to hydrogen's low energy density. Infrastructure development, including hydrogen refuelling stations at ports, is critical for implementing hydrogen-powered vessels.

4 Economic and Market Implications

The transition to hydrogen requires significant economic considerations. While hydrogen production and usage costs are currently higher than conventional fossil fuels, economies of scale, technological advancements, and policy incentives are expected to reduce costs. Developing a solid hydrogen market infrastructure, supported by public and private investments, is essential for market readiness. Future projections indicate a growing market for hydrogen in maritime transport, with potential cost competitiveness achieved through global and national-based efforts. The economic viability of hydrogen as a maritime fuel depends on several factors, such as production costs, market readiness, and future projections.

Current production costs for green hydrogen are remarkably high, but ongoing advancements in electrolysis technology and economies of scale are expected to reduce costs significantly. Blue hydrogen offers a transitional solution with lower costs compared to green hydrogen. Hydrogen production costs will decrease to under €2 per kg by 2050. Globally, costs are projected to drop by approximately 35% by 2030 and 49% by 2050, compared to 2020. By 2030, hydrogen production costs could reach around €3 per kg. By 2050, costs are expected to fall well below €3 per kg, making hydrogen a more economically viable energy source in the future [30]. Hydrogen production from renewable sources could become cost-competitive between 2030 and 2050 if capital costs decrease significantly. By 2050, under typical conditions, the minimum production cost could drop to $1.5 per kilogram, with the potential to fall below $1 per kilogram in more optimistic scenarios, particularly in regions with abundant renewable resources. This price reduction would position hydrogen as an economically viable and sustainable energy solution, enhancing its attractiveness for large-scale adoption across various sectors, including maritime shipping and transportation [31]. Another study shows that to make hydrogen produced from fossil fuels with carbon capture and storage (CCS) competitive with hydrogen produced from fossil fuels without CCS, carbon prices would need to be in the range of $22–46 per ton of CO2e. There are indications that electrolysis powered by renewable energy could soon become cheaper than fossil fuel-based hydrogen with CCS. National strategies prioritizing fossil fuel-based hydrogen may conflict with decarbonization goals and risk creating stranded assets in the future hydrogen economy [32]. Investments in hydrogen infrastructure, such as refuelling stations and bunkering facilities, are critical for market readiness [32, 33]. Public and private partnerships are vital in developing the necessary infrastructure and reducing market barriers. Furthermore, an established global bunkering net for ships is essential for competition with conventional fuels.

As a future projection, hydrogen demand will significantly increase, with the maritime sector being a significant contributor. Hydrogen's potential to achieve cost competitiveness with conventional fuels depends on coordinated global efforts, supportive policies, and technological advancements. Stricter regulations on emissions, increased investment in hydrogen infrastructure, and technological advancements will drive the adoption of hydrogen in the maritime sector. Developing hydrogen hubs and corridors and integrating production, storage, and distribution will facilitate market growth. Several economic models and scenarios suggest that hydrogen could achieve price parity with conventional marine fuels by the 2030s with sufficient investment and policy support. Government incentives, such as subsidies, tax credits, and grants, can accelerate the development of hydrogen infrastructure and technology. Lastly, through pilot projects in the maritime industry, acquiring technical, economic, and operational experience will lead the way in removing barriers to fuel cells and hydrogen fuel by advancing their adoption.

5 Discussion

Although hydrogen is a promising option to ensure emission-free shipping, several challenges remain regarding hydrogen's widespread adoption in the maritime industry. Technological barriers include the need for advanced storage solutions and developing high-efficiency fuel cells. Economic barriers involve high production and infrastructure costs, necessitating substantial investments. Regulatory barriers include the lack of standardized global regulations and the need for international cooperation. Strategies to overcome these barriers include increased funding for research and development and public–private partnerships with the coherence of global standards. The scope and possible strategies to overcome these barriers can be discussed as follows in Table 3.

6 Conclusion

The future of hydrogen in the maritime industry is promising, with ongoing technological advancements, policy support, and market development. Predictions for hydrogen legislation indicate a trend towards stricter regulations on emissions and increased support for clean energy solutions. Policy recommendations to accelerate hydrogen adoption include enhancing international collaboration, providing financial incentives, investing in infrastructure, and promoting innovation through research and development.

The maritime industry's transition to hydrogen is critical to global decarbonization efforts. In the long term, widespread hydrogen adoption in the maritime sector could have transformative implications for the industry and global trade. Hydrogen-powered vessels, with their potential to significantly reduce emissions, will play a central role in shaping a more sustainable global shipping network. As nations move towards decarbonization, maritime routes and logistics chains could evolve, with ports adapting to hydrogen refuelling and storage infrastructure, reshaping global trade dynamics. While significant progress has been made in legislation, technology, and market development, continued efforts are required to overcome existing challenges. By fostering collaboration, innovation, and

Table 3 Scope and overcome strategies for the main barriers of hydrogen in maritime

Main Barriers	Scope and Overcome Strategies
Technological	• Advances in fuel cell technology, storage solutions, and production methods are needed to overcome current limitations. Continued research and development are essential to improve efficiency, durability, and safety – Increased funding for R&D to drive technological advancements in hydrogen production, storage, and utilization. Collaborative research initiatives between academia, industry, and government can accelerate innovation and commercialization. Focus on improving efficiency, reducing costs, and enhancing safety to overcome technological barriers
Economical	• High production and infrastructure costs require substantial investments and policy incentives to become economically viable. Financial support from governments and international organizations is crucial to eliminate the risk investments and encourage private sector participation – Collaboration between governments and industry to share risks and benefits, facilitate infrastructure development, and promote hydrogen adoption. Public–private partnerships can also support pilot projects and demonstrations to showcase hydrogen's viability in maritime applications – Financial incentives, such as subsidies, tax credits, and grants, to encourage investment in hydrogen infrastructure and technology. Policy measures to internalize the environmental costs of conventional fuels, such as carbon pricing, can also improve hydrogen's competitiveness. Support mechanisms for early adopters can reduce risks and accelerate market development
Regulatory	• The lack of standardized global regulations and certification processes for hydrogen fuel cells and storage systems poses significant challenges. International cooperation is essential to develop and harmonize these standards, ensuring safety and interoperability – Harmonization of regulations and standards through international cooperation, enabling the global adoption of hydrogen technologies. IMO and regional organizations can facilitate dialogue and coordination among stakeholders. Strengthen collaboration between countries to develop and harmonize hydrogen technology regulations, standards, and certification processes. International cooperation can also support joint research and development initiatives

investment, the maritime industry can achieve a sustainable and clean energy future, with hydrogen playing a central role. The transition to hydrogen is critical to the maritime industry's efforts to achieve sustainability and reduce environmental impact. While significant progress has been made, continued efforts are required to overcome existing challenges and barriers. Furthermore, hydrogen's role in mitigating environmental damage could lead to stricter regulatory frameworks for shipping emissions and influence global trade patterns. The growing demand for green hydrogen in maritime applications could stimulate international collaborations, potentially creating new hydrogen trade routes and supply chains, particularly from regions with abundant renewable energy resources to those with high demand. Ultimately, hydrogen adoption will help the maritime industry meet decarbonization targets and reshape the future of global trade, driving economic and environmental sustainability worldwide.

References

1. Inal, O. B. (2024). Decarbonization of shipping: Hydrogen and fuel cells legislation in the maritime industry. *Brodogradnja, 75*(2), 75205. https://doi.org/10.21278/brod75205
2. Zincir, B. (2023). Slow steaming application for short-sea shipping to comply with the CII regulation. *Brodogradnja, 74*(2), 21–38. https://doi.org/10.21278/brod74202
3. Inal, O. B., & Deniz, C. (2020). Assessment of fuel cell types for ships: Based on multi-criteria decision analysis. *Journal of Cleaner Production, 265*, 121734. https://doi.org/10.1016/j.jclepro.2020.121734
4. Inal, O. B., Zincir, B., & Dere, C. (2022). Hydrogen as maritime transportation fuel: A pathway for decarbonization. In *Greener and scalable E-fuels for decarbonization of transport* (Vol. 67–110). Springer. https://doi.org/10.1007/978-981-16-8344-2_4
5. IMO (International Maritime Organization). (2020). *The Fourth IMO GHG Study -Reduction of GHG Emissions from Ships*. IMO.
6. Sevim, C., & Zincir, B. (2022). Biodiesel and renewable diesel as a drop-in fuel for decarbonized maritime transportation. In *Energy, environment, and sustainability* (Vol. 319–345). Springer. https://doi.org/10.1007/978-981-16-8414-2_10
7. IMO (International Maritime Organization). (2021). *EEDI—Energy efficiency design index*. IMO. https://www.imo.org/en/OurWork/Environment/Pages/Improving%20the%20energy%20efficiency%20of%20ships.aspx
8. Dnv GL. (2021) *EEXI, energy efficiency existing ship index*. https://www.dnv.com/maritime/insights/topics/eexi/index.html
9. Dnv GL (2021) *CII—Carbon intensity indicator*. https://www.dnv.com/maritime/insights/topics/CII-carbon-intensityindicator/answers-to-frequent-questions.html
10. Welaya, Y. M. A., El Gohary, M. M., & Ammar, N. R. (2011). A comparison between fuel cells and other alternatives for marine electric power generation. *International Journal of Naval Architecture and Ocean Engineering, 32*, 141–149. https://doi.org/10.3744/JNAOE.2011.3.2.141
11. Chiong, M. C., Kang, H. S., Shaharuddin, N. M. R., Mat, S., Quen, L. K., Ten, K. H., & Ong, M. C. (2021). Challenges and opportunities of marine propulsion with alternative fuels. *Renewable and Sustainable Energy Reviews, 149*, 111397. https://doi.org/10.1016/j.rser.2021.111397
12. Zincir, B., & Deniz, C. (2018). Assessment of alternative fuels from the aspect of shipboard safety. *Journal of ETA Maritime Science, 6*(3), 199–214. https://doi.org/10.5505/jems.2018.71676
13. Inal, O. B., Dere, C., & Deniz, C. (2021). Onboard hydrogen storage for ships: An overview. In *5th international hydrogen technologies congress, 26–28 May, Nigde, Turkey*.
14. Rivard, E., Trudeau, M., & Zaghib, K. (2019). Hydrogen storage for mobility: A review. *Materials, 12*(12), 1973. https://doi.org/10.3390/ma12121973
15. Incer-Valverde, J., Korayem, A., Tsatsaronis, G., & Morosuk, T. (2023). "Colors" of hydrogen: Definitions and carbon intensity. *Energy Conversion and Management, 291*, 117294.
16. Ajanovic, A., Sayer, M., & Haas, R. (2022). The economics and the environmental benignity of different colors of hydrogen. *International Journal of Hydrogen Energy, 47*(57), 24136–24154.
17. Inal and Senol. (2024). SWOT-AHP analysis of different Colours of hydrogen for Decarbonization of shipping. *International Journal of Environment and Geoinformatics (IJEGEO), 11*(2), 029–037. https://doi.org/10.30897/ijegeo.1480096
18. Raucci, C., Calleya, J., de la Fuente, S. S., & Pawling, R. (2015). *International conference on shipping in changing climates 12* (Vol. Hydrogen on board ship: A first analysis of key parameters and implications).
19. O'Malley, K., Ordaz, G., Adams, J., Randolph, K., Ahn, C. C., & Stetson, N. T. (2015). Applied hydrogen storage Research and Development: A perspective from the U.S. Department of Energy. *Journal of Alloys and Compounds, 645*(S1), S419–S422. https://doi.org/10.1016/j.jallcom.2014.12.090

20. Hoecke, V., Laurens, L. L., Campe, R., Perreault, P., Verbruggen, S. W., & Lenaerts, S. (2021). Challenges in the use of hydrogen for maritime applications. *Energy & Environmental Science, 14*, 815. https://doi.org/10.1039/d0ee01545h
21. Mench, M. M. (2008). *Fuel cell engines*. Wiley. https://doi.org/10.1002/9780470209769
22. De-Troya, J. J., Álvarez, C., Fernández-Garrido, C., & Carral, L. (2016). Analysing the possibilities of using fuel cells in ships. *International Journal of Hydrogen Energy, 41*(4), 2853–2866. https://doi.org/10.1016/j.ijhydene.2015.11.145
23. van Biert, L., Godjevac, M., Visser, K., & Aravind, P. V. (2016). A review of fuel cell systems for maritime applications. *Journal of Power Sources, 327*(X), 345–364. https://doi.org/10.1016/j.jpowsour.2016.07.007
24. Wang, Z., Li, M., Zhao, F., Ji, Y., & Han, F. (2024). Status and prospects in technical standards of hydrogen-powered ships for advancing maritime zero-carbon transformation. *International Journal of Hydrogen Energy, 62*, 925–946. https://doi.org/10.1016/j.ijhydene.2024.03.083
25. Wang, Z., Dong, B., Yin, J., Li, M., Ji, Y., & Han, F. (2023). Towards a marine green power system architecture: Integrating hydrogen and ammonia as zero-carbon fuels for sustainable shipping. *International Journal of Hydrogen Energy., 50*, 1069. https://doi.org/10.1016/j.ijhydene.2023.10.207
26. Tarkowski, M. (2021). Towards a more sustainable transport future—The cases of ferry shipping electrification in Denmark, Netherland, Norway and Sweden. In *Innovations and traditions for sustainable development* (pp. 177–191). Springer.
27. IMO. (2017). *Global industry alliance launched to support low carbon shipping*. IMO. https://www.imo.org/en/MediaCentre/PressBriefings/Pages/17-GIA-GloMeep-launch.aspx
28. Pramuanjaroenkij, A., & Kakaç, S. (2023). The fuel cell electric vehicles: The highlight review. *International Journal of Hydrogen Energy, 48*(25), 9401–9425. https://doi.org/10.1016/j.ijhydene.2022.11.103
29. European Union. (2021). *European hydrogen strategy*. https://energy.ec.europa.eu/topics/energy-systemsintegration/hydrogen_en.
30. Frieden, F., & Leker, J. (2024). Future costs of hydrogen: A quantitative review. Sustainable. *Energy & Fuels., 8*, 1806.
31. Brändle, G., Schönfisch, M., & Schulte, S. (2021). Estimating long-term global supply costs for low-carbon hydrogen. *Applied Energy, 302*, 117481.
32. Longden, T., Beck, F. J., Jotzo, F., Andrews, R., & Prasad, M. (2022). 'Clean'hydrogen?–comparing the emissions and costs of fossil fuel versus renewable electricity based hydrogen. *Applied Energy, 306*, 118145.
33. Pan, G., Gu, W., Hu, Q., Wang, J., Teng, F., & Strbac, G. (2021). Cost and low-carbon competitiveness of electrolytic hydrogen in China. *Energy & Environmental Science, 14*(9), 4868–4881.

Future Pathways and Technological Advances of Fuel Cell and Hydrogen Technologies for Ships

Mauricio Latapí and Levent Bilgili

Abbreviations

AFC	Alkaline Fuel Cell
DAFC	Direct Alcohol Fuel Cell
DMFC	Direct Methanol Fuel Cell
EC)	European Commission
EEA	European Economy Area
EEDI	Efficiency Design Index
EEXI	Efficiency Existing Ship Index
EMS	Energy Management System
EU	European Union
FCH JU	Fuel Cells and Hydrogen Joint Undertaking
GE	General Electric
GHG	Greenhouse gas
HFC	Hydrogen Fuel Cell
ICE	Internal Combustion Engine
IMO	International Maritime Organization
LCA	Life Cycle Assessment
MCFC	Molten Carbonate Fuel Cells
NASA	National Aeronautics and Space Administration
PEFC	Polymer Electrolyte Fuel Cell
PEMFC	Proton Exchange Membrane Fuel Cells

M. Latapí (✉)
Mount Royal University, Calgary, AB, Canada
e-mail: malatapi@mtroyal.ca

L. Bilgili
Bandirma Onyedi Eylul University, Balikesir, Turkey
e-mail: lbilgili@bandirma.edu.tr

C. O. Colpan et al. (eds.), *Fuel Cell and Hydrogen Technologies in Maritime Transportation*, https://doi.org/10.1007/978-3-031-86110-9_13

RoPax	Roll-on/Roll-off passenger
SOFC	Solid Oxide Fuel Cells
UK	United Kingdom
UN	United Nations
USA	United States of America

1 Introduction

In 2018, the International Maritime Organization (IMO) adopted its "Initial Strategy on Reduction of GHG Emissions from Ships", indicating that the shipping industry must reduce its emissions in the coming decades [1]. This was followed by the adoption of the "2023 IMO Strategy on Reduction of GHG Emissions from Ships" (hereon IMO's 2023 GHG Strategy), which states that international shipping should aim to reach net-zero greenhouse gas (GHG) emissions by or around 2050 [2]. To do so, the shipping industry is expected to transition to zero-emission technologies and energy sources. A promising solution comes from fuel cells with low and zero emissions (depending on their type and fuel input).

While fuel cells have been used and commercialized for other modes of transportation, such as fuel cell vehicles and even for smaller applications such as forklifts [see ref. 3], their adoption for shipping has been limited to pilot and demonstration projects. However, in recent years, the funding toward developing and testing hydrogen fuel cells (HFCs) for shipping has increased significantly. A key supporter has been the European Union (EU) which has provided direct funding for accelerating the uptake of HFCs. As a result, various studies have been conducted to evaluate the feasibility of HFCs, mainly focusing on proton exchange membrane fuel cells (PEMFCs) and solid oxide fuel cells (SOFCs).

Despite the increasing funding and growing interest in HFCs, the technology has not become commercially viable for shipping due to a series of barriers that include their high costs, lack of available fuel, and challenges in the handling and storage of hydrogen on board [3–5]. Nevertheless, the level of ambition set by IMO's 2023 GHG Strategy can be expected to translate into an accelerated uptake of zero-emission technologies in the coming decades. Given this context, this chapter explores two research questions:

- What is the outlook for the adoption of hydrogen fuel cells for shipping in the coming decades?
- How can hydrogen fuel cells be used as energy sources for shipping?

To answer these questions, this chapter provides an overview of HFCs, their status and development, their potential use for shipping as the main energy source or as part of hybrid systems, and their limitations. At the same time, the overview provides an outlook for their adoption, with a particular focus on the EU, where shipping companies are expected to become early adopters. The chapter is divided into six sections. Section 1 provides the introduction to the topic and is followed by Sect. 2, which includes the background of HFCs going from their origins to the most recent and ongoing projects. Section 3 provides a path forward for HFCs identifying two potential options: HFCs as the main energy source and as part of hybrid systems. Then, Section 4 identifies fuel cells that use other fuels besides hydrogen and is followed by Sect. 5 that presents the discussion for the chapter. Section 6 concludes the chapter.

2 Background

2.1 Historical Origin

While the notion of using fuel cells for the shipping industry is relatively new, their history can be traced back to the 1830s [6, 7]. However, it was not until 1933 that the technology reached a level of practical application when Thomas Francis Bacon used electrochemical processes to convert hydrogen and oxygen into electricity, effectively creating the first hydrogen fuel cell [6, 7]. After that and during the Second World War, Bacon's work focused on developing fuel cells that could be used for maritime applications, specifically for submarines for the Royal Navy of the United Kingdom [6]. However, their adoption remained limited and did not reach a significant uptake [8].

Nevertheless, Bacon's seminal work provided the fundamentals for the aerospace industry of the United States of America to develop fuel cells for the space program of the National Aeronautics and Space Administration (NASA) [9]. Specifically, General Electric (GE) developed the first polymer electrolyte fuel cell (PEFC) to be used in NASA's Gemini Program, and later, the company Pratt and Whitney developed the alkaline fuel cell (AFC), which would provide the electrical energy for the Apollo modules [6, 9]. Since then, HFCs have been key for the aerospace industry and are expected to remain significant for future missions [10].

While HFCs have reached technological maturity in the aerospace industry and are frequently used in space missions, their adoption in the maritime sector has been limited. In the case of submarine applications, the space and volume needed, along with the complexity of their use on board, have been the main constraints for their adoption at a large scale [8]. Similar factors have limited the outlook for their adoption for vessels and shipping in general. However, for the shipping industry, the main barriers[1] have been the high costs of technology and green hydrogen [3, 11], the lack of infrastructure along with the lack of available fuel [12–14], as well as operational challenges, mainly in relation to the storage and bunkering of

[1] See Latapí et al. [5] for a comprehensive analysis of the drivers and barriers for the large-scale adoption of HFCs in shipping.

hydrogen [4, 15, 16]. Nevertheless, the interest in using HFCs for shipping has increased significantly during the late 2010s and early 2020s.

2.2 *Pilot and Demonstration Projects*

Since the late 1990s and early 2000s, there has been a growing interest in reducing the GHG emissions from shipping. As a result, a variety of technologies have been considered and tested. In the case of HFCs, the first pilot test took place in Germany in 2000 when the companies ZeTek Power, TTIB, and Etaing Gmbh launched a small 22-person ferry boat named "Hydra" which used hydrogen for a 5-kW AFC to provide electricity for running the propeller [8, 17, 18]. The interest in fuel cells for shipping increased and the EU began funding projects for developing and testing PEMFCs. This was the case of the zero emission ships (Zemships), an EU-funded project that in 2008 gathered nine industry and academic partners to design a prototype of an inland tourist ship propelled by two 48-kW PEMFCs that received hydrogen from twelve pressurized tanks with 50 kg of hydrogen each [8, 19, 20].

Since then, a series of pilot tests and demonstration projects for HFCs followed. Some of the most relevant include the Hydrogenesis Passenger Ferry (2013) [see: 20, 21], the HySeas III project, which developed a prototype of a roll-on/roll-off passenger (RoPax) ferry (2018) [see: 22], the Jules Verne 2 passenger ferry (2018) [see: 23], and the MF Hydra (2023) [see: 24].

Notably, most of these projects have taken place in Europe (including countries in the EU, the European Economy Area (EEA), and the UK), mainly as a result of direct funding for developing fuel cells. This support has been evident with the establishment of the Fuel Cells and Hydrogen Joint Undertaking (FCH JU), which partnered the European Commission (EC) with the fuel cell and hydrogen industry and the academia to support the research, development, and demonstration stages for the use of HFCs in Europe [25]. In 2021, the FCH JU was replaced by the Clean Hydrogen Partnership with the aim of accelerating the development of hydrogen technologies and end-use applications through funding and partnerships [26]. Some of the most recent and relevant projects funded by the Clean Hydrogen Partnership, directly related to the use of HFCs in shipping, are included in Table 1.

3 The Path Forward for Hydrogen Fuel Cells

While the number of pilot and demonstration projects using HFCs for shipping has increased significantly in the last decade, the technology has not reached the level of technological maturity to be commercially viable without governmental funding [see: 27, 28]. However, the international interest in the transition toward zero-emission shipping is expected to materialize as additional support in the coming years. In the case of HFCs, European companies are expected to become early

Table 1 Projects funded by the Clean Hydrogen Partnership related to the use of HFCs in shipping

Name	Objective	Stage of the H_2 value chain	Timeframe	Total funding in euros	Country
H2MARINE	Design and test two PEMFCs to generate 250–300 kW. The project is focused on the marine applications of these fuel cells	End-use	Jan 1, 2024–Jun 30, 2027	€7,499,172	Greece
DelHyVEHR	Accelerate the technological maturity phase for large-scale liquid hydrogen refueling stations	Bunkering	Jan 1, 2024–Dec 31, 2026	€5,064,971	France
TRIERES	Establish a hydrogen valley including the production of green hydrogen and its direct use for one short-distance ferry	Production, bunkering, and end-use	Jul 1, 2023–Apr 30, 2028	€10,492,431	Greece
LH_2CRAFT	Design and develop long-term and long-distance storage for transporting liquid hydrogen on board vessels. This includes the preliminary design for a ship that can transport liquid hydrogen at a large scale	Transportation	Jun 1, 2023–may 31, 2027	€5,627,596	Greece
RH2IWER	Set up the demonstration phase for six commercially operated vessels using HFCs	End-use	Mar 1, 2023–Aug 31, 2027	€20,531,971	Finland
StasHH	Develop and test a new standard for fuel cell modules. The project is focused on heavy-duty applications including shipping	End-use	Jan 1, 2021–Dec 31, 2023	€15,167,500	Norway
HYShip	Develop a maritime supply chain for liquid hydrogen including the production of green H_2 and the design and testing of a 2 MW fuel cell liquid hydrogen ship	Production, bunkering, and end-use	Jan 1, 2021–Dec 31, 2025	€10,796,600	Norway

(continued)

Table 1 (continued)

Name	Objective	Stage of the H_2 value chain	Timeframe	Total funding in euros	Country
e-SHyIPS	Develop a preliminary standardization plan for updating IMO's IGF code for the use of hydrogen in passenger vessels	End-use	Jan 1, 2021–Dec 31, 2024	€2,500,000	Italy

Source: This information was obtained from the repository of projects of the Clean Hydrogen Partnership on March 18, 2024 from the website: https://www.clean-hydrogen.europa.eu/projects-dashboard/projects-repository_en?f%5B0%5D=oe_project_title%3Aship

adopters given the extensive support in Europe toward the hydrogen value chain including the development, testing, and demonstration of hydrogen-based technologies and fuels as well as direct funding for HFCs.

Nevertheless, the physicochemical characteristics of hydrogen constitute significant challenges for its bunkering, handling, and storage on board. At the same time, the energy output and size of HFCs represent limitations for their adoption for long-distance travel and larger ships. Given these aspects, HFCs are considered to be useful mainly for short-distance voyages and smaller vessels, such as ferries. At the same time, not all types of fuel cells are suitable for being used in shipping. As Fu et al. [29] point out, the three most promising types of hydrogen fuel cells for maritime applications are PEMFCs, SOFCs, and molten carbonate fuel cells (MCFCs). Their characteristics are summarized in Table 2.

The characteristics, technological maturity, and direct funding for the development of PEMFCs for shipping suggest that there could be a significant advancement in their adoption for niche applications and specific maritime segments between 2030 and 2050 [38]. For instance, a variety of studies see potential in HFCs, in particular PEMFCs, to become the main energy source for small vessels with short routes [see: 39, 40]. However, the latest research points toward the adoption of hybrid or combined systems in which fuel cells are only one of the key elements [see: 41, 42].

3.1 HFCs as the Main Energy Source

HFCs, particularly PEMFCs, have been evaluated as the main energy source for small vessels. For instance, Cavo et al. [43] and Xie et al. [44] analyzed the potential for using PEMFCs alongside batteries. For this kind of system to be efficient, the vessel's energy management is optimized by considering HFCs as the main energy source and the battery banks as the energy storage system to provide flexibility for dynamic and variable loads [45]. Using this combination with the appropriate energy

Table 2 The main characteristics of PEMFCs, SOFCs, and MCFCs

Characteristics	PEMFC	SOFC	MCFC
Electrical efficiency[a, b, c, d]	50–60%	50–60%	50–55%
Cost in USD/kW[a, c, e]	>1500	>4500	>1300
Relative cost among fuel cells[a, c, e]	Low	High	Low
Temperature of operation[f, c, d]	65–85 °C	500–1000 °C	600–700 °C
Power capacity[c, d]	≤ 120 kW	≤ 10 MW	120 kW to 10 MW
Start-up[f, b]	Quick	Slow	Slow
Partial load efficiency[a]	High	High	Medium to high
Tolerance to load variations[a, g]	Medium to high	Low	Low to medium
Lifetime[a, c, d,h]	Low to medium	Low	Medium
Size[c, d]	Small	Medium	Large
Noise/vibration[a]	Low	Low	Low to medium
Fuel input[b, c]	High purity hydrogen	High fuel flexibility including methane, methanol, ethanol, propane, ammonia, diesel, etc.	Hydrogen, methanol, and hydrocarbons
Sensitivity to fuel impurities[g, i]	High	Low	Low
Maturity[a, b, g]	Medium to high	Low to medium	Low

Sources: The information used for this table comes from [a]IRENA [30]
[b]Baldi et al. [31]
[c]Xing et al. [32]
[d]Li et al. [35]
[e]Ryu et al. [34]
[f]Fu et al. [29]
[g]DNV [33]
[h]Cui et al. [37]
[i]Contreras et al. [36]
Note: See Xing et al. [32] for details regarding the system components for each type of fuel cell

management system (EMS) allows to limit the energy fluctuations through the fuel cells which increases their efficiency and lifespan [46]. While the use of HFCs as the main energy source has been found technically feasible [see: 47], the uncertainties regarding hydrogen availability and supply for shipping, along with the high costs of the fuel and the technology, represent significant barriers to their wider adoption [5].

3.2 *Hybrid Systems*

In the case of hybrid systems that include HFCs, some potential options consider their use alongside renewable energy sources such as wind and solar energy [48] as well as heat recovery systems involving steam turbines [49]. However, some of these hybrid systems are expected to include a fossil fuel component, such as a diesel generator or an internal combustion engine (ICE). Nevertheless, their adoption is projected to reduce fossil fuel consumption, GHG emissions, as well as noise pollution and vibration at a significant scale [50, 51].

Hybrid systems involving HFCs have the potential to result in low to zero emissions for the vessel from a tank-to-wake perspective while increasing the ship's overall energy efficiency. For instance, Oh et al. [49] indicate that a hybrid system that uses SOFCs with hydrogen and a heat recovery system can reach a 22.5% increase in energy efficiency and a 17.5% decrease in fuel consumption compared to only using the SOFCs. However, this kind of system is mainly suitable for short routes. In the case of ocean-going vessels with longer routes, the literature points out hybrid systems that include fossil-fuel elements. For example, the modeling conducted by Lee et al. [52] indicates that a tanker on a Middle East–East Asia route using a hybrid system consisting of five small generators complemented with SOFCs and a steam boiler can reduce its tank-to-wake emissions up to 8.8% in comparison to its conventional system of five generators and a steam boiler. Given these aspects, hybrid systems can be expected to have wider adoption than HFCs being used as main energy sources (see Fig. 1 for a simplified diagram of a potential hybrid system).

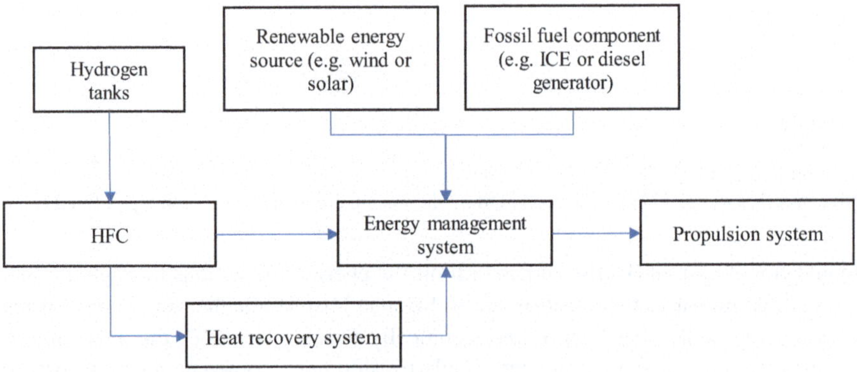

Fig. 1 Simplified diagram of a potential hybrid system (the arrows indicate the energy flow through the system)

4 Fuel Cells that Use Other Fuels Besides Hydrogen

While HFCs have promising potential for being used in short-distance voyages and small vessels, there are still significant challenges to their adoption at a large scale. As a result, fuel cells that use another kind of fuel besides hydrogen can become more readily available and more attractive for the maritime sector. Some options include AFCs, direct methanol fuel cells (DMFC), direct alcohol fuel cells (DAFC), as well as MCFCs, and SOFCs. It is relevant to point out that AFCs, MCFCs, and SOFCs are able to work with other kinds of fuels besides hydrogen, which gives them more flexibility in comparison to PEMFCs [40].

4.1 Alkaline Fuel Cells (AFCs)

AFCs are considered to be one of the first fuel cells developed, and although their general power output is expected to be between 1 and 5 kW, some experimental samples have reached 200 kW [53]. AFCs use alkaline as electrolyte and nickel and silver as anode and cathode, respectively [53]. It is relevant to consider that the power produced by AFCs may be slightly higher than by SOFCs.

4.2 Direct Methanol and Direct Alcohol Fuel Cells

These fuel cells may use ethanol or methanol as fuels [40], and their efficiency is generally considered low [54]. However, there are several advantages of using DMFCs including sufficient fuel supply, simple structure, quick refueling, stable performance, and low exhaust emissions. Furthermore, methanol is one of the most promising fuels in maritime decarbonization, which means that DMFCs may find a significant place among other fuel cells for the shipping industry [55].

4.3 Molten Carbonate Fuel Cells (MCFCs)

MCFCs operate between 600 and 700 °C, resulting in high ion conductivity. While the electrolyte is a mixture of a molten alkali, metals carbonate, nickel, and nickel oxide can be used as electrodes at the anode and cathode, respectively [53]. MCFCs mainly use natural gas, waste gas, and biomass gas as primary fuel [40], but they can also use ammonia [56], making them very flexible in terms of fuel intake.

One of the first studies on using MCFCs on board stated that the carbon dioxide (CO_2) emissions were reduced by 10–30% compared with conventional diesel engines [57], with similar results being reached in a 2023 study (reduction of

20%) [see ref. 58]. When MCFCs are integrated with $LiNiMnCoO_2$ battery cells, they demonstrate the best environmental and economic performance, reducing CO_2 emissions by up to 49.75% and electricity production cost by 0.181 USD/kWh [59]. MCFCs are also investigated in terms of safety due to various external conditions and as it is a newly adopted mechanism for ships, it is indicated that a detailed cost–benefit analysis is needed with regard to their safety for onboard operations [60].

4.4 *Solid Oxide Fuel Cells (SOFCs)*

The operating temperature of SOFCs may reach 1000 °C, resulting in high efficiency rates. SOFCs use solid-state yttrium-stabilized zirconia as the electrolyte, while nickel alloy and lanthanum strontium manganite are used as anode and cathode, respectively [53]. Some of the main advantages of SOFCs include their installation flexibility, reduced noise and vibration, clean exhaust emissions, and good performance under variable external conditions [61]. Besides hydrogen, SOFCs can use natural gas, liquefied petroleum gas, or ammonia as primary energy sources [40]. As a result, SOFCs are a promising and viable technology that can use ammonia instead of hydrogen and still have a relatively good environmental performance [62, 63]. However, using hydrogen is considered more energy efficient and would lead to zero emissions at the exhaust [64].

SOFCs (like other fuel cells) are a relatively new technology and have gained importance after the adoption of the GHG reduction strategies adopted by the IMO in 2018 and 2023 [see: refs. 2, 65]. A SOFC is an efficient, available, noiseless energy source with low maintenance effort and dependability, in addition to providing long-term durability. For instance, an ammonia SOFC can produce zero CO_2 emission from a tank-to-wake perspective. However, the system's total cost (fuel price, installation of the propulsion system, fuel storage system, and overall CAPEX) is higher than internal combustion systems [56].

SOFCs can also be used integrated with internal combustion engines, and a recent study indicated that nitrogen oxides (NO_x) and CO_2 emissions may be reduced by 30% and 12%, respectively, when this combined system is used [66]. Another study on integrating SOFCs and internal combustion engines showed that their use can reduce the energy efficiency design index (EEDI) and carbon intensity indicator (CII) by 15.6% and 17.5%, respectively [67]. Utilizing SOFCs is also considered compatible with the energy efficiency existing ships index (EEXI) [68]. Given these aspects, SOFCs and PEMFCs are the most promising hydrogen fuel cells for maritime operations [69].

5 Discussion

The IMO 2023 GHG Strategy sets a new level of ambition for the shipping industry [2]. However, transitioning toward zero emissions requires the development and commercialization of new technologies. One of the most promising options is the adoption of HFCs. In particular, PEMFCs seem to be the best suited for early adoption considering their characteristics (e.g., low relative cost, low temperatures of operation, quick start-up, and small size), their technological maturity (medium to high), and the funding that is being directed toward their development and testing (mainly in the EU through the Clean Hydrogen Partnership).

Notably, the European funding toward HFCs seems to focus on regional areas, including the Nordic (e.g., Norway) and the Mediterranean countries (e.g., Greece). This supports the notion that HFCs might become readily available and viable for specific regions and niche shipping industry segments. This also suggests that the wider adoption of HFCs could potentially start with the ferry segment of Nordic and Mediterranean regions.

The increasing interest in the transition of the shipping industry toward zero emissions has resulted in the evaluation of HFCs from two key perspectives:

1. Using HFCs as the main energy source with a particular focus on reaching zero emissions. For this system to be feasible, the literature suggests that HFCs would need to be complemented with battery banks that can serve as energy storage and provide flexibility during dynamic and variable loads [see refs. 43, 44, 45]. This kind of system would most likely use PEMFCs and be suitable for small vessels with short routes such as ferries. However, this would be limited by the availability of green hydrogen and renewable energy along the ship's routes for it to be considered zero emissions from a complete life cycle assessment (LCA), which is an innovative approach for the reduction of emissions from the maritime sector and is planned to be used extensively in the next decades by the IMO. Furthermore, the energy system and energy storage would use a large space on board, which might not appeal to ship owners and operators. Nevertheless, the development of new storage units and the research being conducted for PEMFCs might make the technology attractive and suitable for their adoption in the coming decade.
2. Using HFCs as part of a hybrid system with the goal of optimizing energy efficiency and reducing GHG emissions. A hybrid system can provide flexibility for early adopters during the transition considering that it can include fossil fuel components that operate along with HFCs and heat recovery systems [see refs 49, 52]. This kind of hybrid system would most likely use PEMFCs or SOFCs along with other energy sources and be suitable for small and medium vessels with longer routes than the previous option. An advantage of a hybrid system with SOFCs is that it could function with hydrogen-based fuels such as ammonia, which might be more readily available than green hydrogen. These aspects can make a hybrid system attractive for ship owners and operators by providing them with flexibility and less uncertainty regarding fuel and energy availability along their routes.

While both options are considered technically feasible, practical aspects such as fuel supply, costs, lack of regulations and guidelines, and the need for developing or retrofitting infrastructure might become defining factors for the adoption of HFCs. This suggests that the further development of the technology and its potential market penetration and commercialization might be inextricably driven by policies. This has been pointed out by recent studies that indicate that zero-emission technologies for shipping still face market failures and barriers that require policy intervention [see refs. 28, 70]. Given these aspects, the adoption of hybrid systems seems to be the most likely option for the early stages of the transition beginning at a small and regional scale, mainly in Nordic Europe and the Mediterranean.

6 Conclusions

The shipping industry is expected to reduce its GHG emissions in the coming decades with the goal of reaching net-zero emissions by or around mid-century. To do so, new technologies need to be developed, tested, and commercialized at a large scale. A promising option comes from HFCs which can be used as the main energy source for vessels (e.g., PEMFCs complemented with battery tanks for energy storage) or can also be adopted as an element of a hybrid system (e.g., PEMFCs or SOFCs in combination with other energy sources as well as heat recovery systems). In particular, PEMFCs seem to be the most suitable for early adopters, given their technological maturity, low relative costs, and the direct funding being provided for their development. For instance, the financial support from the EU through the Clean Hydrogen Partnership suggests that the main interest in the adoption of HFCs resides in the Nordic European and Mediterranean countries.

This chapter provides an overview of the technological advancements of HFCs and their potential adoption for the shipping industry. The analysis indicates that HFCs represent a feasible alternative for shipping for niche segments and at a regional scale—for example, for ferries with short routes. Given these aspects and the information presented, this chapter can be useful for policymakers and key stakeholders to better understand the status of HFCs, their potential, and their limitations. It can help bring attention to the benefits of low or zero-emission technologies for the shipping industry.

References

1. International Maritime Organization. *Historic Background: IMO and the UNFCCC policy framework*. 2019 February 22 2023]; https://www.imo.org/en/OurWork/Environment/Pages/Historic%20Background%20GHG.aspx
2. International Maritime Organization. (2023). *2023 IMO strategy on reduction of GHG emissions from ships*. International Maritime Organization.

3. Ajanovic, A., & Haas, R. (2021). Prospects and impediments for hydrogen and fuel cell vehicles in the transport sector. *International Journal of Hydrogen Energy, 46*(16), 10049–10058.
4. Van Hoecke, L., et al. (2020). Challenges in the use of hydrogen as a maritime fuel. *Energy & Environmental Science, 14*, 815.
5. Latapí, M., Davíðsdóttir, B., & Jóhannsdóttir, L. (2023). Drivers and barriers for the large-scale adoption of hydrogen fuel cells by Nordic shipping companies. *International Journal of Hydrogen Energy, 48*, 6099.
6. Andújar, J. M., & Segura, F. (2009). Fuel cells: History and updating. A walk along two centuries. *Renewable and Sustainable Energy Reviews, 13*(9), 2309–2322.
7. Ortiz-Rivera, E. I., Reyes-Hernandez, A. L., & Febo, R. A. (2007). *Understanding the history of fuel cells*. In *IEEE conference on the history of electric power* (p. 2007). IEEE.
8. Wang, X., Zhu, J., & Han, M. (2023). Industrial development status and prospects of the marine fuel cell: A review. *Journal of Marine Science and Engineering, 11*(2), 238.
9. Perry, M. L., & Fuller, T. F. (2002). A historical perspective of fuel cell technology in the 20th century. *Journal of the Electrochemical Society, 149*(7), S59.
10. FuelCellsWork. *How fuel cells are used in space travel*. 2022 [cited 2024 February 19 2024]; https://fuelcellsworks.com/news/how-fuel-cells-are-used-in-space-travel/
11. Prussi, M., et al. (2021). Potential and limiting factors in the use of alternative fuels in the European maritime sector. *Journal of Cleaner Production, 291*, 125849.
12. European Commission. (2020). In E. Commission (Ed.), *State of the art on alternative fuels transport Systems in the European Union: Update 2020*. European Commission.
13. Bach, H., et al. (2020). Implementing maritime battery-electric and hydrogen solutions: A technological innovation systems analysis. *Transp Res D Transp Environ, **87***, 87.
14. Fuel Cells and hydrogen 2nd joint undertaking. (2019). *Hydrogen roadmap Europe: A sustainable pathway for the European energy transition* (Vol. 2019, p. 13). European Commission.
15. Xing, H., et al. (2021). Alternative fuel options for low carbon maritime transportation: Pathways to 2050. *Journal of Cleaner Production, 297*, 126651.
16. Zhan, Z. (2022). Current status and future development of fuel cell ships in China. *Journal of Physics Conference Series, 2160*(1), 012061.
17. die Brennstoffzelle. *Brennstoffzellen-Boot Hydra*. [cited 2024 February 22, 2024]; http://www.diebrennstoffzelle.de/h2projekte/mobil/hydra.shtml
18. Fuell Cells. (2000). *Fuel cell technology update—August 1, 2000*. In *Fuel cells*.
19. Pflegerl, M. *Proton Motor liefert Antrieb für erstes Fahrgastschiff mit Brennstoffzellen*. 2007 [cited 2024 February 22, 2024]; https://web.archive.org/web/20071217212759/http://www.wattgehtab.com/index.php/content/view/1860/25/.
20. *FCS Alterwasser—ZemShips*. n.d. [cited 2024 February 22, 2024]; https://www.elizabethqueenseaswann.com/HISTORY/LH2_Ships_Ferries_Yachts_Hydrogen_Projects/Zemships_FCS_Alterwasser100Passenger_Fuel_Cell_Zero_Emission.html
21. Ship Technology. *Hydrogenesis passenger ferry*. 2013 [cited 2024 February 22, 2024]; https://www.ship-technology.com/projects/hydrogenesis-passenger-ferry/?cf-view
22. HySeas III., HySeas III. (2019). [cited 2024 February 22, 2024]; https://www.hyseas3.eu/
23. Energy-Observer. *Jules Verne 2, a new hydrogen-powered passenger vessel*. 2018 [cited 2024 February 22, 2024]; https://www.energy-observer.org/innovations/jules-verne-2-navibus-nantes-hydrogen-propulsion
24. Ballard. *World's first liquid-powered hydrogen ship, MF Hydra, is powered by Ballard's fuel cells*. 2023 [cited 2024 February 22, 2024]; https://blog.ballard.com/marine/worlds-first-liquid-powered-hydrogen-ship-mf-hydra-is-powered-by-ballards-fuel-cells
25. Fuel Cells and Hydrogen Joint Undertaking. *Who we are*. [cited 2024 February 23, 2024]; https://www.fch.europa.eu/page/who-we-are
26. Clean Hydrogen Partnership. *Who we are*. [cited 2024 February 23, 2024]; https://www.clean-hydrogen.europa.eu/about-us/who-we-are_en
27. DNV, Maritime forecast to 2050: Energy transition outlook 2021, DNV, Editor. 2021, : DNV.
28. Latapí, M., et al. (2024). Hydrogen fuel cells in shipping: A policy case study of Denmark, Norway, and Sweden. *Marine Policy, 163*, 106109.

29. Fu, Z., et al. (2023). Fuel cell and hydrogen in maritime application: A review on aspects of technology, cost and regulations. *Sustainable Energy Technologies and Assessments, 57*, 103181.
30. IRENA. (2022). *Global hydrogen trade to meet the 1.5°C climate goal: Part II—Technology review of hydrogen carriers*. International Renewable Energy Agency.
31. Baldi, F., et al. (2020). The role of solid oxide fuel cells in future ship energy systems. *Energy, 194*, 116811.
32. Xing, H., et al. (2021). Fuel cell power systems for maritime applications: Progress and perspectives. *Sustainability, 13*(3), 1213.
33. DNV. (2021). Energy transition outlook 2021. In *Technology progress report*. DNV.
34. Ryu, J.-Y., et al. (2020). Thermo-economic assessment of molten carbonate fuel cell hybrid system combined between individual sCO_2 power cycle and district heating. *Applied Thermal Engineering, 169*, 114911.
35. Li, J.-C., et al. (2024). A review on the research progress and application of compressed hydrogen in the marine hydrogen fuel cell power system. *Heliyon, 10*, e25304.
36. Contreras, R. R., Almarza, J., & Rincón, L. (2021). Molten carbonate fuel cells: A technological perspective and review. *Energy Sources, Part A: Recovery, Utilization, and Environmental Effects*, 1–15.
37. Cui, C., et al. (2021). Review of molten carbonate-based direct carbon fuel cells. *Materials for Renewable and Sustainable Energy, 10*(2), 12.
38. Global Maritime Forum. (2023). *The shipping industry's fuel choices on the path to net zero*. Global Maritime Forum.
39. Inal, O. B., et al. (2024). Hydrogen fuel cell as an electric generator: A case study for a general cargo ship. *Journal of Marine Science and Engineering, 12*(3), 432.
40. Wang, Z., et al. (2024). Analysis and evaluation of fuel cell technologies for sustainable ship power: Energy efficiency and environmental impact. *Energy Conversion and Management: X, 21*, 100482.
41. Dall'Armi, C., Pivetta, D., & Taccani, R. (2023). Hybrid PEM fuel cell power plants Fuelled by hydrogen for improving sustainability in shipping: State of the art and review on active projects. *Energies, 16*(4), 2022.
42. Cao, W., et al. (2023). Energy management strategy considering energy storage system degradation for hydrogen fuel cell ship. *Polish Maritime Research, 30*(2), 95–104.
43. Cavo, M., et al. (2023). An advanced control method for fuel cells—metal hydrides thermal management on the first Italian hydrogen propulsion ship. *International Journal of Hydrogen Energy, 48*(54), 20923–20934.
44. Xie, P., et al. (2024). A two-layer energy management system for a hybrid electrical passenger ship with multi-PEM fuel cell stack. *International Journal of Hydrogen Energy, 50*, 1005–1019.
45. Ganjian, M., et al. (2024). Optimal design strategy for fuel cell-based hybrid power system of all-electric ships. *International Journal of Hydrogen Energy, 50*, 1558–1571.
46. Wu, J., et al. (2024). Design and performance evaluation of power system for unmanned ship based on proton exchange membrane fuel cell. *International Journal of Hydrogen Energy, 59*, 730–741.
47. Hansson, J., et al. *Hydrogen HOPE for the Nordics: Shipping as a frontrunner*. Nordic Energy Research.
48. Dolatabadi, S. H., Ölçer, A. I., & Vakili, S. (2023). The application of hybrid energy system (hydrogen fuel cell, wind, and solar) in shipping. *Renewable Energy Focus, 46*, 197–206.
49. Oh, D., Cho, D.-S., & Kim, T.-W. (2023). Design and evaluation of hybrid propulsion ship powered by fuel cell and bottoming cycle. *International Journal of Hydrogen Energy, 48*(22), 8273–8285.
50. Ghenai, C., et al. (2019). Hybrid solar PV/PEM fuel cell/diesel generator power system for cruise ship: A case study in Stockholm, Sweden. *Case Studies in Thermal Engineering, 14*, 100497.

51. Gianni, M., Pietra, A., & Taccani, R. (2021). *Outlook of future implementation of PEMFC and SOFC onboard cruise ships*. In *E3S web of conferences* (Vol. 238). EDP Sciences.
52. Lee, J. I., Yoon, B. Y., & Cha, S. W. (2023). Analysis of solid oxide fuel cell hybrid power system in marine application for CO_2 reduction. *Energy Reports, 9*, 3072–3081.
53. Inal, O. B., & Deniz, C. (2020). Assessment of fuel cell types for ships: Based on multi-criteria decision analysis. *Journal of Cleaner Production, 265*, 265.
54. van Biert, L., et al. (2016). *A review of fuel cell systems for maritime applications. Journal of Power Sources*, 327, 345–364.
55. Fang, S., et al. (2024). Comprehensive energy conversion efficiency analysis of micro direct methanol fuel cell stack based on polarization theory. *Energy, 287*, 129670.
56. Wu, S., Miao, B., & Chan, S. H. (2022). Feasibility assessment of a container ship applying ammonia cracker-integrated solid oxide fuel cell technology. *International Journal of Hydrogen Energy, 47*(63), 27166–27176.
57. Dimopoulos, G. G., Stefanatos, I. C., & Kakalis, N. M. P. (2016). Exergy analysis and optimisation of a marine molten carbonate fuel cell system in simple and combined cycle configuration. *Energy Conversion and Management, 107*, 10–21.
58. Bosio, B., et al. (2023). Process analysis of a molten carbonate fuel cell on-board application to reduce vessel CO_2 emissions. *Chemical Engineering and Processing - Process Intensification, 190*, 109415.
59. Korkmaz, S. A., et al. (2023). Environmental and economic analyses of fuel cell and battery-based hybrid systems utilized as auxiliary power units on a chemical tanker vessel. *International Journal of Hydrogen Energy, 48*(60), 23279–23295.
60. Ahn, J., et al. (2017). Fuzzy-based failure mode and effect analysis (FMEA) of a hybrid molten carbonate fuel cell (MCFC) and gas turbine system for marine propulsion. *Journal of Power Sources, 364*, 226–233.
61. van Veldhuizen, B. N., et al. (2023). Experimental evaluation of a solid oxide fuel cell system exposed to inclinations and accelerations by ship motions. *Journal of Power Sources, 585*, 233634.
62. Rathore, S. S., et al. (2021). Direct ammonia solid-oxide fuel cells: A review of progress and prospects. *International Journal of Hydrogen Energy, 46*, 35365–35384.
63. Fuerte, A., et al. (2009). Ammonia as efficient fuel for SOFC. *Journal of Power Sources, 192*(1), 170–174.
64. Ryu, B. R., Duong, P. A., & Kang, H. (2023). Comparative analysis of the thermodynamic performances of solid oxide fuel cell–gas turbine integrated systems for marine vessels using ammonia and hydrogen as fuels. *International Journal of Naval Architecture and Ocean Engineering, 15*, 15.
65. International maritime organization. (2018). *Initial IMO strategy on reduction of GHG emissions from ships*. International Maritime Organization.
66. Sapra, H., et al. (2021). Integration of solid oxide fuel cell and internal combustion engine for maritime applications. *Applied Energy, 281*, 115854.
67. Qu, J., et al. (2023). Assessment of a methanol-fueled integrated hybrid power system of solid oxide fuel cell and low-speed two-stroke engine for maritime application. *Applied Thermal Engineering, 230*, 120735.
68. Yuksel, O. (2023). A comprehensive feasibility analysis of dual-fuel engines and solid oxide fuel cells on a tanker ship considering environmental, economic, and regulation aspects. *Sustainable Production and Consumption, 42*, 106–124.
69. Benet, Á., et al. (2022). Knowledge gaps in fuel cell-based maritime hybrid power plants and alternative fuels. *Journal of Power Sources, 548*, 232066.
70. Getting to Zero Coalition. (2021). *Policy options for closing the competitiveness gap between fossil and zero-emission fuels in shipping*. Getting to Zero Coalition.

Index

C. O. Colpan et al. (eds.), *Fuel Cell and Hydrogen Technologies in Maritime Transportation*, https://doi.org/10.1007/978-3-031-86110-9